OPTICAL SYSTEM DESIGN

Optical System Design

Robert E. Fischer
President, OPTICS 1, Incorporated

Biljana Tadic-Galeb
Senior Staff Optical Engineer,
OPTICS 1, Incorporated

With contributions by
Rick Plympton
Bob Weiderhold
Chanda Bartlett Walker
Ranko Galeb
Michael Newell

McGraw-Hill

New York · San Francisco · Washington, D.C. · Auckland · Bogotá
Caracas · Lisbon · London · Madrid · Mexico City · Milan
Montreal · New Delhi · San Juan · Singapore
Sydney · Tokyo · Toronto

Library of Congress Cataloging-in-Publication Data

Fischer, Robert Edward
 Optical systems design / Robert E. Fischer, Biljana Tadic-Galeb.
  p.   cm.
 Includes bibliographical references and index.
 ISBN 0-07-134916-2
  1. Imaging systems—Design and construction.   2. Optical instruments—Design and construction.   3. Electrooptical devices—Design and construction.   I. Tadic-Galeb, Biljana.II. Title.

TK8315 .F57 2000
681'.4—dc21   00-032892

McGraw-Hill

*A Division of The **McGraw·Hill** Companies*

2 3 4 5 6 7 8 9 0   PBT/PBT   0 2 1

ISBN 0-07-134916-2

The sponsoring editor for this book was Steve Chapman, the editing supervisor was Gerry Fahey, and the production supervisor was Pamela Pelton. It was set in Vendome ICG by Deirdre Sheean of McGraw-Hill's Professional Book Group composition unit, Hightstown, N.J.

Printed and bound by Phoenix Book Tech.

McGraw-Hill books are available at special quantity discounts to use as premiums and sales promotions, or for use in corporate training programs. For more information, please write to the Director of Special Sales, Professional Publishing, McGraw-Hill, Two Penn Plaza, New York, NY 10121-2298. Or contact your local bookstore.

 This book is printed on recycled, acid-free paper containing a minimum of 50% recycled, de-inked fiber.

We dedicate this book to our families who have tolerated many lonely evenings and weekends as we were writing, computing, and otherwise getting our material together. We also would like to acknowledge the many colleagues with whom we have both worked with over the years who have helped to make our field of optical design so fascinating, challenging, and rewarding. We also appreciate the efforts of our colleagues at OPTICS 1, especially Michael Newell and Mike Couture, who helped review the manuscript. Also thanks to our dear friend and colleague Max Reidl for his valuable support. Also, the support, encouragement, and friendship of our colleague Warren Smith whose books are classics in their own right. We also want to pay a special thanks to our professors from whom we learned optical design and engineering: R. E. Hopkins, R. Kingslake, W. L. Hyde, H. H. Hopkins, and J. McDonald.

CONTENTS

Contents

Contents

PREFACE

The design of imaging optical systems is an engineering discipline which has been practiced and written about for many years. In many ways, optical design is both a science and an art, and for this reason it is a technology that can cause problems if it is not done properly. Furthermore, most books on the subject tend to be complex and difficult to follow and to understand. With this book, we hope to bring the understanding of our discipline to everyone.

We are all aware of cameras, binoculars, and other optical systems and instruments. With the advent of the new millennium, the field of optics and photonics will see a tremendous surge in applications. This will be fueled by a closer association with electronics in devices such as digital cameras, enhanced machine vision systems, MEMS, and microoptical systems for telecommunications and other related applications, many of which have yet to be invented.

With this surge in the applications of optics, the educational process of training experienced optical designers and engineers becomes extremely important if not critical.

We realize that it is difficult to be an expert in everything. We also realize that optical manufacturing, polarization, and optical coatings are extremely important subjects that need to be covered in this book. We are honored to have guest chapters written by experts in their fields: Rick Plympton and Bob Weiderhold on optical manufacturing (Chapter 17), Chanda Barlett Walker on polarization in optical systems (Chapter 18), and Ranko Galeb on thin films and optical coatings (Chapter 19). We also appreciate all of the hard work by our colleague Michael Newell who read the manuscript in great depth and gave us many constructive suggestions. Also Max Riedl, Mike Couture, Jim Contreras, Dave Kappel, and Steve McClain all helped in the review.

Our goal in writing this book is to teach optical design and engineering in a fully unintimidating way, using clear and easy-to-understand graphics and explanations. Many authors feel an obligation to include complex mathematical derivations. We have taken a very different approach. We will make this book clear and easy to understand, with the goal that you will learn the subject matter with a combination of complete graphics, easy-to-follow explanations, and just enough math to be useful, but not too much math to make the book hard to follow or difficult to understand.

Optical System Design is largely based on the firm foundation of the short course by the same title taught by Bob Fischer to hundreds of students over the past 15 years. The course has been honed, polished, and expanded over the years. It is available on CD-ROM and videotape, and finally, via this book. Typical comments have been:

> *This course was just what I had hoped it would be. It condensed the vast optical world into the key elements necessary for a broad understanding of the subject and an excellent foundation for future study. Good Job!*

> *Excellent presentation! This is an invaluable course for those who are engaged in optical systems efforts and have a minimum training in optics.*

> *A fast paced, well-prepared study, presented by a hands-on-instructor.*

> *I learned what I came to learn, thanks.*

> *Excellent! Wonderful presentation and technique. Material was well covered.*

> *Excellent in explaining and answering questions. Very useful rules of thumb, great presentation. Thanks!*

> *Very professional and excellent presentation.*

This book is for everyone from program managers to seasoned optical designers and engineers, mechanical engineers, electrical engineers, and others. You will find that it is like reading *Gulliver's Travels*. We all read *Gulliver's Travels* in elementary school, some of us again in high school, and some scholars wrote their PhD theses on the book. *Gulliver's Travels* can be read at multiple levels, just like this book.

Robert E. Fischer
Biljana Tadic-Galeb

OPTICAL
SYSTEM
DESIGN

Basic Optics and Optical System Specifications

This chapter will discuss what a lens or mirror system does and how we specify an optical system. You will find that properly and completely specifying a lens system early in the design cycle is an imperative ingredient required to design a good system.

The Purpose of an Imaging Optical System

The purpose of virtually all image-forming optical systems is to resolve a specified minimum-sized object over a desired field of view. The *field of view* is expressed as the spatial or angular extent in object space, and the minimum-sized object is the smallest *resolution element* which is required to identify or otherwise understand the image. The word "spatial" as used here simply refers to the linear extent of the field of view in the plane of the object. The field of view can be expressed as an angle or alternatively as a lateral size at a specified distance. For example, the field of view might be expressed as $10 \times 10°$, or alternatively as 350×350 m at a distance of 2 km, both of which mean the same thing.

A good example of a resolution element is the dot pattern in a dot matrix printer. The capital letter E has three horizontal bars, and hence five vertical resolution elements are required to resolve the letter. Horizontally, we would require three resolution elements. Thus, the minimum number of resolution elements required to resolve capital letters is in the vicinity of five vertical by three horizontal. Figure 1.1 is an example of this. Note that the capital letter B and the number 8 cannot be distinguished in a 3×5 matrix, and the 5×7 matrix of dots will do just fine. This applies to telescopes, microscopes, infrared systems, camera lenses, and any other form of image-forming optics. The generally accepted guideline is that approximately three resolution elements or 1.5 line pairs over the object's spatial extent are required to acquire an object. Approximately eight resolution elements or four line pairs are required to recognize the object and 14 resolution elements or seven line pairs are required to identify the object.

There is an important rule of thumb, which says that this smallest desired resolution element should be matched in size to the minimum detector element or pixel in a pixelated charged-coupled device (CCD) or complementary metal-oxide semiconductor (CMOS)—type sensor. While

Figure 1.1
Illustration of Number of Resolution Elements Required to Resolve or Distinguish Alphanumerics

5 x 7
resolution elements

3 x 5
resolution elements

not rigorous, this is an excellent guideline to follow for an optimum match between the optics and the sensor. This will become especially clear when we learn about the Nyquist Frequency in Chap. 21, where we show a digital camera design example. In addition, the aperture of the system and transmittance of the optics must be sufficient for the desired sensitivity of the sensor or detector. The detector can be the human eye, a CCD chip, or film in your 35-mm camera. If we do not have enough photons to record the imagery, then what good is the imagery?

The preceding parameters relate to the optical system performance. In addition, the design *form* or *configuration* of the optical system must be capable of meeting this required level of performance. For example, most of us will agree that we simply cannot use a single magnifying glass element to perform optical microlithography where submicron line-width imagery is required, or even lenses designed for 35-mm photography for that matter. The form or configuration of the system includes the number of lens or mirror elements along with their relative position and shape within the system. We discuss design configurations in Chap. 8 in detail.

Furthermore, we often encounter special requirements, such as cold stop efficiency, in infrared systems, scanning systems, and others. These will be addressed later in this book.

Finally, the system design must be producible, meet defined packaging and environmental requirements, weight and cost guidelines, and satisfy other system specifications.

How to Specify Your Optical System: Basic Parameters

Consider the lens shown in Fig. 1.2 where light from infinity enters the lens over its *clear aperture diameter.* If we follow the solid ray, we see that

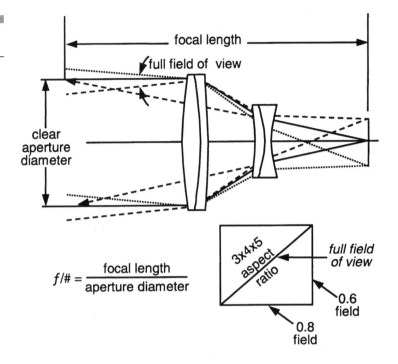

Figure 1.2
Typical Specifications

it is redirected by each of the lens element groups and components until it comes to focus at the image. If we now extend this ray *backwards* from the image towards the front of the system as if it were not bent or refracted by the lens groups, it intersects the entering ray at a distance from the image called the *focal length.* The final imaging cone reaching the image at its center is defined by its *f/number* or *f/#*, where

$$f/\text{number} = \frac{\text{focal length}}{\text{clear aperture diameter}}$$

You may come across two other similar terms, *effective focal length* and *equivalent focal length,* both of which are often abbreviated EFL. The effective focal length is simply the focal length of a lens or a group of lenses. Equivalent focal length is very much the same; it is the overall focal length of a group of lens elements, some or all of which may be separated from one another.

The lens is used over a *full field of view,* which is expressed as an angle, or alternatively as a linear distance on the object plane. It is important to express the total or full field of view rather than a subset of the field of view. This is an extremely critical point to remember. For example, assume we have a CCD camera lens covering a sensor with a $3 \times 4 \times 5$ aspect ratio. We could specify the *horizontal* field of view, which is often done in video technology and cinematography. However, if we do this, we would be ignoring the full diagonal of the field of view. If you do specify a field of view less than the full or total field, you absolutely must indicate this. For example, it is quite appropriate to specify the field of view as $\pm 10°$. This means, of course, that the total or full diagonal field of view is 20°. Above all, do not simply say "field of view 10°" as the designer will be forced to guess what you really mean!

System specifications should include a defined *spectral range* or wavelength band over which the system will be used. A visible system, for example, generally covers the spectral range from approximately 450 nm to 650 nm. It is important to specify from three to five specific wavelengths and their corresponding relative weights or importance factors for each wavelength. If your sensor has little sensitivity, say, in the blue, then the image quality or performance of the optics can be more degraded in the blue without perceptible performance degradation. In effect, the spectral weights represent an importance factor across the wavelength band where the sensor is responsive. If we have a net spectral sensitivity curve, as in Fig. 1.3, we first select five representative wavelengths distributed over the band, $\lambda_1 = 450$ nm through $\lambda_5 = 650$ nm, as shown. The circular data points represent the relative sensitivity at the specific wavelengths, and the relative weights are now the normalized area or integral within each band from band 1 through band 5, respectively. Note that the weights are not the ordinate of the curve at each wavelength as you might first expect but rather the integral within each band. Table 1.1 shows the data for this example.

Even if your spectral band is narrow, you must work with its bandwidth and derive the relative weightings. You may find some cases where you think the spectral characteristics suggest a monochromatic situation but in reality, there is a finite bandwidth. Pressure-broadened spectral lines emitted by high-pressure arc lamps exhibit this characteristic. Designing such a system monochromatically could produce a disastrous result. In most cases, laser-based systems only need to be designed at the specific laser wavelength.

System *packaging constraints* are important to set at the outset of a design effort, if at all possible. These include length, diameter, weight, distance or clearance from the last surface to the image, location and space

Figure 1.3
Example of Spectral
Sensitivity Curve

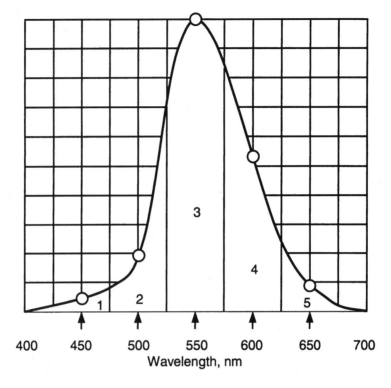

TABLE 1.1

Example of
Spectral Sensitiv
ity and Relative
Wavelength
Weights

Wavelength, nm	Relative sensitivity	Relative weight
450	0.05	0.08
500	0.2	0.33
550	1.0	1.0
600	0.53	0.55
650	0.09	0.16

for fold mirrors, filters, and/or other components critical to the system operation.

Sets of specifications often neglected until it is too late are the *environmental parameters* such as *thermal soak* conditions (temperature range) that the system will encounter. Also, we may have *radial thermal gradients*, which are changes in temperature from the optical axis outward; *diame-*

tral thermal gradients, which are thermal gradients across the diameter of the system in a nonaxially symmetrical profile; and *axial gradients,* which are thermal gradients from the front to the rear of the system. You may also be provided with a set of *operational* specifications and a set of *storage* specifications with respect to temperature.

System *transmittance,* or throughput, as well as *relative illumination,* or brightness, uniformity over the image format are also often specified.

One of the most important specifications is the optical performance or *image quality.* The following list contains some of the more common ways of specifying the image quality, along with simple definitions. Each of these will be discussed in more detail in Chap. 10.

- *Modulation transfer function (MTF).* The modulation (think of the word contrast) versus the number of line pairs per millimeter in the image
- *RMS blur diameter.* The diameter of a circle containing approximately 68 percent of the energy imaged from a point source
- *Encircled energy (or ensquared energy).* The diameter of a circle (or side of a square such as a pixel) containing a given percent of energy, for example, 80 percent
- *Root-mean-square (rms) wavefront error.* The rms departure of the real wavefront from a perfect wavefront
- *Other.* Depending on the functional requirements of the system, there may be other performance requirements relating to image quality, for example, point spread function (PSF), control of specific aberrations, etc.

The most fundamental set of first-order specifications are *focal length, clear aperture diameter* (more properly called *entrance pupil diameter,* as will be explained in Chap. 2), and *f/number.* As we know, the *f*/number is the focal length divided by the clear aperture diameter (the entrance pupil diameter). There is, however, another important and related quantity called the *numerical aperture* (NA), which is often used. The numerical aperture is simply the sine of the half cone angle of the limiting edge ray coming to the axial image, or the sine of the half cone angle coming from the axial object point, as shown in Fig. 1.4. Why would we want or need yet another term to remember? The reason is that the definition of focal length is based on light from infinity entering the system. What if we have a so-called finite conjugate system where neither the

Figure 1.4
Numerical Aperture
and f/#

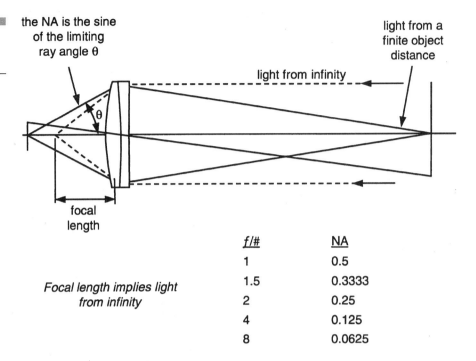

the NA is the sine
of the limiting
ray angle θ

light from a
finite object
distance

light from infinity

θ

focal
length

*Focal length implies light
from infinity*

f/#	NA
1	0.5
1.5	0.3333
2	0.25
4	0.125
8	0.0625

object nor the image is at infinity? The traditional definition of focal length and $f/\#$ would be misleading since the system really is not being used with collimated light input. Numerical aperture is the answer. The *numerical aperture* is simply the sine of the image cone half angle, *regardless of where the object is located.* We can also talk about the numerical aperture at the object, which is the sine of the half cone angle from the optical axis to the limiting marginal ray emanating from the center of the object. Microscope objectives are routinely specified in terms of numerical aperture. Some microscope objectives reimage the object at a finite distance, and some have collimated light exiting the objective. These latter objectives are called *infinity corrected* objectives, and they require a "tube lens" to focus the image into the focal plane of the eyepiece or alternatively onto the CCD or other sensor.

As noted earlier, the definition of focal length implies light from infinity. And similarly, f/number is focal length divided by the clear aperture diameter. Thus, f/number is also based on light from infinity. Two terms commonly encountered in finite conjugate systems are "f/number at used conjugate" and "working f/number." These terms define the equivalent f/number, even though the object is not at infinity. The f/number at used conjugate is $1/(2 \cdot \text{NA})$, and this is valid whether the object is at infinity or at a finite distance.

It is important at the outset of a design project to compile a specification for the desired system and its performance. The following is a candidate list of specifications:

Optical system basic operational and performance specifications and requirements

Basic system parameters:
    Object distance
    Image distance
    Object to image total track
    Focal length
    f/number (or numerical aperture)
    Entrance pupil diameter
    Wavelength band
    Wavelengths and weights for 3 or 5 λs
    Full field of view
    Magnification (if finite conjugate)
    Zoom ratio (if zoom system)
    Image surface size and shape
    Detector type
Optical performance:
    Transmission
    Relative illumination (vignetting)
    Encircled energy
    MTF as a function of line pairs/mm
    Distortion
    Field curvature
Lens system:
    Number of elements
    Glass versus plastic
    Aspheric surfaces
    Diffractive surfaces
    Coatings
Sensor:
    Sensor type
    Full diagonal
    Number of pixels (horizontal)
    Number of pixels (vertical)
    Pixel pitch (horizontal)
    Pixel pitch (vertical)
    Nyquist frequency at sensor, line pairs/mm
Packaging:
    Object to image total track
    Entrance and exit pupil location and size
    Back focal distance
    Maximum diameter

Optical system basic
operational and
performance
specifications and
requirements
(Continued)

Maximum length	_____
Weight	_____
Environmental:	
Thermal soak range to perform over	_____
Thermal soak range to survive over	_____
Vibration	_____
Shock	_____
Other (condensation, humidity, sealing, etc.)	_____
Illumination:	
Source type	_____
Power, in watts	_____
Radiometry issues, source:	
Relative illumination	_____
Illumination method	_____
Veiling glare and ghost images	_____
Radiometry issues, imaging:	
Transmission	_____
Relative illumination	_____
Stray light attenuation	_____
Schedule and cost:	
Number of systems required	_____
Initial delivery date	_____
Target cost goal	_____

Basic Definition of Terms

There is a term called *first-order optics*. In first-order optics the bending or refraction of a lens or lens group happens at a specific plane rather than at each lens surface. In first-order optics, there are no aberrations of any kind and the imagery is perfect, by definition.

Let us first look at the simple case of a perfect thin positive lens often called a *paraxial lens*. The limiting aperture that blocks the rays beyond the lens clear aperture is called the *aperture stop*. The rays coming from an infinitely distant object that passes through the lens clear aperture focus in the image plane. A paraxial positive lens is shown in Fig. 1.5. The rays coming from an infinitely distant point on the optical axis approach the lens as the bundle parallel to the optical axis. The ray that goes along the optical axis passes through the lens without bending. However, as we move away from the axis, rays are bent more and more as we approach the edge of the clear aperture. The ray that goes through the edge of the aperture parallel to the optical axis is called the *marginal ray*. All of the rays parallel to the optical axis focus at a point on the optical axis in the focal plane. The rays that are coming from a nonaxial object point form an angle with the optical axis. One of these rays is called a *chief ray,* and it goes through the center of the lens (center of the *aperture stop*) without bending.

A common first-order representation of an optical system is shown in Fig. 1.6. What we have here is the representation of any optical system, yes, any optical system! It can be a telescope, a microscope, a submarine periscope, or any other imaging optical system.

Figure 1.5
Paraxial Positive Lens

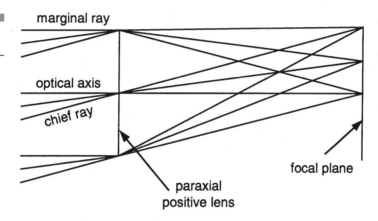

Figure 1.6
Cardinal Points of an
Optical System

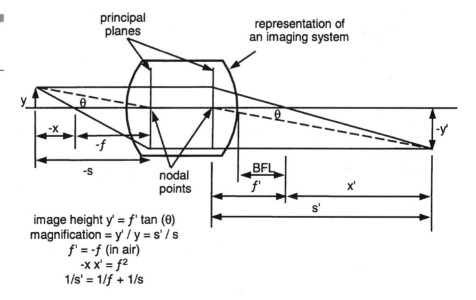

image height $y' = f' \tan(\theta)$
magnification $= y' / y = s' / s$
$f' = -f$ (in air)
$-x\,x' = f^2$
$1/s' = 1/f + 1/s$

The easiest way to imagine what we have here is to think of having a shoebox with a 2-in-diameter hole in each end and inside is some arbitrary optical system (or perhaps nothing at all!). If we send a laser beam into the shoebox through the center of the left-hand hole normal to the hole, it will likely exit through the center of the hole at the other end of the shoebox. The line going through the center of each of the holes is the optical axis.

If we now send the laser beam into the shoebox displaced nearly 1 in vertically, it may exit the shoebox on the other end exactly the same and parallel to how it entered, in which case there is probably nothing in the shoebox. Alternately, the laser beam may exit the shoebox either descending or ascending (going downhill or uphill). If the laser beam is descending, it will cross the optical axis somewhere to the right of the shoebox, as shown in Fig. 1.6. If we connect the entering laser beam with the exiting laser beam, they will intersect at a location called the *second principal plane*. This is sometimes called the *equivalent refracting surface* because this is the location where all of the rays appear to bend about. In a high-peformance lens, this equivalent refracting surface is spherical in shape and is centered at the image. The distance from the second principal plane to the plane where the ray intersects the optical axis is the *focal length*.

If we now send a laser beam into the hole on the right parallel to the optical axis and in a direction from right to left, it will exit either ascending or descending (as previously), and we can once again locate the principal plane, this time the *first principal plane*, and determine the focal length. Interestingly, the focal length of a lens system used in air is identical whether light enters from the left or the right. Figure 1.7a shows a telephoto lens whose focal length is labeled. Recall that we can compute the focal length by extending the marginal ray back from the image until it intersects the incoming ray, and this distance is the focal length. In the telephoto lens the focal length is longer than the physical length of the lens, as shown. Now consider Fig. 1.7b, where we have taken the telephoto lens and simply reversed it with no changes to radii or other lens parameters. Once again, the intersection of the incoming marginal ray with the ray extending forward from the image is the focal length. The construction in Fig. 1.7b shows clearly that the focal lengths are identical with the lens in either orientation!

The center of the principal planes (where the principal planes cross the optical axis) are called the *nodal points*, and for a system used in air, these points lie on the principal planes. These nodal points have the unique property that light directed at the front nodal point will exit

Figure 1.7
The Identical Lens Showing How the Focal Length Is Identical When the Lens Is Reversed

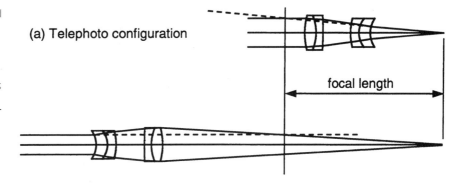

(a) Telephoto configuration

focal length

(b) Inverse telephoto configuration

the lens from the second nodal point at exactly the same angle with respect to the optical axis. This, too, we can demonstrate with our laser beam and shoebox.

So far, we have not talked about an object or an image at all. We can describe or represent a cone of light leaving an object (at the height, Y, in Fig. 1.6.) as including the ray parallel to the optical axis, the ray aimed at the front nodal point, and lastly the ray leaving the object and passing through the focal point on the left side of the lens. All three of these rays (or laser beams) will come together once again to the right of the lens a distance, Y', from the optical axis, as shown. We will not bore you with the derivation, but rest assured that it does happen this way.

What is interesting about this little example is that our shoebox could contain virtually any kind of optical system, and all of the preceding will hold true. In the case where the laser beam entering parallel to the optical axis exits perhaps at a different distance from the axis but parallel to the axis, we then have what is called an *afocal lens* such as a laser beam expander, an astronomical telescope, or perhaps a binocular. An afocal lens has an infinite focal length, meaning that both the object and the image are at infinity.

Useful First-Order Relationships

As discussed earlier, in first-order optics, lenses can be represented by planes where all of the bending or refraction takes place. Aberrations are

Figure 1.8
Newton's Equation

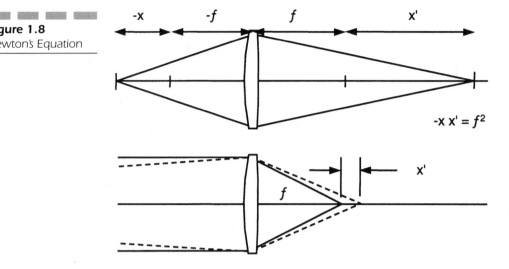

Figure 1.8
Newton's Equation

nonexistent in first-order optics, and the imagery is by definition absolutely perfect. There are a series of first-order relationships or equations, which come in very handy in one's everyday work, and we will discuss the most useful ones here.

Consider the simple lens system shown in Fig. 1.8. Newton's equation says:

$$(-x)(x') = f^2$$

where x is the distance from the focal point on the front side of the lens to the object, and x' is the distance from the rear focal point to the image. Note that x is negative according to the sign convention, since the distance from the image to the object is in a direction to the left. This is an interesting equation in that, at first glance, it seems to be of marginal use. However, consider the example where we need to determine how far to refocus a 50-mm focal length lens for an object at a distance of 25 m. The result is 0.1 mm, and this is, in all likelihood, a very reliable and accurate answer. We must always remember, however, that first-order optics is an approximation and assumes no aberrations whatsoever. For small angles and large f/#s the results are generally reliable; however, as the angles of incidence on surfaces increase, the results become less reliable. Consider Fig. 1.9a where we show how light proceeds through a three-element lens known as a Cooke triplet, with the object at infinity. If we were to use Newton's equation to determine how far to refocus the lens for a relatively close object distance, as shown in Fig. 1.9*b*, the resuting amount of refocusing may not be reliable. This is because the ray heights and angles of incidence are different from the infinite object condition, especially at the outer positive elements, as shown in Fig. 1.9*c*, which is an overlay of the infinite and close object distance layouts. These different ray heights and angles of incidence will cause aberrations, and the net effect is that the result determined by Newton's equation might not be reliable for predicting where best focus is located Consider a typical *f*/5 50-mm focal length Cooke triplet lens used at an object distance of 0.5 m. Newton's equation gives a required refocusing of 2.59 mm from infinity focus, versus 3.02 mm based on optimum image quality, a difference of 0.43 mm. However, for a 10-m object distance, the difference between Newton's equation and best focus reduces to 0.0008 mm, which is negligible.

The important message here is to use first-order optics with caution. If you have any question as to its accuracy in your situation, you really should perform a computer analysis of what you are modeling. If you

Figure 1.9
Light Imaging through a Cooke Triplet for Two Object Distances

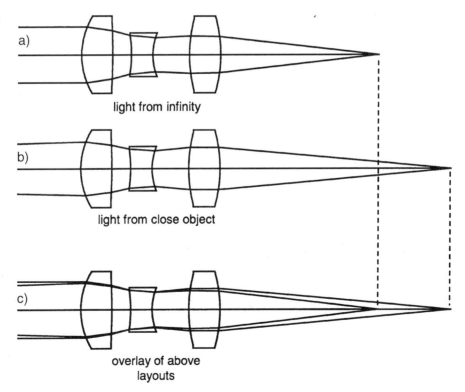

a)

light from infinity

b)

light from close object

c)

overlay of above layouts

then find that your first-order analysis is sufficiently accurate, continue to use it with confidence in similar situations. However, if you find inaccuracies, you may need to work with real rays in your computer model.

Another useful and commonly used equation is

$$\frac{1}{s'} = \frac{1}{f} + \frac{1}{s}$$

where s and s' are the object and image distances, respectively, as shown in Fig. 1.10.

Consider now the basic definitions of magnification from an object to an image. In Fig. 1.11, we show how *lateral magnification* is defined. Lateral implies in the plane of the object or the image, and lateral magnification is therefore the image height, y' divided by the object height, y. It is also the image distance, s', divided by the object distance, s.

There is another form of magnification: the *longitudinal magnification*. This is the magnification *along the optical axis*. This may be a difficult concept to visualize because the image is always in a given plane. Think of longitudinal magnification this way: if we move the object a distance, d, we need to move the image, d', where d'/d is the longitudinal magnification. It can be shown that the longitudinal magnification is the square of the lateral magnification, as shown in Fig. 1.12. Thus, if the lateral magnification is $10\times$, the longitudinal magnification is $100\times$. A good

Figure 1.10
Basic Relationship of
Object and Image:
The "Lens Makers"
Equation

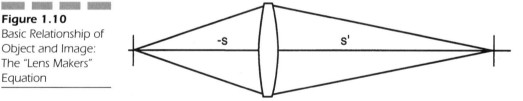

Figure 1.11
Lateral Magnification

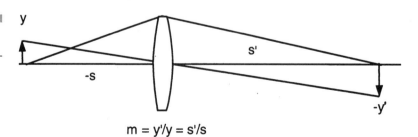

$$m = y'/y = s'/s$$

Figure 1.12
Longitudinal
Magnification

longitudinal magnification = d'/d = (lateral magnification)^2 = (y'/y)^2

example is in the use of an overhead projector where the viewgraph is in the order of 250 mm wide and the screen is in the order of 1 m wide, giving a lateral magnification of 4×. If we were to move the viewgraph 25 mm toward the lens, we would need to move the screen outward by 16 × 25 = 400 mm.

As a further example of the concept, consider Fig. 1.13 where we show a two-mirror reflective system called a *Cassegrain*. Let us assume that the large or *primary mirror* is 250 mm in diameter and is ƒ/1. Also, assume that the final image is ƒ/20. The small, or *secondary,* mirror is, in effect, magnifying the image, which would be formed by the primary mirror by 20× in lateral magnification. Thus, the longitudinal magnification is 400×, which is the square of the lateral magnification. Now let us move the secondary mirror 0.1 mm toward the primary mirror. How far does the final image move? The answer is 0.1 × 400 = 40 mm to the right. This is a very large amount and it illustrates just how potent the longitudinal magnification really can be.

While we are on the subject, how can we easily determine which way the image moves if we move the secondary mirror to the right as discussed previously? Indeed there is an easy way to answer this question (and similar questions). The approach to follow when presented by a question of this kind is to consider moving the component a very large amount, perhaps even to its limit, and ask "what happens?" For example, if we move the secondary mirror to a position approaching the image formed by the primary, clearly the final image will coincide with the secondary mirror surface when it reaches the image formed by the primary. This means that the final image will move in the same direction as the secondary mirror motion. In addition, if you take the secondary and move it a large amount toward the primary, eventually the light will reflect back to the primary when the rays are incident normal to

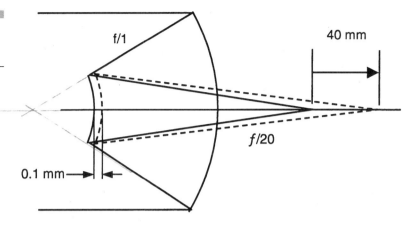

Figure 1.13
Cassegrain Reflective
System

the secondary mirror surface. Moreover, at some intermediate position, the light will exit to the right collimated or parallel. The secret here, and for many other similar questions, is to *take the change to the limit*. Take it to a large enough magnitude so that the direction of the result becomes fully clear and unambiguous.

Figure 1.14 shows how the *optical power* of a single lens element is defined. The optical power is given by the Greek letter, Φ, and Φ is the reciprocal focal length or 1 divided by the focal length. In optics, we use a lot of reciprocal relationships. Power = Φ = 1/(focal length), and curvature = (1/radius) is another.

If we know the radii of the two surfaces, r_1 and r_2, and the refractive index, n, we find that

$$\Phi = \frac{1}{\text{focal length}} = (n-1)\left(\frac{1}{r_1} - \frac{1}{r_2}\right)$$

In addition, if we have two thin lenses separated by an air space of thickness d, we find that

$$\Phi = \frac{1}{\text{focal length}} = \Phi_a + \Phi_b - d(\Phi_a\Phi_b)$$

One very important constant in the optical system is the *optical invariant* or *Lagrange invariant* or *Helmholtz invariant*. It has a constant value throughout the entire system, on all surfaces and in the spaces between them. The optical invariant defines the system throughput. The basic characteristic of an optical system is known when the two main rays are traced through the system: the *marginal ray* going from the center of the object through the edge of the aperture stop, and the *chief* or *principal ray*

Figure 1.14
Optical Power and
Focal Length of a
Single Lens Element

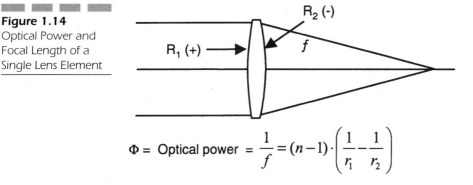

Figure 1.14
Optical Power and
Focal Length of a
Single Lens Element

$$\Phi = \text{Optical power} = \frac{1}{f} = (n-1)\cdot\left(\frac{1}{r_1} - \frac{1}{r_2}\right)$$

going from the edge of the object through the center of the aperture stop. These rays are shown in Fig. 1.15. The optical invariant defines the relationship between the angles and heights of these two rays through the system, and in any location in the optical system it is given as

$$I = y_p\,n\,u - y\,n\,u_p$$

where the subscript p refers to the principal ray, no subscript refers to the marginal ray, and n is the refractive index.

The optical invariant, I, once computed for a given system, remains constant everywhere within the system. When this formula is used to calculate the optical invariant in the object plane and in the image plane where the marginal ray height is zero, then we get the commonly used form of the optical invariant

$$I = hnu = h'n'u'$$

where h, n, and u are the height of the object, the index of refraction, and angle of the marginal ray in the object plane, and h', n', and u' are the corresponding values in the image space. Although this relationship is strictly valid only in the paraxial approximation, it is often used with sufficient accuracy in the form

$$nh \sin u = n'h' \sin u'$$

From this form of optical invariant we can derive the magnification of the system $M = h'/h$ as

$$M = \frac{n \sin u}{n' \sin u'}$$

Figure 1.15
The Optical Invariant

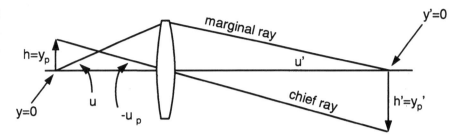

In simple terms these relationships tell us that if the optical system magnifies or increases the object M times, the viewing angle will be decreased M times.

In systems analysis, the specification of the optical invariant has a significant importance. In the radiometry and photometry of an optical system, the total light flux collected from a uniformly radiating object is proportional to I^2 of the system, commonly known as *etendue*, where I is the optical invariant. For example, if the optical system is some kind of a projection system that uses a light source, then the projection system with its optical invariant defines the light throughput. It is useful to compare the optical invariant of the light source with the invariant of the system to see how much light can be coupled into the system. It is not necessarily true that the choice of a higher-power light source results in a brighter image. It can happen that the light-source optical invariant is significantly larger than the system optical invariant, and a lot of light is stopped by the system. The implications of the optical invariant and etendue on radiometry and photometry will be discussed in more depth in Chap. 14.

The *magnification* of a visual optical system is generally defined as the ratio of the angles subtended by the object *with* or looking through the optical system to the angle subtended by the object *without* the optical system or looking at the object directly with unaided vision. In visual optical systems where the human eye is the detector, a nominal viewing distance without the optical system when the magnification is defined as unity is 250 mm. The reason that unity magnification is defined at a distance of 250 mm is that this is the closest distance that most people with good vision can focus. As you get closer to the object, it subtends a larger angle and hence looks bigger or magnified.

This general definition of magnification takes different forms for different types of optical systems. Let us look first at the case of a microscope objective with a CCD camera, as shown in Fig. 1.16. The image from the CCD is brought to the monitor and the observer is located at the distance, D, from the monitor. The question is what is the magnification of this system. In the first step, a microscope objective images the object with the height, y, onto the CCD camera, with the magnification

$$\frac{y'}{y}$$

where y' is the image height at the CCD camera. In the next step, the image from the CCD is brought to the monitor with the magnification

$$\frac{y''}{y'}$$

where y'' is the image height at the monitor. In the third step, the observer watches the monitor from the distance, D, with the magnification

$$\frac{250 \text{ mm}}{D}$$

Overall, the magnification of this system is

$$M = \frac{y'}{y} \frac{y''}{y'} \frac{250}{D}$$

$$M = \frac{y''}{y} \frac{250}{D}$$

The second example is a magnifier or an eyepiece, as shown in Fig. 1.17. The object with height h at a distance of 250 mm is seen to subtend

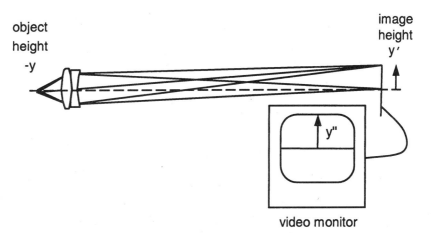

Figure 1.16
Magnification of a Microscope

object height -y

image height y'

y"

video monitor

Figure 1.17
Magnification
of a Magnifier
or Eyepiece

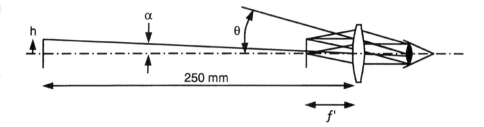

an angle, α. When the same object is located in the first focal plane of the eyepiece, the eye sees the same object at an angle, θ, where

$$\alpha = \frac{h}{250} \qquad \theta = \frac{h}{f}$$

Therefore, magnification M is given by

$$M = \frac{\theta}{\alpha} = \frac{250}{f}$$

The next example is the visual microscope shown in Fig. 1.18. A microscope objective is a short focal length lens, which forms a highly magnified image of the object. A visual microscope includes an eyepiece which has its front focal plane coincident with the objective image plane. The image formed by the objective is seen through the eyepiece, which has its magnification defined as

$$M_e = \frac{250}{f}$$

where f is the focal length of the eyepiece. The magnification of the microscope is the product of the magnification of the objective times the magnification of the eyepiece. Thus

$$M_m = M_o M_e$$

A visual telescope is shown in Fig. 1.19. A distant object is seen at an angle, α, without the telescope and at an angle, θ, with the telescope. The angular magnification of the telescope is

$$M = \frac{\theta}{\alpha}$$

Using the similarity of triangles, it can also be shown that the telescope magnification is

Figure 1.18
Magnification of a
Visual Microscope

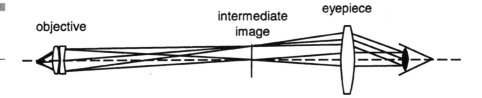

Figure 1.19
Magnification of a
Visual Telescope

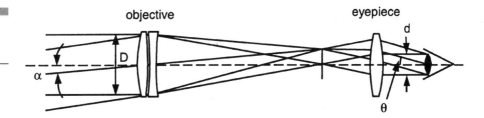

$$M = \frac{f_o}{f_e} = \frac{D}{d}$$

where f_o is the focal length of the objective, f_e is the focal length of the eyepiece, D is the diameter of the entrance pupil, and d is the diameter of the exit pupil.

There are several useful first-order relationships regarding plane parallel plates in an optical system. The first relates to what happens in an optical system when a wedge is added to a plane parallel plate. If a ray, as shown in Fig. 1.20, goes through the wedged piece of material of index of refraction n and a small wedge angle α, the ray deviates from its direction of incidence by the angle, θ, according to

$$\theta \approx (n - 1)\alpha$$

The angle of deviation depends on the wavelength of light, since the index of refraction is dependent on the wavelength. It is important to understand how the wedge can affect the performance of the optical system. When a parallel beam of white light goes through the wedge, the light is dispersed into a rainbow of colors, but the rays of the individual wavelengths remain parallel. Therefore, the formula that gives the angle of deviation through the wedge is used to quickly determine the allowable wedge in protective windows in front of the optical system. However, if the wedge is placed into a converging beam, not only will

Figure 1.20
Light Deviation
through Wedged
Material

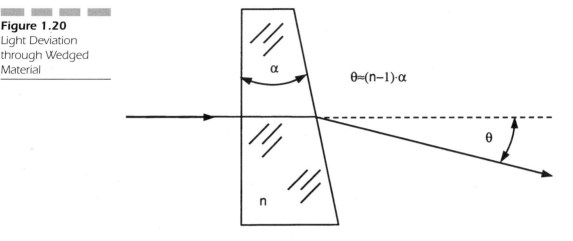

$$\theta \approx (n-1) \cdot \alpha$$

Figure 1.21
Focus Shift
Introduced by a
Plane Parallel Plate

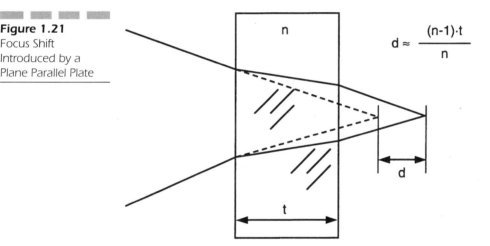

$$d \approx \frac{(n-1) \cdot t}{n}$$

the different colors be focused at different distances from the optical axis, but also the individual colors will be blurred. There is a term called *boresight error,* which means the difference between where you think the optical system is looking and where it really is looking. A wedged window with a wedge angle, α, will cause a system to have a boresight error of angle θ.

A plane parallel plate in a converging beam moves the image plane further along the optical axis, as shown in Fig. 1.21. If the thickness of the plate is t, the image displacement, d, along the optical axis is

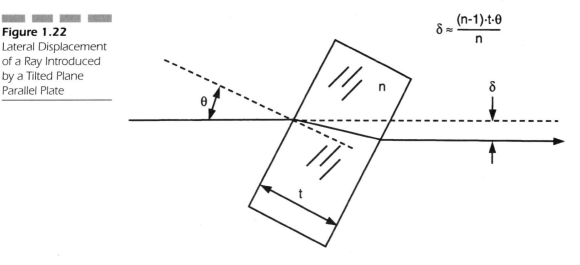

Figure 1.22
Lateral Displacement
of a Ray Introduced
by a Tilted Plane
Parallel Plate

$$\delta \approx \frac{(n-1)\cdot t\cdot\theta}{n}$$

$$d = (n - 1)\frac{t}{n}$$

When a plane parallel plate is tilted in the optical system, as in Fig. 1.22, then the ray incident at an angle, θ, is displaced laterally by the amount, δ, given by

$$\delta = (n - 1)\,\frac{t\theta}{n}$$

Note that if we look through a telescope at an infinitely distant object and we put a tilted plane parallel plate in front of the telescope, there will be no change in the image. One would think that color fringes would be seen because the different wavelengths are displaced differently. However, because the parallel bundle of rays going through the tilted plate is only laterally displaced, it remains parallel to itself after transmission through the plate, and therefore there is no color fringing. If there is a wedge in the plate, however, chromatic dispersion will, of course, cause the appearance of color fringing.

2

Stops and Pupils
and
Other Basic
Principles

The Role of the Aperture Stop

In an optical system, there are apertures which are usually circular that limit the bundles of rays which go through the optical system. In Fig. 2.1 a classical three-element form of lens known as a Cooke triplet is shown as an example. Take the time to compare the exaggerated layout (Figs. 2.1*a* and *b*) with an actual computer optimized design (Fig. 2.1*c*). From each point in the object only a given group or bundle of rays will go through the optical system. The *chief ray,* or *principal ray,* is the central ray in this bundle of rays. The *aperture stop* is the surface in the system where all of the chief rays from different points in the object cross the optical axis and appear to pivot about. There is usually an iris or a fixed mechanical diaphragm or aperture in the lens at this location. If your lens has an adjustable iris at the stop, its primary purpose is to change the brightness of the image. The chief ray is, for the most part, a mathematical convenience; however, there definitely is a degree of symmetry that makes its use valuable. We generally refer to the specific height of the chief ray on the image as the image height.

Entrance and Exit Pupils

The *entrance pupil* is the image of the aperture stop when viewed from the front of the lens, as shown in Fig. 2.1. Indeed, if you take any telescope, such as a binocular, and illuminate it from the back and look into the optics from the front, you will see a bright disk which is formed, in most cases, at the objective lens at the front of the binocular. In the opposite case, if you illuminate the system from the front, there will be a bright disk formed behind the eyepiece. The image of the aperture stop in the image space is called the *exit pupil.* If you were to write your initial with a grease pencil on the front of the objective lens and locate a business card at the exit pupil, you would see a clear image of the initial on the card.

There is another way to describe entrance and exit pupils. If the chief ray entering the lens is extended without bending or refracting by the lens elements, it will cross the optical axis at the entrance pupil. This is shown in Figs. 2.1*a* and *b* where only the chief ray and the pupil locations are shown for clarity. Clearly, it is the image of the aperture stop, since the chief ray crosses the optical axis at the aperture stop. In a

Figure 2.1
Aperture Stop and
Pupils in an Optical
System

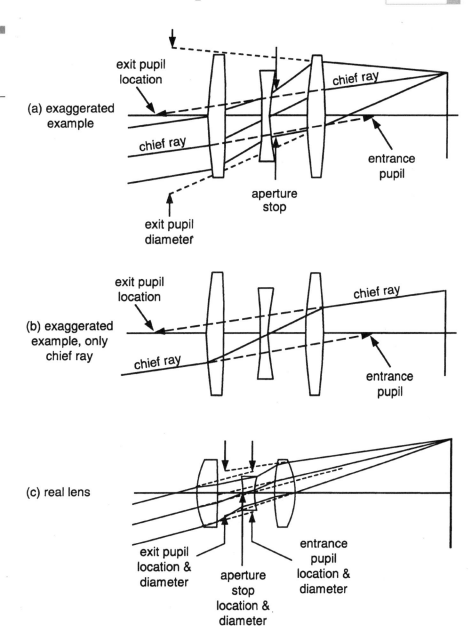

(a) exaggerated
    example

exit pupil
location

chief ray

chief ray

entrance
pupil

aperture
stop

exit pupil
diameter

(b) exaggerated
    example, only
    chief ray

exit pupil
location

chief ray

chief ray

entrance
pupil

(c) real lens

exit pupil
location &
diameter

aperture
stop
location &
diameter

entrance
pupil
location &
diameter

similar way, the exit pupil will be at the location where the chief ray appears to have crossed the optical axis. The location of the exit pupil can be obtained if the chief ray that exits the optical system is extended backwards until it crosses the optical axis. Both definitions are synonymous, and it will be valuable to become familiar with each.

Let us assume that we have an optical system with a lot of optical components or elements, each of them having a known clear aperture diameter. There are also a few mechanical diaphragms in the system. The question is, which of all these apertures is the aperture stop? In order to answer this question, we have to image each aperture into object space. The aperture whose image is seen from the object plane at the smallest angle is the aperture stop. It is the limiting aperture within the lens.

There are many systems such as stand-alone camera lenses, where the location of the entrance and exit pupils are generally not important. The exit pupil location of a camera lens will, of course, dictate the angle of incidence of the off-axis light onto the sensor. However, the specific pupil locations are generally not functionally critical. When multiple groups of lenses are used together, then the pupil locations become very important since the exit pupil of one group must match the entrance pupil location of the following group. This will be discussed later in this chapter.

Vignetting

The position of the aperture stop and the entrance and exit pupils is very important in optical systems. Two main reasons will be mentioned here. The first reason is that the correction of aberrations and image quality very much depends on the position of the pupils. This will be discussed in detail later in the book. The second reason is that the amount of light or throughput through the optical system is defined by the pupils and the size of all elements in the optical system. If ray bundles from all points in the field of view fill the aperture stop entirely and are not truncated or clipped by apertures fore or aft of the stop, then there is no *vignetting* in the system.

For a typical lens, light enters the lens on axis (the center of the field of view) through an aperture of diameter D in Fig. 2.2, and focuses down to the center of the field of view. As we go off axis to the maximum field of view, we are now entering the lens at an angle. In order to allow the rays from the entire diameter, D, to proceed through the lens, in which case the aperture stop will be filled with the ray bundle from the edge of the field, the rays at the edge of the pupil have to go through points *A* and *B*. At these positions, A and B, the rays undergo severe bending which means that they contribute significantly to the image aberrations of the system, as will be discussed in Chaps. 3 and 5. At the same time, mounting of the lenses with larger diameters is more expensive. Further, the lens will be heavier and thicker. So why don't we truncate the aperture in the plane of Fig. 2.2 to 0.7 *D*? We will lose approximately 30% of the energy at the edge of the field of view compared to the center of the field; however, the positive elements in our Cooke triplet example will be smaller in diameter, which means that they can also be thinner and the housing can be smaller and lighter in weight. Telescopes, projectors, and other visual optical systems can have vignetting of about 30 to 40%, and the eye can generally "tolerate" this amount of vignetting. When we say that the eye can tolerate 30 to 40% vignetting, what we mean is that a slowly varying brightness over an image of this magnitude is generally not noticed. A good example is in overhead viewgraph and slide projectors where this amount of brightness falloff is common, yet unless it is pointed out, most people simply will not notice. If the film in a 35-mm camera has a large dynamic range, then this magnitude of vignetting is also acceptable in photography.

In Fig. 2.3 a triplet lens example is shown first in its original form without vignetting (Fig. 2.3*a*). In the next step, the elements are sized for 40% vignetting, but with the rays traced as if there is no vignetting (Fig. 2.3*b*). In the last step, the lens is shown with the vignetted bundle of rays at the edge of the field (Fig. 2.3*c*).

Although vignetting is acceptable and often desirable in visible optical systems, it can be devastating in thermal infrared optical systems because of image anomalies, as will be discussed in Chap. 12. One must also be very careful when specifying vignetting in laser systems, as will be discussed in Chap. 11.

Figure 2.2
Vignetting

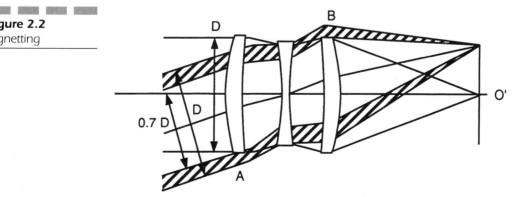

Figure 2.3
Example of
Vignetting

(a) Basic design *f*/5 ±20
degree field, no
vignetting

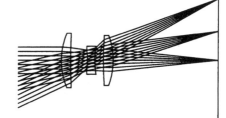

(b) Elements sized for
40% vignetting

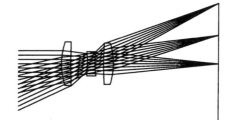

(c) Final design with 40%
vignetting

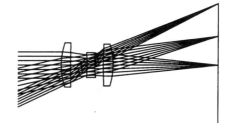

Figure 2.4
Matching of Pupils

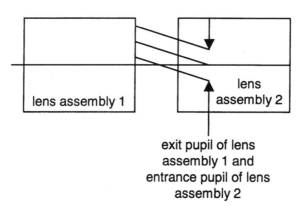

lens assembly 1

lens
assembly 2

exit pupil of lens
assembly 1 and
entrance pupil of lens
assembly 2

When a system is designed using off-the-shelf components with a combination of two or more modules or lens assemblies, it is very important to know the positions of the entrance and exit pupils of these modules. The exit pupil of the first module *must* coincide with the entrance pupil of the second module, etc. This is shown in Fig. 2.4.

There can be a very serious pupil-matching problem when using off-the-shelf (or even custom) zoom lenses as modules in optical systems. Zoom lenses have a given size and position of their pupils which change as a function of zoom position or focal length. It is very easy to make a mistake when the exit pupil of the first module is matched to the entrance pupil of the second module for only one zoom position. When the pupils move with respect to one another through zoom and do not image from one to another, we can lose the entire image.

Diffraction,
Aberrations,
and
Image Quality

What Image Quality Is All About

Image quality is never perfect! While it would be very nice if the image of a point object could be formed as a perfect point image, in reality we find that image quality is degraded by either geometrical aberrations and/or diffraction. Figure 3.1 illustrates the situation. The top part of the figure shows a hypothetical lens where you can see that all of the rays do not come to a common focus along the optical axis. Rather, the rays entering the lens at its outer periphery cross the optical axis progressively closer to the lens than those rays entering the lens closer to the optical axis. This is one of the most common and fundamental aberrations, and it is known as *spherical aberration.* Geometrical aberrations are due to the failure of the lens or optical system to form a perfect geometrical image. These aberrations are fully predictable to many decimal places using standard well-known ray trace equations.

If there were no geometrical aberrations of any kind, the image of a point source from infinity is called an *Airy disk.* The profile of the Airy disk looks like a small gaussian intensity function surrounded by low-intensity rings of energy, as shown in Fig. 3.1, exaggerated.

Figure 3.1
Image Quality, Geo-
metrical Aberrations
(Top) and Diffraction
Limited (Bottom)

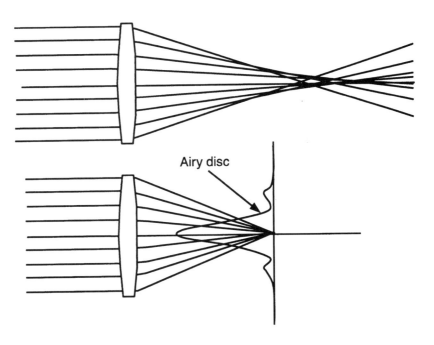

Airy disc

If we have a lens system in which the geometrical aberrations are significantly larger than the theoretical diffraction pattern or blur, then we will see an image dominated by the effect of these geometrical aberrations. If, on the other hand, the geometrical aberrations are much smaller than the diffraction pattern or blur, then we will see an image dominated by the effect of the Airy disk. If we have a situation where the blur diameter from the geometrical aberration is approximately the same size as the theoretical diffraction blur, we will see a somewhat degraded diffraction pattern or Airy disk. Figure 3.1, while exaggerated, does show a situation where the resulting image would, in fact, be a somewhat degraded Airy disk.

What Are Geometrical Aberrations and Where Do They Come From?

In the previous section, we have shown the distinction between geometrical aberrations and diffraction. The bottom line is that imagery formed by lenses with spherical surfaces simply is not perfect! We use spherical surfaces primarily because of their ease of manufacture. In Fig. 3.2, we show how a large number of elements can be ground and polished using a common or single tool. The elements are typically mounted to what is called a *block*. Clearly, the smaller the elements and the shallower the radius, the more elements can be mounted on a given block. The upper tool is typically a spherical steel tool. The grinding and polishing operation consists of a rotation about the vertical axis of the blocked elements along with a swinging motion of the tool from left to right, as indicated by the arrows. The nature of a sphere is *that the rate of change of slope is constant everywhere on a sphere*, and because of this mathematical definition, the tool and lens surfaces will only be in perfect contact with one another over the full range of motions involved when both are perfectly spherical. Due to asymmetries in the process, the entire surface areas of the elements and tool are not in contact the same period of time. Hence this process is not perfect. However, the lens surfaces are driven to a near-spherical shape in reasonable time by a skilled optician. This is the reason we use spherical surfaces for most lenses. Chapter 17 discusses optical component manufacturing in more detail.

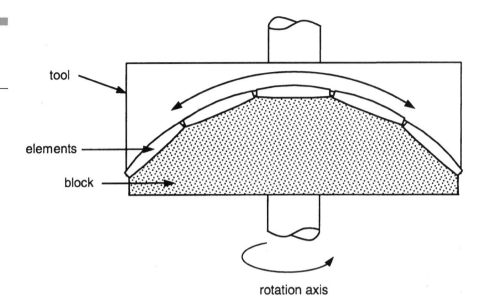

We will discuss the use of nonspherical or aspheric surfaces in Chap. 8.

Earlier we said that geometrical aberrations are due entirely to the failure of the lens or optical system to form a perfect geometrical image. Maxwell formulated three conditions that have to be met for the lens to form a perfect geometrical image:

1. All the rays from object point O after passing through the lens, must go through the image point O'.

2. Every segment of the object plane normal to the optical axis that contains the point O must be imaged as a segment of a plane normal to the optical axis, which contains O'.

3. The image height, h', must be a constant multiple of the object height, h, no matter where O is located in the object plane.

Violation of the first condition results in the image degradation, or image aberrations. Violation of the second condition results in the presence of image curvature, and violation of the third condition in image distortion. A different way to express the first condition is that all the rays from the object point, O, must have the same *optical path length* (OPL) to the image point, O'.

$$\text{OPL} = \int_{O(x,y)}^{O'(x',y')} n(s)\,ds$$

where $n(s)$ is the index of refraction at each point along the ray path, s.

The lenses that meet the first Maxwell condition are called stigmatic. Perfect stigmatic lenses are generally stigmatic only for one pair of conjugate on-axis points. If the lens shown in Fig. 3.3 is to be stigmatic not only for the points, O and O', but also for the points, P and P', it must satisfy the *Herschel condition*

$$n \, dz \sin^2 \frac{u}{2} = n' \, dz' \sin^2 \frac{u'}{2}$$

If the same lens is to be stigmatic at the off-axis conjugate points, Q and Q', it must satisfy the *Abbe sine condition*

$$n \, dy \sin u = n' \, dy' \sin u'$$

Generally, these two conditions cannot be met exactly and simultaneously. However, if the angles u and u' are sufficiently small, and we can substitute the sine of the angle with the angle itself

$$\sin u \approx u \quad \text{and} \quad \sin u' \approx u'$$

then both the Herschel and Abbe sine condition are satisfied. We say that the lens works in the *paraxial region,* and it behaves like a perfect stigmatic lens. The other common definition of paraxial optics is that paraxial rays are rays "infinitely close to the optical axis." This is a fine and correct definition; however, it can become difficult to understand when we consider tracing a paraxial ray through the edge of a lens system, a long way from the optical axis. This creates a dilemma since rays traced through the edge of the system are hardly infinitely close to the optical axis! This is why the first definition of paraxial optics, i.e., using the small-angle approximation to the ray tracing equations, as would be the case for rays infinitely close to the optical axis, is easier to understand. Consider Fig. 3.4a, where we show how the rays are

Figure 3.3
Definition of Paraxial
Lens

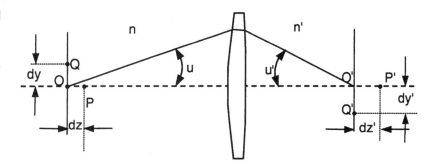

refracted at the interface between two optical media, according to Snell's law.

$$n \sin \theta = n' \sin \theta'$$

In Table 3.1 we show just how a real ray, according to Snell's law, and a paraxial ray, using the small-angle approximation $\sin \theta \sim \theta$, refract or bend after refraction from a spherical surface at a glass-air interface (index of refraction of glass $n = 1.5$). These data use the nomenclature of Fig. 3.4b. Note that the difference in angle between the paraxial and the real rays define the resulting image blur. For angles of incidence θ of 10° or less we see that the real refracted ray is descending within 0.1° of the paraxial ray (0.0981° difference at a 10° angle of incidence). However, as the angles of incidence increase, the difference between the real and the paraxial descending angles increases quite significantly. This is where aberrations come from.

Along with this understanding, it is evident that in order to keep aberrations small, it is desirable if not mandatory to keep the angles of

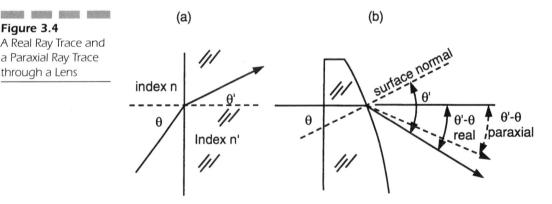

Figure 3.4

A Real Ray Trace and a Paraxial Ray Trace through a Lens

TABLE 3.1

Paraxial Approximation Versus Real Ray Angles of Refraction

θ	$\theta'-\theta$ real (degrees)	$\theta'-\theta$ paraxial (degrees)	Difference (degrees)
1	0.5001	0.5	0
10	5.0981	5	0.0981
20	10.8659	10	0.8659
30	18.5904	15	3.5904
40	34.6186	20	14.6186

incidence as small as possible on the various surfaces within your system.

What Is Diffraction?

Diffraction is a phenomena or effect resulting from the interaction of light (which of course is electromagnetic radiation) with the sharp limiting edge or aperture of an optical system. While we could very easily fill the next few pages with integral signs and Bessel functions, it is not the intention of the authors to provide this level of detail. Rather, the following explanation is easy to follow and should provide a sufficient level of understanding of the causes of diffraction and resultant observable effects.

Imagine a swimming pool at 3 o'clock in the morning with no wind present; the water is like a sheet of glass. Imagine throwing a large rock into one end of the pool. Water waves will emanate outward from where the rock has entered the water as concentric, ever-expanding circles. Before proceeding onward, note that the physics of water waves is virtually identical to the physics of electromagnetic radiation, and all of the derivations are quite analogous.

Now let us proceed to the other end of the swimming pool. If the pool is large enough, the water waves will be nearly straight and parallel to one another. In reality, of course, they are going to be curved and centered about the point where the rock entered the water. For this discussion, consider the water waves as straight. Let us now immerse a 1- by 2-m sheet of plywood partially into the pool, as shown at the top of Fig. 3.5. What you will see to a reasonable extent above the edge of the board at the top of Fig. 3.5 is that the water waves will continue to propagate left to right undisturbed. Below the edge where the major part of the board is located, you will see to the right of the board virtually no disturbance in the water. To the right of the intersection of the upper edge of the board and the water waves, you will see little curlicues traveling or emanating outward from the edge of the board. These curlicues are, in reality, *diffraction* of water waves. We sincerely hope that none of our readers would think there would be a sharp step in the water to the right of the edge...if you did think this was the case, we urge you to try this little experiment in your backyard swimming pool.

The peaks of the water waves are called *wavefronts*. Perpendicular to the wavefronts are the *rays*. While we rarely if ever talk about "water

Figure 3.5
Diffraction Effects

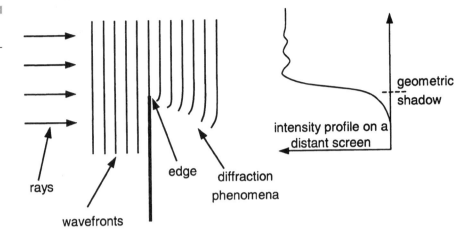

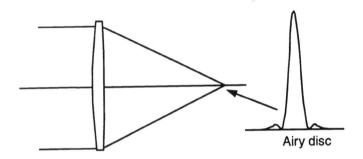

rays," we certainly do talk all the time about light rays. The important point here is that the rays are perpendicular or normal to the wavefronts and the wavefronts are perpendicular to the rays. Throughout your reading of this book, we would like you to understand the difference between ray optics and wave optics.

Now, back to our little example. If instead of our swimming pool example we were to have parallel or collimated light incident upon the edge of a razor blade, we would have diffraction of the electromagnetic radiation much in the same way as we showed diffraction of water waves. On a distant screen or card you will not see a very sharp edge or step function, but rather an intensity gradient with slight variations in intensity occurring in a similar fashion to the curlicues of the water waves.

The previous explanation represents our attempt to illustrate how diffraction occurs without the lengthy and messy mathematical derivations. If we now have a typical lens system, it is easy to understand that there will, by definition, be a limiting edge or aperture at which the light effectively stops or is blocked. This edge, which in many cases is the aperture stop of your system, wraps around the optical axis generally in a symmetrical, circular fashion, and the resulting diffraction pattern acquires a rotationally symmetric shape known as the "Airy disk."

It is important to note that diffraction occurs perpendicular to an edge. Since a circular aperture, in effect, wraps around in a full 360°, the resulting diffraction pattern (the Airy disk) is a rotationally symmetrical blur. However, if your aperture were a triangular shape as shown in Fig. 3.6a, the resulting diffraction pattern would be star shaped with three spikes as shown in Fig. 3.6c. The reason there are three notable spikes is that the diffraction spreading has occurred perpendicular to the three straight edges of the aperture, as shown in Fig. 3.6 b. Note that the relative length of the spikes is proportional to the length of the edge.

Diffraction-Limited Performance

As discussed previously, if the geometrical aberrations are significantly smaller than the diffraction blur, the image is, in effect, well represented by the Airy disk. This form of optics is called *diffraction-limited optics*. Understanding the limits of diffraction-limited optics becomes extremely important, especially with today's extremely demanding levels of performance. Figure 3.7 shows two very important principles: (1) the physical

Figure 3.6
Diffraction from a Triangular Aperture

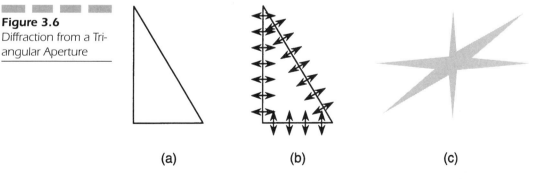

(a) (b) (c)

diameter of the Airy disk and (2) the angular diameter or subtense of the Airy disk. It can be shown that:

$$\text{Physical diameter of the Airy disk} = 2.44\,\lambda\ f/\text{number}$$

Shown in the top part of Fig. 3.7 are three different lenses, all of diameter D. One lens focuses images fairly close to the lens, the second has a somewhat longer focal length, and the third a still longer focal length. For these three lenses, all of which have the same entrance pupil diameter D, the $f/\#$ increases in proportion to the increase in focal length. From the equation, we see that the Airy disk increases in diameter in direct proportion to the $f/\#$ increase and thus in proportion to the focal length as well. A very useful rule of thumb to remember is: *The Airy disk diameter in the visible part of the spectrum is approximately equal to the $f/\#$ expressed in microns.*

Figure 3.7
A Clarifying Illustration of "Diffraction-Limited" Imagery

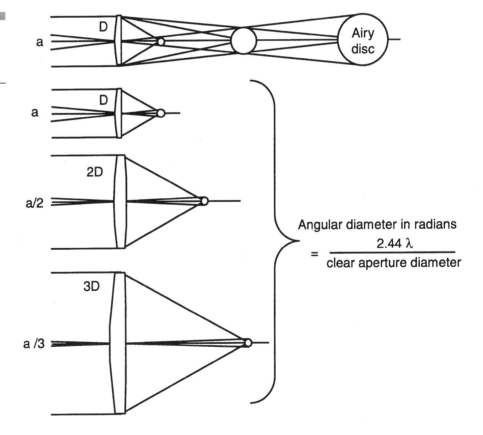

This is easy to see if you consider a wavelength of 0.5 µm, which would be approximately the center of the visible spectrum. In this case, the physical diameter of the Airy disk

$$D = 2.44 \times 0.5 \, f/\# = 1.22 \, f/\#$$

which is approximately equal to the f/number itself expressed in microns!

We now show three separate lenses of diameter D, $2D$, and $3D$, all of identically the same $f/\#$. What this means is that the Airy disk or diffraction blur will be identical in all three cases. You can see quite clearly that for each lens the focal length increases in proportion to the increase in diameter (since the $f/\#$ is identical). What this means is that the angular subtense of the Airy disk also decreases in proportion to the diameter or the increase in focal length. The resulting relationship becomes

$$\text{Angular diameter of the Airy disk} = \frac{2.44 \, \lambda}{\text{clear aperture diameter}}$$

The angular diameter is expressed in radians if the wavelength and the clear aperture diameter are in the same units.

Note that in all of the preceding discussion, the diameter of the Airy disk is assumed to be the diameter of the first dark ring in the diffraction pattern.

Derivation of System Specifications

There is a broad term "systems analysis" which generally refers to the task of deriving the basic optical system parameters based on the functional system performance requirements. We can apply what we learned earlier to perform a simple, yet noteworthy systems analysis example.

Consider, for example, an optical system used in the long-wave infrared (LWIR) which operates in the 8- to 12-µm spectral band. Our task is to derive the system $f/\#$ and clear aperture diameter. Let us assume that the detector is mercury cadmium telluride or (HgCdTe) with a 50-µm pixel pitch and further assume that we need to resolve 0.25 mrad in object space. These values are typical for an LWIR system such as a forward looking infrared (FLIR).

Earlier in this chapter, we discussed that as a rule of thumb the smallest resolvable image blur should be matched to the pixel size of the detector (sensor), i.e., smallest element size. Thus, we would require that the diffraction blur or Airy disk should be approximately the same diameter as our 50-μm pixel. Recall that the diameter of the Airy disk, $D = 2.44 \lambda f/\#$, and we can solve for the $f/\#$ to produce a 50-μm Airy disk diameter, and the result is $f/2.2$ at $\lambda = 10$ μm. Before we continue, it is interesting to note that for diffraction-limited optics, an $f/2.2$ system that is 6 mm in diameter will have exactly the same diffraction blur diameter as an $f/2.2$ system that is 3 m in diameter, and that is 50 μm!

For a given diffraction blur diameter, as the focal length increases, the angular subtense of the Airy disk decreases proportionally. We can use the relationship that the angular diameter of the Airy disk = 2.44 λ/(clear aperture diameter) to solve for the clear aperture diameter required so that the 50-μm Airy disk subtends 0.25 mrad in object space, and the result is a 100-mm-diameter clear aperture.

Figure 3.8 shows parametrically how the $f/\#$ and clear aperture diameters relate to the diffraction-limited image blur or Airy disk diameter and the angular subtense of the diffraction blur. This illustrates how we can quickly and easily take the most basic system functional requirements and derive the system $f/\#$ and clear aperture diameters. Do keep in mind that this assumes diffraction-limited optics. Further, it is based on the criteria that the Airy disk is matched to the pixel pitch. These are generally good assumptions to work with, and as your system needs become better understood, you may need to revise the results.

As we begin to learn more about image formation, it is important to understand just how light bends or refracts when passing through an air-glass or glass-air interface. As shown in Fig. 3.9, the *refractive index of a material* = n, where n is the ratio of the velocity of light in a vacuum to the velocity of light in the denser material such as glass. Since the light or electromagnetic radiation slows down in the denser material, the refractive index is always greater than unity. For optical glass the refractive index ranges from about $n = 1.5$ to $n = 1.85$.

According to the Snell's law, $n \sin \theta = n' \sin \theta'$. With air on the input side of the interface, the equation reduces to $\sin \theta = n' \sin \theta'$. For small angles the equation further reduces to $\theta = n'\theta'$. We will use this result later!

We discussed earlier how light could be represented by either rays or wavefronts, where the two are orthogonal to one another. We will be using both representations throughout this book and we hope that you will become "bilingual" or fluent with both representations. To help understand these concepts, we show in Fig. 3.9 how a light ray, as well as a series of wavefronts, is incident on an air-glass interface and how the

Figure 3.8
Example of Systems
Analysis

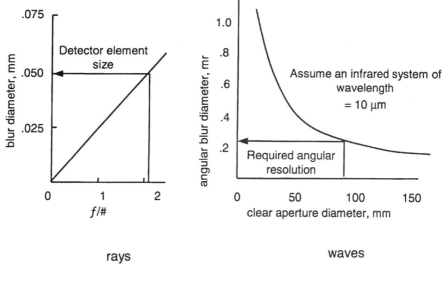

Figure 3.9
Bending of Light at
an Optical Surface

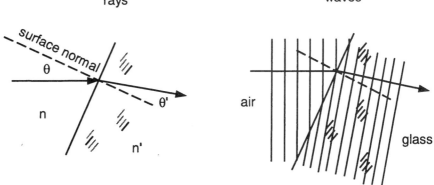

light bends or refracts. From ray optics, we can simply use Snell's law to determine the angle of refraction. Consider how we represent the same thing using wave optics. The wavefronts are traveling from left to right, with their peaks separated by the wavelength of light. As the wavefront enters the denser medium such as glass, its velocity is reduced by $1/n$, with the result being that the wavefronts are closer together. There is a fundamental law of physics, which says that the wavefronts must be continuous at the interface between the media. Considering the velocity reduction along with the wavefront continuity requirement, we can see how the entire wavefront is rotated around in a clockwise direction as it proceeds into the denser medium. Interestingly, you can use this construction to rederive Snell's law!

The Concept of Optical Path Difference

Optical Path Difference (OPD) and The Rayleigh Criteria

OPD is an extremely useful measure of the performance of an imaging optical system. If the wavefronts proceeding to a given point image are spherical, concentric, and centered at the point image for a given field of view, then the imagery will be geometrically perfect, or *diffraction limited*. As shown earlier, the image will then be a perfect Airy disk. This is, in effect, the reverse of our earlier example where we threw a rock into a pool of water to illustrate the wave nature of light and diffraction. If we think of the water waves traveling in reverse to where the rock entered the water, we will emulate light imaging to a point image. By definition, the wavefronts will be perfectly spherical, concentric, and centered where the rock entered the water. Recall also that rays are perpendicular to the wavefronts. It is thus clear that if the wavefronts are spherical, concentric, and centered at a point in the image, then the rays will all come to that same point as defined by the center of curvature of the wavefronts. As we learned earlier, diffraction at the limiting edge of the pupil will create an Airy disk, which is the reason why we do not have a perfect point image.

Consider Fig. 4.1 where we show a hypothetical lens with a perfectly spherical reference wavefront and a real wavefront. The real wavefront departs from sphericity due to aberrations induced by the lens. The *optical path difference* is the difference between the real wavefront and a spherical reference wavefront, which is usually selected to be a near best fit to the aberrated wavefront.

One of the reasons the OPD is so valuable a parameter is evident from the Rayleigh criteria. Lord Rayleigh (real name William Strutt, a Nobel Prize winner for discovering the gas argon) showed that

> An optical instrument would not fall seriously short of the performance possible with an absolutely perfect system if the distance between the longest and shortest paths leading to a selected focus did not exceed one-quarter of a wavelength.

What the Rayleigh criteria say is that if the OPD is less than or equal to one-quarter of a wave (one-quarter of the wavelength of the light), then the performance will be almost indistinguishable from perfect. If this is the case, then the imagery of a point object will be very nearly a

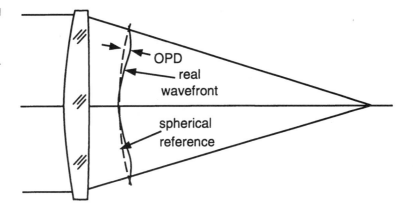

Figure 4.1
Optical Path Difference (OPD)

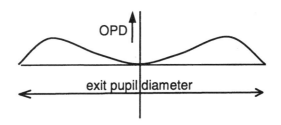

perfect Airy disk. This is a *very* useful tool, and as will become evident, its validity is quite broadly applicable. It is important to note, however, that it is not 100% infallible, and should only be used as a guide or relative measure of the level of optical performance.

Figure 4.2 shows the appearance of the image of a point source, which is known as a *point-spread function* (PSF), for optical path differences of 0 waves (a perfect Airy disk), 0.25 wave, 0.5 wave, and 1.0 wave. Note that the 0.25 wave imagery is qualitatively nearly indistinguishable from the perfect Airy disk. The character of the central maximum is maintained, and the first bright ring is fully intact. As soon as we go to 0.5 wave and above, the imagery is clearly degraded from perfect. Figure 4.3 shows perspective views of the point spread function for the same values of the OPD; however, these data are for spherical aberration rather than coma. Here, too, we can conclude that the 0.25 wave imagery is nearly indistinguishable from perfect. We do see a drop in peak intensity; however, the overall character of the pattern is very similar to the perfect Airy disk.

Figure 4.2
Image of a Point
Source with Different
Amounts of Peak-to-
Valley Optical Path
Difference Due to
Coma

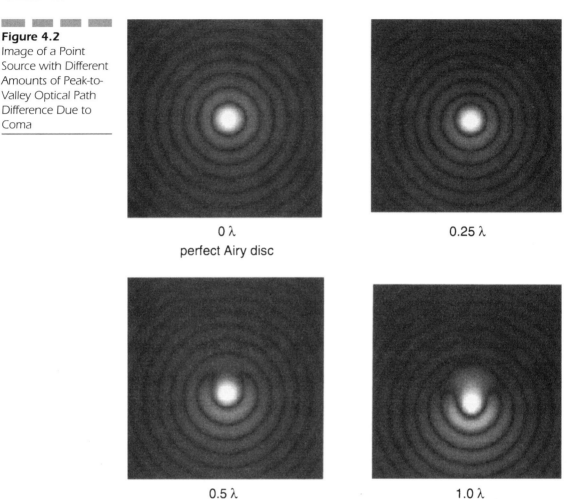

0 λ
perfect Airy disc

0.25 λ

0.5 λ

1.0 λ

Peak-to-Valley and RMS Wavefront Error

The OPD as shown here is known as *peak-to-valley* (P-V) optical path difference. Peak-to-valley is the total difference between the portion of the wavefront closest to the image (leading, or ahead of the reference wavefront) and the farthest lagging portion of the wavefront (lagging, or behind the reference wavefront). Figure 4.4 shows this as the separation between the two dashed reference spheres.

Figure 4.3
Image of a Point
Source with Different
Amounts of Peak-to-
Valley Optical Path
Difference Due to
Spherical Aberration

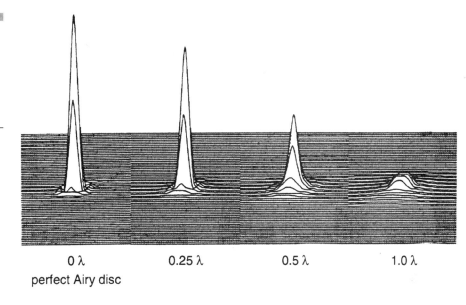

0 λ   0.25 λ   0.5 λ   1.0 λ
perfect Airy disc

Figure 4.4
Peak-to-Valley and
rms Wavefront Error

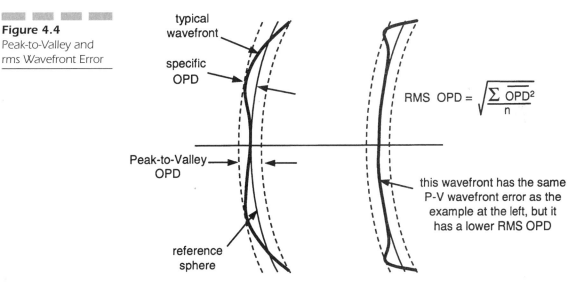

typical
wavefront

specific
OPD

$$RMS\ OPD = \sqrt{\frac{\sum \overline{OPD^2}}{n}}$$

Peak-to-Valley
OPD

this wavefront has the same
P-V wavefront error as the
example at the left, but it
has a lower RMS OPD

reference
sphere

There is another term, and that is *rms wavefront error*. The definition of rms wavefront error is shown in Fig. 4.4 as the square root of the sum of the squares of the OPDs as measured from a best-fit reference spherical wavefront over the total wavefront area. The rms wavefront error represents more of an averaging over the wavefront than the P-V wavefront error. The example shown on the right-hand side of Fig. 4.4 has the same P-V OPD as the left-hand side; however, the rms would be lower. This is because most of the wavefront error, or wavefront distortion, is at the outer periphery of the aperture, and over most of the area of the wavefront, the wavefront is nearly perfect.

Consider, for example, a large telescope mirror 3 m in diameter. In order to assure near-diffraction-limited performance, let us assume that the P-V wavefront error is specified as 0.125 wave on the surface. This is the P-V departure from the ideal or perfect surface profile. In reflection, the wavefront departure will be double this value, or 0.25 wave which just meets the Rayleigh criteria. Now let's further assume that the optical shop produces a mirror which has a P-V surface departure from the nominal of 0.02 wave, with the exception of a small depression the size of a pencil eraser 0.5 wave deep. The mirror is clearly out of spec as the reflected wavefront will have a P-V error of 1.0 wave, which is four times the Rayleigh criteria. However, the area of this small depression in the surface would be 0.0004% of the total mirror area, an almost negligible amount. This will have virtually zero effect on the optical performance of our telescope, and if the scattering from such an error were of concern, we could simply paint the 6-mm-diameter depression with a flat black paint. While there will still be some scattering from the mirror/paint interface, this, too, will be extremely negligible in all but the most demanding applications (such as with space telescopes).

The rms wavefront error typically ranges from approximately one-fifth to one-third of the P-V error. This ratio is highly dependent on the correlation of the wavefront, where the correlation is the inverse number of bumps over the surface. For a given number of bumps, a lower correlation has greater surface slope errors and conversely if we assume a ratio of 5× between P-V and rms, the Rayleigh criteria of 0.25 wave P-V equates to 0.05 wave rms.

Figure 4.5 shows an exaggerated illustration of just how an aberrated wavefront proceeds to an image. This figure reminds us of several key points such as the fact that rays are perpendicular to the wavefront. The peak-to-valley OPD is the maximum deviation from the real wavefront

Figure 4.5
OPD Showing Wave-
fronts and Ray Paths

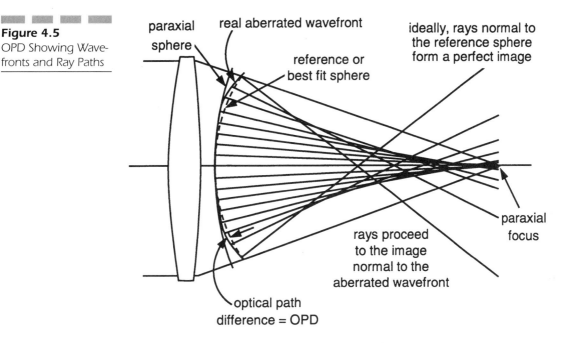

and a spherical reference wavefront, which best fits the real wavefront. While the figure is quite exaggerated, it is drawn to scale and the various factors we have learned about all apply.

The Wave Aberration Polynomial

The optical path difference, or the wave aberration function, can be mathematically expressed in the form of a polynomial for rotationally symmetric optical systems.

A single ray proceeding from a given point in the object through an optical system is defined by the coordinates in the object plane and its coordinates in the pupil of the system. The wave aberration function can be expressed as a Taylor expansion polynomial in field and pupil coordinates. The wave aberration polynomial can be simplified by using the symmetrical properties of the optical system. In its final form the wave aberration polynomial has two quadratic terms which are not the intrinsic aberrations—they present a focal shift, five terms of the fourth power of the field and pupil coordinates which are primary aberrations, and sixth, eighth, and tenth, etc., power terms which are higher-order aberrations.

In order to obtain the coefficients in the wave aberration polynomial, it is sufficient to trace a small number of rays and then fit the data to the polynomial. To obtain the higher-order aberrations, it is necessary to do finite ray tracing, but the primary ray aberrations can be calculated by a paraxial ray trace. In optical systems with moderate to small apertures and fields, primary aberrations dominate. The wave aberration polynomial, W, or OPD, is of the form

$$W = W_{020} \, r^2 + W_{111} \, h \, r \cos \theta + W_{040} \, r^4 + W_{131} \, h \, r^3 \cos \theta + W_{222} \, h \, r^2 (\cos \theta)^2$$
$$+ \; W_{220} \, h^2 \, r^2 + W_{311} \, h^3 \, r \cos \theta + ...(\text{higher-order terms})$$

where h is the height of the object and r and θ are polar ray coordinates in the pupil (see Fig. 4.6).

It can be shown that the ray coordinates in the image plane relative to the perfect image coordinates are proportional to the partial derivatives of the wave aberration polynomial, i.e.

$$\partial y' \sim - \frac{\partial W}{\partial y} \qquad \partial x' \sim - \frac{\partial W}{\partial x}$$

This means that if the OPD or the wave aberration polynomial is known, the ray intersections in the image plane or spot diagrams can be easily calculated. The exponent of the pupil radius term is higher by one in the wave aberration polynomial than in the ray-intercept equations. Thus, for example, third-order spherical aberration affects the image blur diameter in proportion to the cube of the radius of the pupil, whereas the optical path difference is proportional to the fourth power of the pupil radius.

Depth of Focus

As we now know, if the optical path difference is less than or equal to $^1/_4\lambda$, our system meets the Rayleigh criteria and the system imagery is nearly indistinguishable from perfect. This result can be effectively used to determine just how much defocus is tolerable to maintain diffraction-limited performance.

Figure 4.6
Nomenclature for
Wave Aberration
Polynomial

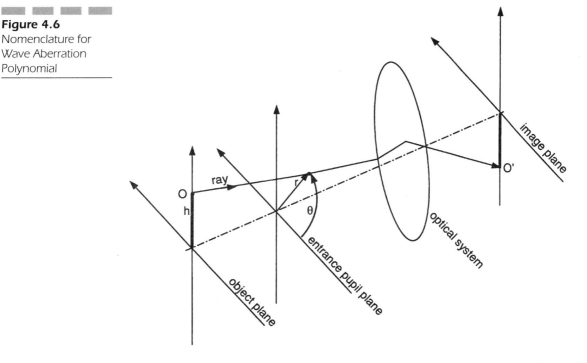

Consider an otherwise perfect optical system, as shown in Fig. 4.7. The solid line in the upper part of Fig. 4.7 represents the nominally perfect spherical wavefront proceeding to the nominal image plane. If we now locate a compass point displaced fore and aft of the nominal image plane and draw two circles which touch the nominal wavefront on the optical axis, these circles will depart from the nominal wavefront along the limiting marginal ray by an amount which is, in effect, the optical path difference. We now adjust the compass point until this displacement from the nominal wavefront is ±0.25 wave. This yields the image plane locations which correspond to one-quarter wave of defocus. The depth of focus which corresponds to an OPD of $\pm^{1}/_{4}\lambda$ is

$$\delta = \pm\lambda/(2\ n \sin^2 \theta) = \pm 2\ \lambda\ (f/\#)^2$$

An extremely useful rule of thumb is that the depth of focus in the visible is approximately $\pm(f/\#)^2$, in micrometers. Thus, for an $f/4$ lens in the visible the depth of focus is approximately ±16 µm. For an $f/2$ system the depth of focus is approximately ±4 µm, and so on. In the lower

Figure 4.7
Depth of Focus

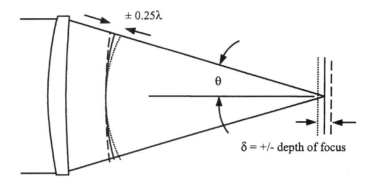

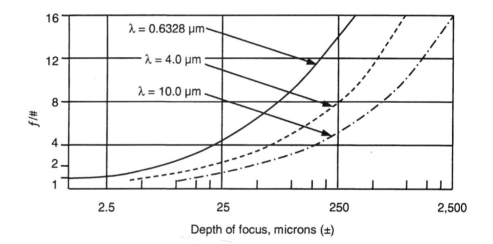

Depth of focus, microns (±)

portion of Fig. 4.7, we show the depth of focus for systems in the visible, the medium-wave infrared (3 to 5 µm), and the long-wave infrared (8 to 12 µm), respectively. This is shown as a function of $f/\#$. It will become very apparent that as the $f/\#$ and wavelength increase, the depth of focus increases as well. This increase is linear with wavelength and quadratic with $f/\#$.

Do keep in mind that this assumes an otherwise perfect system. If your lens system has some inherent aberrations and/or wavefront errors due to design or manufacturing errors, then it will not be nominally perfect to begin with and you may not be able to allow a full one-quarter wave of depth of focus in image location or defocus before you degrade the performance by the quarter-wave limit.

There is another term, "depth of field," which is often confused with depth of focus. Both terms are defined here, which should dispel any further confusion.

Depth of focus. The amount of image defocus which corresponds to being out of focus by one-quarter wave. This means that the optical path difference between the real wavefront leaving the exit pupil at its outer periphery and a reference wavefront centered at the nominal image plane is one-quarter of the wavelength of light.

Depth of field. This is a term used mostly in photography. What it means is that if you focus a camera at a given distance or range, how much further from the camera and closer to the camera than this distance will objects be in acceptable focus. This is analogous to depth of focus; however, it is not as stringent, and it is directly related to how acceptable the image looks to the eye.

Review of Specific Geometrical Aberrations and How to Get Rid of Them

As discussed in Chap. 3, aberrations are the failure of the optical system to produce a perfect or point image from a point object. The geometry of focusing light using spherical surfaces is simply not perfect, and spherical surfaces are used primarily due to their inherent ease of manufacturing. Many lenses can be ground and polished at the same time, as was shown in Fig. 3.2. Lenses are blocked together on the rotating part of the machine called a "block." The top part, which is called a tool, has the desired radius of curvature, and it moves back and forth as the block rotates, forming spherical surfaces of the same radius on all lenses.

As was discussed, paraxial optics applies Snell's law using the small angle approximation where the sine of the angles of incidence on surfaces is equal to the angle, in radians. In paraxial optics, there are no aberrations whatsoever, and by definition, the image of a point object is a perfect point image. Aberrations occur because in a real system the angles of incidence are nearly always so large that the paraxial approximation is invalid and this causes the rays not to converge to a single point image.

As will be discussed in Chap. 7, the use of nonspherical, or aspheric, surfaces can often help significantly in minimizing, if not eliminating, aberrations. It is important to note that the use of aspherics does not automatically guarantee that the aberrations will be zero; in fact, for the most part, this will not happen. Their use is yet another technique for minimizing and balancing aberrations. There are techniques for manufacturing aspheric surfaces or aspheric lenses such as injection molding of plastic lenses, compression molding of glass, or diamond-turning aspheric surfaces in plastic, some crystals, or metal. Aspheric surfaces are used for additional aberration correction, but for the most part, spherical surfaces are used in optical systems. Aspheric optical components are often expensive, such as diamond-turned surfaces and glass-molded lenses, and not sufficiently accurate, or unstable with a change of temperature such as plastic lenses.

There is a class of small lenses used in optical storage applications such as CD read lenses where aspherics are mandatory. These lenses are about the diameter of an aspirin tablet and are compression molded or manufactured by other techniques. In addition, many of today's digital cameras contain very small lenses (less than 6 to 8 mm in diameter) and glass aspherics are becoming more common in these application areas.

The index of refraction of the glass and other transmitting materials is used for making lenses changes with the wavelength of light, a phenomena called *dispersion*. The result is aberrations which change as a function of the wavelength. These aberrations are called *chromatic aberrations*. The image of a point is a superposition of the images for the entire wavelength band or spectral range, each of them blurred with the presence of monochromatic aberrations.

With a well-chosen combination of optical parameters such as lens shapes, number of optical elements, and different optical materials, aberrations in real optical systems with large ray angles can be reduced to a minimum or may be able to be eliminated to the level of the diffraction limit.

Spherical Aberration

If light is incident on the single lens shown in Fig. 5.1, rays that are infinitely close to the optical axis will come to focus at the paraxial image position. As the ray height above the optical axis at the lens increases, the rays in image space cross the axis or focus closer and closer to the lens. This variation of focus position with aperture is called *spherical aberration*.

The magnitude of this spherical aberration depends on the height of the ray in the entrance pupil. The amount of spherical aberration is proportional to the cube of ray height incident onto the lens. If the spherical aberration is measured along the optical axis, it is called *longitudinal spherical aberration*. More often, it is measured as a lateral or transverse aberration, and it represents the image blur radius. For a given focal length lens, a lens with twice the diameter will have eight times larger image blur. For a given focal length and aperture of a single lens, spherical aberration is a function of the object distance and bending (shape) of the lens.

Also shown in Fig. 5.1 are lenses of different bending. The meaning of the term "lens bending" is that the focal length and hence the power of the lens is maintained while changing the radii of both surfaces. This would be the same as physically bending a lens made of flexible plastic. Spherical aberration is highly dependent on the relative lens bending, as will be discussed later.

Another powerful method of controlling spherical aberration is by splitting the optical power into more elements, as shown in the lower portion of Fig. 5.1. By splitting the optical power among several elements, the angles of incidence on each surface can be reduced, resulting

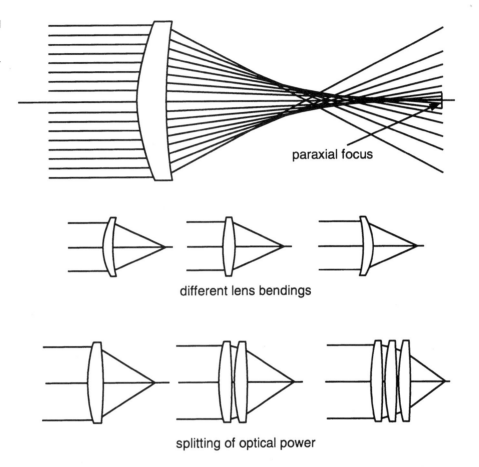

Figure 5.1
Spherical Aberration

paraxial focus

different lens bendings

splitting of optical power

in reduced aberrations. As we learned earlier in Chap. 3, reducing the angle of incidence results in a smaller deviation between paraxial rays and real rays, and hence reduced aberrations.

Consider Fig. 5.2 where we show a single $f/2$ lens element with an enormous amount of spherical aberration. The lower part of Fig. 5.2 also shows an $f/2$ lens; however, in this case the lens is bent for minimum spherical aberration.

When the object point is at a finite distance, the shape of the lens changes for minimum spherical aberration. In a symmetrical case, when the distance of the object point from the lens is the same as the distance of the image, an equiconvex lens is the bending which produces minimum spherical aberration.

Let's look further into reducing the spherical aberration by splitting a lens into several elements. The resulting lens will perform the same

■■■ ■■ ■■ ■■

Figure 5.2
Spherical Aberration
as a Function of Lens
Bending

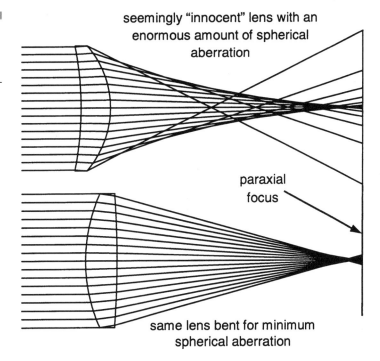

seemingly "innocent" lens with an
enormous amount of spherical
aberration

paraxial
focus

same lens bent for minimum
spherical aberration

function, keeping the total optical power of the elements the same as of the original lens.

We will demonstrate in a simple way how spherical aberration can be reduced by a factor of 2, if the lens is split into two lenses. We will do this in several logical steps:

1. The first step is to start with a lens bent for minimum spherical aberration, and this is shown in Fig. 5.3*a* for a 25-mm-diameter *f*/2 lens of BK7 glass. The residual spherical aberration is 913 μm from real ray tracing.

2. We now scale the lens up by a factor of 2, as shown in Fig. 5.3*b*. The focal length of our new lens is twice as large, the diameter is twice as large, and the spherical aberration is also twice as large. Note that when we scale a lens, all parameters with units of length scale by the same factor such as the radii and thickness. The refractive index is unitless, and thus remains unchanged. The spherical aberration is now doubled to 1826 μm.

3. We now reduce the aperture of this scaled-up lens by a factor of 2, as shown in Fig. 5.3*c*. The spherical aberration reduces by the cube

Figure 5.3
Splitting Optical
Power to Reduce
Spherical Aberration

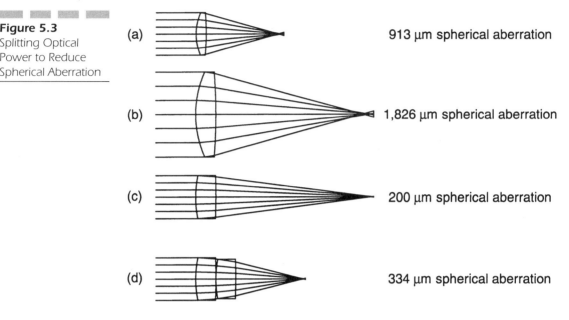

(a)   913 μm spherical aberration

(b)   1,826 μm spherical aberration

(c)   200 μm spherical aberration

(d)   334 μm spherical aberration

of the aperture, which means by a factor of 8, which is 228 μm. Real ray tracing gives 200 μm, which is quite close. Now we have a lens with approximately four times less spherical aberration than the starting point. This new lens has the same aperture as the starting lens, but its focal length is twice as large.

4. Now we add one more identical lens of the same power (same focal length), as in Fig. 5.3d. The spherical aberration is doubled (approximately), but it is twice as small as the aberration of the starting lens. The real ray tracing shows that our final solution has 334 μm of spherical aberration, which is 36% of the starting value with a single element of the same focal length.

The new configuration consists of two lenses performing the same function as the single starting lens, but having one-half of the spherical aberration. The theoretical result of splitting a single lens into multiple lenses is shown in Fig. 5.4. This result shows that if we split an element into four to five elements, the spherical aberration will reduce to about 10 to 15% of the single-element starting point. As it turns out, when we split power in a real lens, the results are significantly better. This is because the light exiting the first element will be converging, and if the second element is now bent for minimum spherical aberration based on converging incident light, the resulting spherical aberration is reduced

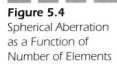

Figure 5.4
Spherical Aberration
as a Function of
Number of Elements

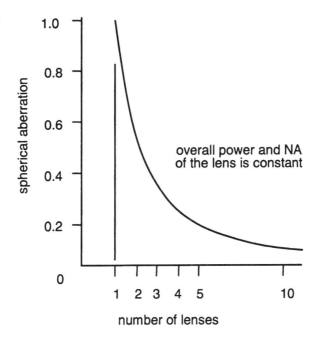

even further. This way, by introducing even more optical elements, the spherical aberration can be reduced significantly. Figure 5.5 shows the situation. Figure 5.5*a* represents a single element bent for minimum spherical aberration. Figure 5.5*b* shows two identical elements as derived previously. Note that the light between the two elements is converging as it enters the second element. Figure 5.5*c* shows how the second element can more optimally "curl" or bend more strongly so as to minimize the angles of incidence onto its surfaces, thereby reducing the spherical aberration from the design in Fig. 5.5*b* with the identical elements.

It is instructive to consider the design of an $f/2$, 100-mm focal length lens for minimum spherical aberration with one, two, three, and four components, and we will do this for glasses with refractive indices ranging from 1.52 to 1.95. We will now plot the peak-to-valley optical path difference for all of these cases in Fig. 5.6.

The results are quite dramatic. Note that for a single $f/2$ element with a 100-mm focal length of BK7 glass (refractive index 1.517) the spherical aberration is approximately 40 to 50 waves P-V. Splitting the element into two elements reduces the OPD to about 6 to 8 waves, and splitting it into three elements further reduces it to about 2 waves. And four elements results in about 0.004 wave, a significant reduction. There is a

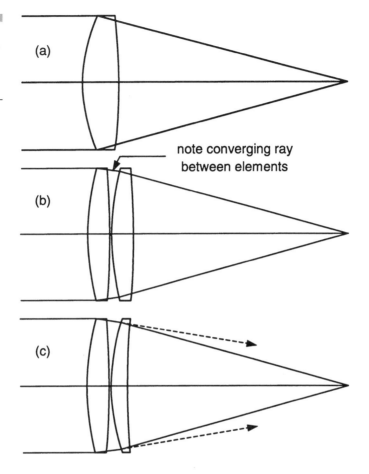

note converging ray
between elements

further reduction in OPD as the refractive index is increased, especially
for three and four elements where nearly six orders of magnitude reduction in OPD is achieved by simply increasing the refractive index from
1.5 to 1.9!

This is for what we have termed the "classical" solution. This is where
each element is bent somewhat more than its predecessor in order to
minimize the angles of incidence and thus the overall spherical aberration. As we will see later, there is a configuration which yields an even
better solution, and we call this the "optimum" configuration. It is characterized by a negatively powered meniscus first element, and it yields
two orders of magnitude less OPD than the classical solution, even at the
lower refractive index region.

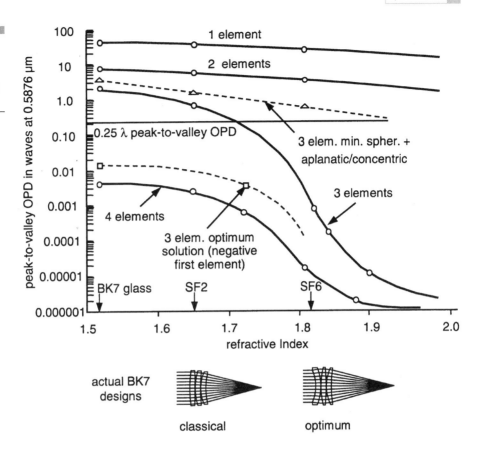

Figure 5.6
Spherical Aberration as a Function of Number of Elements and Refractive Index

Note that the preceding parametric analysis is based on monochromatic light and was computed only on axis. While this is somewhat of an idealized situation, the insight we have gained into aberration reduction is of major significance and it further enhances our understanding of aberrations and where they come from. However, from the analysis thus far we really do not know just why such a dramatic reduction in the aberration is achieved.

Spherical aberration terms in the wave aberration polynomial are the fourth, sixth, eighth, etc., order in terms of the pupil radius. The exponent of the pupil radius term is larger by one in the wave aberration polynomial than in the ray-intercept equations. When we talk about spherical aberration image blur size in the image plane, we talk about third, fifth, seventh, ninth, etc., order in terms of the pupil radius. Let us again look at three component lenses optimized for the smallest spherical aberration, and compare the lenses from low-index glass $n = 1.5$, and then increase the index up to $n = 2$. The spherical aberration as a function of the index of refraction is shown in Fig. 5.7. The contribution to the third-, fifth-, seventh-, and ninth-order spherical aberration is shown for refractive indices ranging from 1.5 to 2.0. Generally, lower orders of aberration have higher values, and they are predominant in the polynomial. As the index increases to somewhere around $n = 1.7$, the fifth-order spherical aberration changes sign and starts to balance the third-order aberration, so that the overall spherical aberration has a significant drop. Although the spherical aberration changes a lot with the change of the index of the components, there is only an imperceptible change in the shape of the lenses. The surfaces become a little shallower, but the overall shape of the lens remains the same.

As a final illustration of what is happening, consider Fig. 5.8 where we show the classical solutions for refractive indices from approximately 1.5 to 1.7. The graphical data is the deviation of the wavefront from perfect as a function of the normalized pupil radius. Note that the OPD decreases from about 2 waves P-V to about 0.25 wave. Figure 5.9 shows the data for refractive indices ranging from 1.8 to 1.95, and we see that the OPD reduces from 0.002 wave to several ten-thousandths of a wave.

Figure 5.7

Third-, Fifth-, Seventh-, and Ninth-Order Spherical Aberration Versus Refractive Index for Three Elements

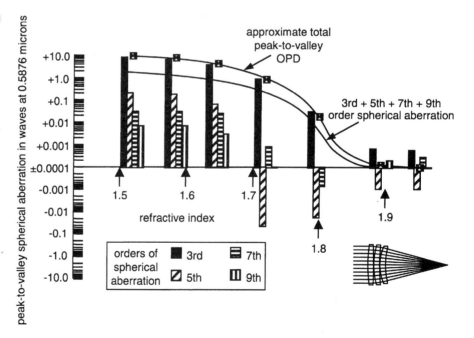

Figure 5.8

Lens Configuration
and Plot of Optical
Path Difference for
Optimized Lenses
of Refractive Index
1.517 to 1.720

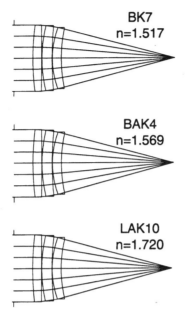

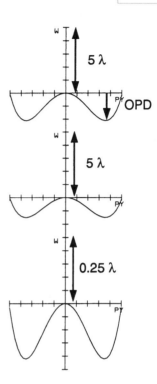

In all of these examples the relative shapes of the elements has remained nearly constant.

We noted earlier that there is a more optimum solution, and we show the three-element "classical" and optimum solutions along with the plot of optical path difference in Fig. 5.10. We are able to reduce the P-V OPD from 2 waves to less than 0.007 wave by changing the configuration. Both of these configurations use BK7 glass. The reduction in aberration is due to balancing of the fifth-order spherical aberration against the third-order, as described previously.

We have discussed orders of aberration in several contexts thus far. By carefully evaluating the plot of optical path difference, you can actually see visually the different orders. Consider Fig. 5.11 where we show again the plots of OPD for the "classical" and optimum designs just discussed. However, here we show each of the orders of spherical aberration. Recall that the exponent is one higher than the transverse ray aberration polynomial, so the third-order spherical aberration is proportional to the fourth-order in OPD, and so on. Of course, focus shift is thus quadratic

Figure 5.9

Lens Configuration and Plot of Optical Path Difference for Optimized Lenses of Refractive Index 1.805 to 1.952

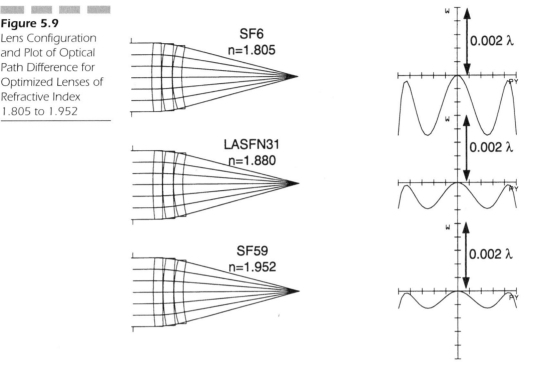

in OPD, as we would expect. Note that each time we see an inflection in the data this is equivalent to another order being added. There will almost, by definition, be higher orders than those shown; however, these data show clearly the presence of the different orders and how they tend to balance each other.

Coma

When we move away from the optical axis in field of view, the image of a point becomes nonrotationally symmetric. In Fig. 5.12 parallel rays come from an infinitely distant point which does not lie on the optical axis of the lens. They enter the lens at an angle, and they are focused by the lens to a certain height from the optical axis, defined by the field angle and the focal length of the lens. If the lens itself limits the bundles of rays from different points in the field, we say that the aperture stop is located on the lens. Rays that go through the center of the

Figure 5.10
"Classical" and Optimum Solutions with
Three Elements for
BK7 Glass

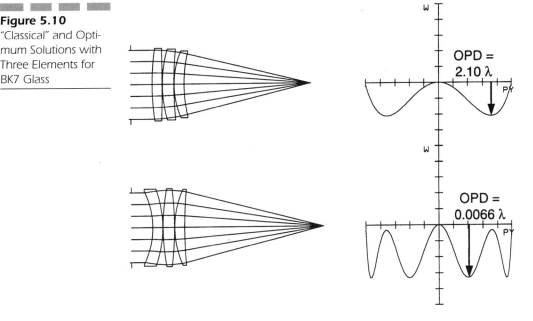

aperture stop are called *chief rays*. There is only one chief ray for each point in the object.

Rays that go through the aperture stop and lie in the plane of the drawing are called *meridional rays*. Rays which do not lie in the meridional plane are called *skew rays*. A plane perpendicular to the drawing in which lies the chief ray called the *sagittal plane*. The meridional and sagittal planes have one common ray, the chief ray.

In an optical system *coma* is defined as the variation of magnification with aperture. Rays that transmit through the lens through different portions of the aperture stop cross the image plane at different heights from the optical axis. In the case of the single positive lens shown in Fig. 5.12, a ray passing through the top and bottom edge portions of the lens converge to a point in the image plane which is further from the optical axis than the point of convergence of other skew rays.

The shape of the image of a point as formed by a system with coma has the shape of a comet. The height of the image is usually defined by the position of the chief ray on the image plane. In the presence of coma, most of the light energy is concentrated in the vicinity of the chief ray. Coma is linearly proportional to the field of view and proportional to the square of the aperture.

Figure 5.11
Illustration of the
Orders of Wavefront
Aberration

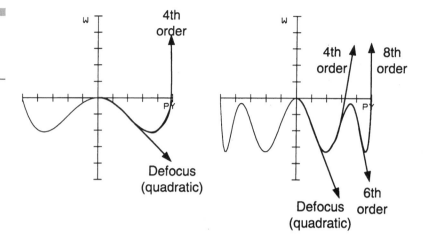

Figure 5.11
Illustration of the
Orders of Wavefront
Aberration

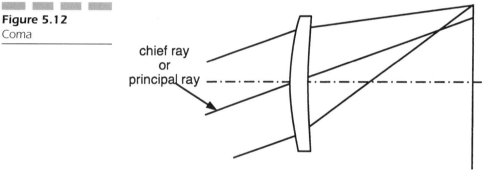

Figure 5.12
Coma

When the aperture stop is not on the lens, moving the position of the stop can control the coma. Having greater ray symmetry on the way through the lens about the aperture stop reduces the coma. Figures 5.13 and 5.14 show the aperture stop to the left and the right of the lens, respectively, and it is clear that the off-axis ray bundles have a higher degree of symmetry and hence significantly reduced coma when the stop is to the right of the lens. This is due to the greater symmetry on the first surface, which results in reduced angles of incidence and hence reduced aberration.

We can best understand coma by looking at Fig. 5.15, which shows the cause of coma. In the top part of the Fig. 5.15, we show a collimated bundle of light incident obliquely onto a convex surface (we will only consider here the first surface of the lens). Note that the entire bundle is displaced from the normal to the surface, which is shown as a dashed line.

Figure 5.13
Coma with Stop in
Front of Lens

aperture
stop

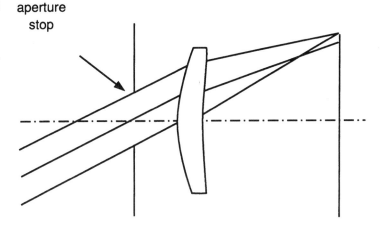

Figure 5.14
Reduced Coma with
Stop Aft of Lens

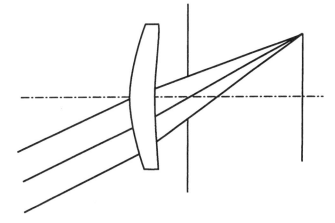

Note also that the upper ray is incident onto the lens surface at a very high angle of incidence with respect to the surface normal, and as we know, this results in a significant ray bending or angle of refraction. The angles of incidence decrease rapidly as we transition to the lower portion of the ray bundle. The coma formation in this situation is quite evident.

Consider now the lower portion of Fig. 5.15, where we show a light bundle whose central or chief ray is normal to the surface. Now we have greatly reduced angles of incidence and, furthermore, the upper and lower limiting rays are symmetrical with each other. The net effect is that there is no coma contribution from this surface whatsoever, and the residual aberration is the same as spherical aberration.

Figure 5.15
Where Does Coma
Come From?

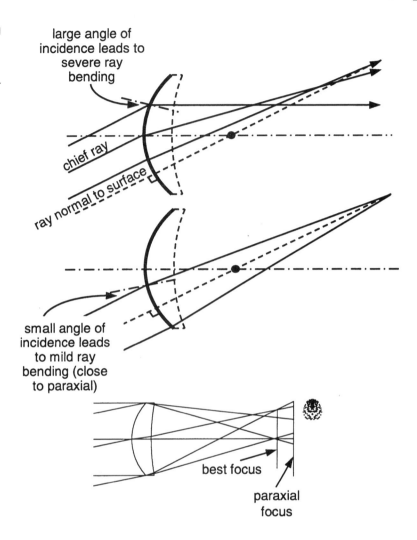

large angle of
incidence leads to
severe ray
bending

chief ray

ray normal to surface

small angle of
incidence leads
to mild ray
bending (close
to paraxial)

best focus

paraxial
focus

Astigmatism

In the presence of astigmatism, rays in the meridional and sagittal
planes are not focused at the same distance from the lens. An astigmatic
image formed by a positive lens is shown in Fig. 5.16.

Rays in the meridional plane focus along the line that is perpendicular to the meridional plane. Rays in the sagittal plane are focused further
away from the lens, along the line perpendicular to the sagittal plane.
Between the astigmatic foci, the image of a point is blurred. It takes the

Figure 5.16
Astigmatism

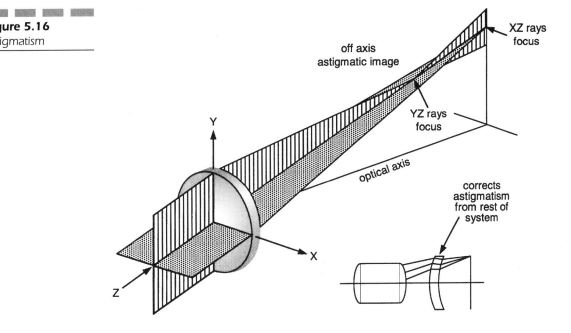

shape of an ellipse or circle. The smallest size of the image blur is half-way between two astigmatic foci when it is circular. Astigmatism is linearly proportional to the lens aperture and to the square of the field angle.

Astigmatism can be controlled by changing the shape of the lens and its distance from the aperture stop, which limits the size and position of the bundle of rays passing through the lens.

A tilted plate in a converging cone of light introduces astigmatism. A weak meniscus lens close to the image plane acts similar to a tilted plate with a tilt angle which changes from zero on axis to a certain angle at the edge of the field. This way, astigmatism created by the meniscus can partly or completely cancel the astigmatism of the rest of the optical system. Also shown in Fig. 5.16 is an example showing just how this works.

Where does astigmatism come from? An oblique cone of light incident on a lens is shown in Fig. 5.17. Assume that this element is immersed somewhere in the middle of an optical system. The area or footprint on the surface of the light cone shown extends over more of the surface in the *y* or tangential direction than in the *x* or sagittal direction. Recall that the rate of change of slope of a sphere is constant

Figure 5.17
Where Does Astigmatism Come From?

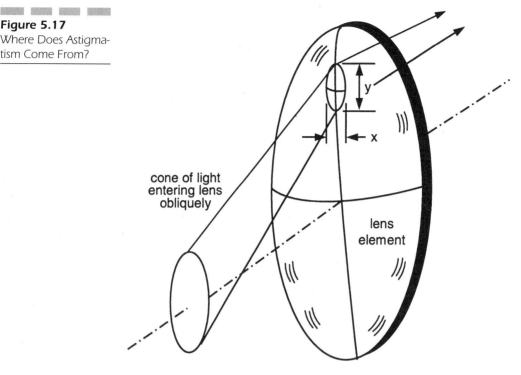

everywhere on the sphere. Thus, the extreme tangential rays see a greater slope change on the surface than the extreme sagittal rays and are hence refracted at greater angles. This causes the tangential ray fan to focus closer to the lens than the sagittal ray fan, and this is astigmatism. As the surface is spherical, we will also find in many cases an off-axis form of spherical aberration called "tangential oblique spherical aberration" which is introduced in the tangential direction.

Field Curvature and the Role of Field Lenses

A positive lens forms an image on a curved surface, as shown in Fig. 5.18. In the absence of astigmatism, a surface on which the image is formed is called the Petzval surface. If a lens has no astigmatism, the sagittal and tangential images coincide with each other, and they coincide with the Petzval surface.

Figure 5.18
Field Curvature

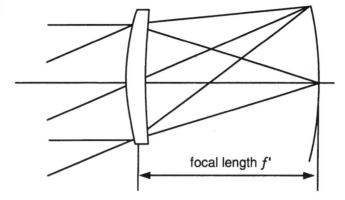

focal length f'

In the presence of astigmatism, both sagittal and tangential image planes are closer to the lens than the Petzval surface, and the tangential image is three times further from the Petzval surface than the sagittal image. The curvature of the Petzval image is inversely proportional to the product of the index of refraction of the lens and its focal length. If there are many components in the optical system, the resulting Petzval curvature is a sum of Petzval curvature contributions from all lenses.

We know that with a 35-mm camera, we can take nice sharp photographs using a flat film. Which method is then used in designing a lens to get a flat image plane? Since the contribution a lens element makes to the Petzval sum is proportional to its optical power, simply splitting of the elements will not change the field curvature. However, positive and negatively powered components can be combined to reduce the field curvature to zero. When negatively powered lenses are added to the system, the resulting power is also reduced.

Fortunately, there is a solution to this problem. The contribution a lens element makes to the system power is proportional to the product of its power and the height of the marginal ray which is the ray going through the edge of the aperture stop. This way, if the position of a negative lens in an optical system is suitably chosen so that its power is substantial, but the height of the marginal ray on the lens is relatively low, its contribution to the overall optical power is relatively low while still having a significant field curvature.

Two examples where a negative component is effectively used to reduce a field curvature are shown in Fig. 5.19. The first example is the Cooke triplet. A negative component is located in the middle between two positive lenses. The marginal ray height on both positive lenses is higher than on the negative lens. However, the power of the negative

Figure 5.19
Negatively Powered
Elements with Small
Value of Y to Flatten
Field

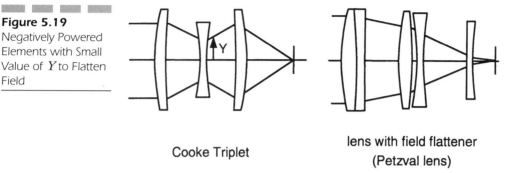

Cooke Triplet

lens with field flattener
(Petzval lens)

lens significantly reduces the field curvature created by the two positive lenses.

The second example is a Petzval lens, where the negative component is located very close to the image plane. Its contribution to the power of the whole lens is very small, since the height of the marginal ray is extremely small when the lens is close to the image. If the lens is placed at the image plane, it does not change the overall power of the system.

Is there any reason to put the lens at the image plane? Yes, indeed both positive- and negative-powered lenses are often located either in the image plane or very close to it. They are called *field lenses*.

The first case when a lens is located at the image plane or just in front of the image plane is when the negative lens is used as a *field flattener* to correct the field curvature and flatten the field. This is common in complex wide field of view, fast (low $f/\#$) lenses.

How does a negative lens flatten the field? Let us imagine the case of a simple positive lens that forms an image a certain distance from the lens. If we add a block of glass between the lens and the image plane, the image will move away from the lens. The image shift is proportional to the thickness of the glass block. In the case of the negative lens in front of the image plane, ray bundles that focus close to the edge of the field of view pass through the part of the lens where the glass thickness is larger than in the center of the lens. This way image points that are closer to the periphery of the image are shifted away from the focusing lens more than the ones in the center of the field. This results in a flatter image plane.

A doublet lens forms a sharp image on a spherical surface shown in Fig. 5.20a. If an achromatic doublet is an objective lens of a telescope, which focuses the image on a reticle, or on a CCD detector, it is desirable to correct the field curvature of the lens. The reticle is usually engraved on a flat piece of glass, and in the case of a CCD detector, the sensitive area is always flat. When the CCD detector is adjusted for best focus, both the center and the edge of the field are slightly blurred, as in Fig. 5.20*b*, and the sharpest image is obtained for the intermediate field. Figure 5.20*c* shows how a field lens in front of the image reduces the field curvature and the image blur at the edge of the field.

The other types of field lenses are positive-powered lenses used in the systems with one or more intermediate images and relay optics. A submarine periscope is an example of such a system. The optics is inside a tube, which is 10 m long or even longer, but the diameter of the optics

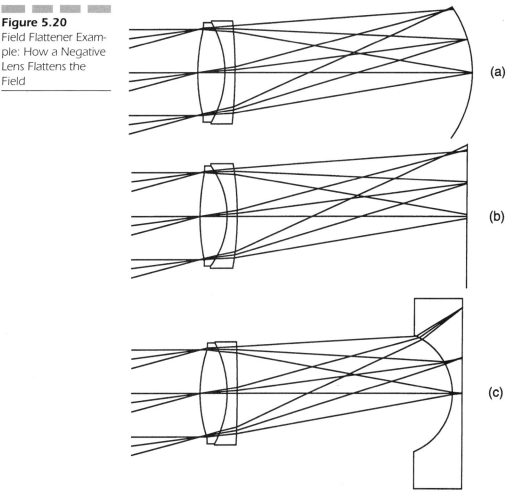

Figure 5.20
Field Flattener Example: How a Negative Lens Flattens the Field

does not exceed 250 to 300 mm. Another example is an endoscope, which is on a much smaller scale, but with a similar ratio of the system length and diameter. Schematically, these systems are shown in Fig. 5.21a, where lenses labeled O are objective lenses and lenses labeled F are field lenses.

What is the function of these positive field lenses? They cannot correct the field curvature; they actually *increase* the field curvature already introduced by the other positive lenses. If we look at the axial beam shown in the schematic drawing, and assume from the beginning that there are no field lenses (only the O lenses are present), the axial beam will be focused at the first intermediate image plane, then relayed with two lenses to the second intermediate plane, and finally relayed with another two lenses to the final image plane. There is no problem

Figure 5.21
Field Lens Example:
How a Positive Lens
Reduces Vignetting

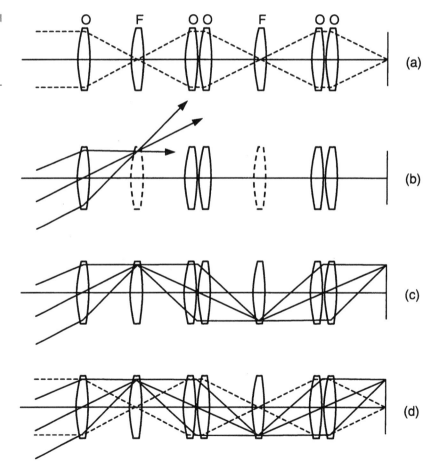

with the axial ray bundle. Now consider the beam entering the optical system at an angle. It will be focused at a certain height from the optical axis in the first intermediate image plane. However, the cone of light is so tilted that almost the whole ray bundle is going to miss the two relay lenses, and it will hit the housing. This is called *vignetting* in the optical system and it reduces the amount of light in the image periphery. If we now add a positive lens in the image plane, it will not do anything to the axial beam, but it will redirect the cone of light coming from the edge of the field into the relay lenses. There will be no vignetting and almost no change in the position and the size of the image. The image brightness is going to be uniform across the field, but the system will have a significant amount of field curvature. The primary purpose of these positive field lenses is to reduce or eliminate the vignetting in the system.

There was a paper in 1980 given by Erhard Glatzel of Carl Zeiss in Germany, where he talked about designing lenses in microlithography that imaged a mask onto a 50-mm-diameter silicon wafer at a 5:1 reduction ratio. Microlithography lenses are the most sophisticated lenses in our industry, since they have to resolve submicrometer structures in the flat image, as shown in Fig. 5.22. Glatzel starts from the basic relation for the Petzval sum. Suppose that only one type of glass is used in the entire lens. In order to correct the field curvature, the sum of the powers of all components in the system has to be zero or very close to zero. At the same time, total power of the system has to be positive. The total power is proportional to the sum of weighted powers of the components, where the weighting factor is the normalized height of the marginal ray on the component. Glatzel first analyzes a so-called planar camera lens similar to a double Gauss-type lens. The positive-powered components are the first two and the last two components. The negatively powered components are located in the center of the lens. This power distribution is very similar to a Cooke triplet lens. Below the lens layout, Glatzel plots the contribution to the power sum as well as the weighted power sum from corresponding lens groups above. Petzval curvature in this first lens is not reduced to zero but has a residual of 0.46.

In order to reduce the Petzval sum or field curvature, Glatzel forces the central elements to have a more negative optical power. The Petzval sum is lower than in the first example but it is still present. Unfortunately, the radii of the central elements have become quite strong or severe, thus increasing the angles of incidence on their surfaces (especially the short-radius inner concave surfaces), therefore introducing their own aberrations.

Figure 5.22
Reduced Field Curvature Lenses as Described by Glatzel of Carl Zeiss

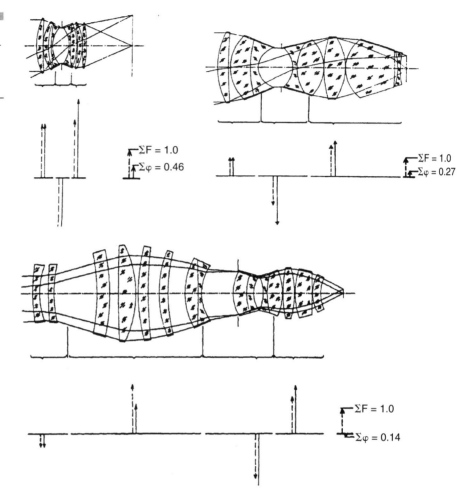

The final step in the design of the flat-field lithography lens is the addition of a negative group in the front of the whole lens. This group has a very important role. Because of its negative power, it slightly reduces field curvature, but it expands the beam and makes it possible for the following positive group to make a high contribution to the total power of the system. The final lens design has -0.14 of field curvature. While this still has a noticeable result, the net aberration of the system is extremely well corrected. This form of lens still represents the state of the art in lithography optics.

While we are discussing this design, we should take a few minutes to look at the design form. After the initial two negative elements in the front, Glatzel uses four larger single elements to take the diverging light and converge it into the next group. He uses four elements in order to split the power and minimize angles of incidence so as to reduce aberrations. However, note that the second element seems to have most of the positive power, also the first element of this group seems to be quite concentric about the diverging light cone, and it seems to have little or no optical power. We point this out in order to suggest that this element may indeed be able to be removed from the design. This is something that you should always be looking for during the optimization of your design in order to, where possible, simplify the design. It is difficult, if not impossible, to make a definitive judgment on this without working with the specific design data and optimization. It is certainly possible that this element may be canceling some higher-order aberration residual.

We show for interest in Fig. 5.23 a 30-element lithography lens from the patent literature. The many lens elements are, of course, used in order to reduce the angles of incidence and thus the residual aberrations. This lens will be extremely difficult to manufacture, assemble, and align due to the large number of surfaces.

Distortion

The only aberration that does not result in image blur is distortion. If all other aberrations in the system, except distortion, are corrected, an object point is imaged onto a perfect image point, which is displaced from its paraxial position. The amount of distortion can be expressed either as a lateral displacement in length units or as a percentage of the paraxial image height. Distortion is defined as

$$\text{Distortion} = \frac{y - y_p}{y_p}$$

where y is the height in the image plane and y_p is the paraxial height.

Figure 5.23
A 30-Element Lithography Lens from the Patent Literature

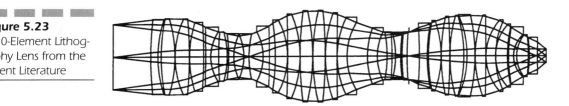

Third-order distortion increases with the cube of the field of view. A distorted image of a rectilinear object is shown in Fig. 5.24. Distortion can be *positive or pincushion distortion* or alternately *negative or barrel distortion.* For a thin lens with the aperture stop on the lens, distortion is equal to zero. The thickness of a lens and its position relative to the aperture stop determines its contribution to the system distortion. An example of a system where a correction of the distortion is a big challenge is a wide field-of-view eyepiece. Its field of view may be as high as 70°, and its aperture stop, which is the pupil of the eye, can be in the order of 20 mm away from the system. If the object and the image are interchanged, the lens that has a barrel distortion in one direction has a pincushion distortion in the opposite direction. This is a very interesting subtlety, and we urge the reader to spend due time thinking about it prior to predicting what a distorted image will look like to make sure your sign of distortion is modeling the real world properly.

Figure 5.25 shows different amounts of negative and positive distortion. Since distortion is a cosmetic-type aberration not affecting

Figure 5.24
Distortion

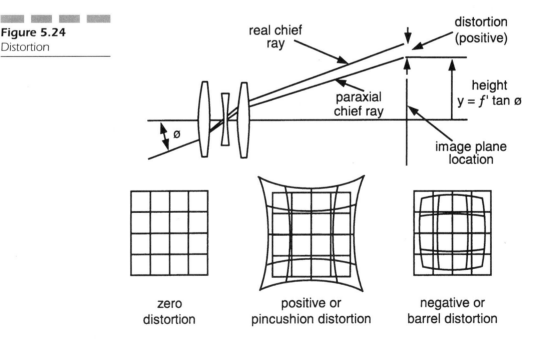

real chief ray

paraxial chief ray

distortion (positive)

height $y = f'\tan ø$

image plane location

ø

zero distortion

positive or pincushion distortion

negative or barrel distortion

Figure 5.25
Illustration of Different Amounts of Negative and Positive Distortion

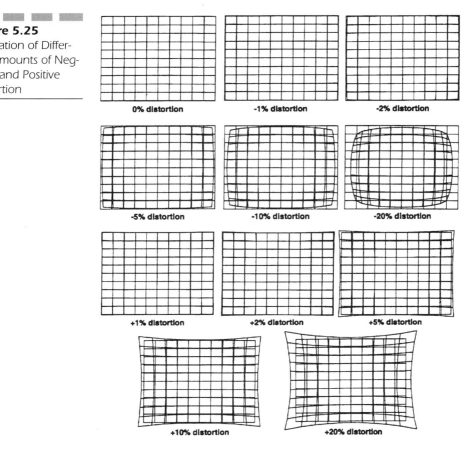

resolution, its appearance is very important, especially in visual systems. Generally, distortion in the order of 2 to 3% is acceptable visually.

Figure 5.26 shows where distortion comes from. In this situation, the aperture stop is located to the left of the lens, and the angle of incidence on the lens by the ray bundle is large enough so that there is a reasonable difference between the paraxial angle of refraction and the real ray angle of refraction. As with spherical aberration, the real rays are refracted more severely than the paraxial rays. In this case, this causes the real image to be pulled inward from the paraxial image thus causing negative or barrel distortion.

Table 5.1 summarizes the aperture and field dependence of the primary aberrations.

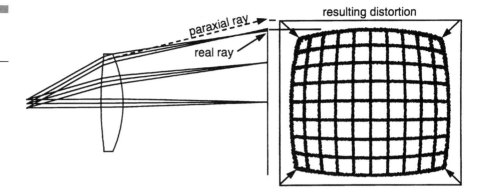

Figure 5.26
Where Does Distortion Come From?

TABLE 5.1

Summary of Third-Order Monochromatic Aberration Dependence on Aperture and Field Angle

Aberration	Aperture Dependence	Field Dependence
Spherical	Cubic	—
Coma	Quadratic	Linear
Astigmatism	Linear	Quadratic
Field curvature	Linear	Quadratic
Distortion	—	Cubic

Axial Color

If white light is incident onto a glass wedge or a prism, it is decomposed into a rainbow. This is called *dispersion*. Blue light is refracted more severely than red light, since the index of refraction is higher for shorter wavelengths than for the longer wavelengths. The properties of lenses also vary with wavelength. White light coming from an axial infinitely distant object, which is incident upon a convergent lens, is shown in Fig. 5.27a. The edge of the lens acts like a *wedge*, refracting or bending the blue light more than the red light. This causes the blue light to focus closer to the lens than the red light. This longitudinal variation of focus with wavelength is called the *axial chromatic aberration* or *axial color*. In the absence of spherical aberration, a system with uncorrected chromatic aberration forms a bright spot surrounded with a purple halo coming from the blue and red light.

Is there a way to correct the axial color? A lens that focuses an infinitely distant object is shown as an example in Fig. 5.27 b. In order to bring the blue and the red to focus together, a positive lens must be split into two lenses made of glasses with different dispersions. The first is a positive lens with low dispersion glass. This type of glass is called a *crown* glass. The second lens has a lower optical power than the first one, so that the total power of the doublet is positive. However, the second lens is made of a high-dispersion glass called a *flint* glass, which means that it spreads light more with color, and it cancels most of the axial chromatic aberration created by the first lens because of its negative power. This doublet is called an *achromatic doublet*.

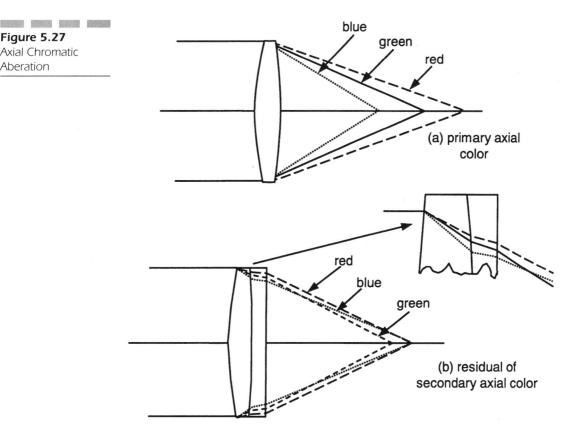

Figure 5.27
Axial Chromatic
Aberation

Lateral Color

When a lens forms an image of an off-axis point at different heights for different wavelengths, the lens has lateral chromatic aberration or chromatic difference of magnification. This aberration is quite common in wide field-of-view systems. A very descriptive name for lateral color is *color fringing* since this is what is seen when looking at an image formed by a lens with lateral color.

Lenses that are further from the aperture stop in a system contribute more to lateral color than the lenses with smaller chief ray heights. Lateral color created by a lens is shown in Fig. 5.28. The chief ray is going through the single-lens element close to its outer periphery. Shorter wavelengths are bent or refracted more severely than the longer wavelengths. The blue image is formed closer to the optical axis than the red image. In wide field-of-view systems, lateral color is often the aberration that is the most difficult to correct. Its correction may require the use of anomalous dispersion glasses, which are often expensive, or diffractive elements.

Visual systems often have lateral color due primarily to the eyepiece, which inherently has a lot of lateral color. If you look, for example, through a pair of binoculars at a sharp bright/dark edge close to and tangent to the edge of the field of view, you will likely see severe lateral color. This is often not a problem, as the user will most often place the object of interest at the center of the field of view. For this reason visual optical systems tend to be somewhat more forgiving than other systems.

Parametric Analysis of Aberrations Introduced by Plane Parallel Plates

Let us assume that we have a diffraction-limited $f/1$ lens. What is the spherical aberration introduced by a plane parallel plate inserted in the converging cone between the lens and the image plane? The spherical aberration introduced by the plate increases as a function of the plate thickness. Figure 5.29 shows how the blur diameter due to the spherical aberration changes as a function of the $f/\#$ of the lens and the plate thickness. The shaded area is the region where the optical system is diffraction limited. For our $f/1$ lens, the Airy disk diameter is 1.5 μm. If a plane parallel plate of glass 50 mm thick is inserted in the $f/1$ cone of

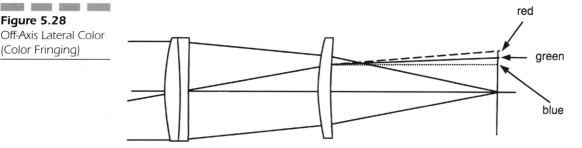

Figure 5.28
Off-Axis Lateral Color
(Color Fringing)

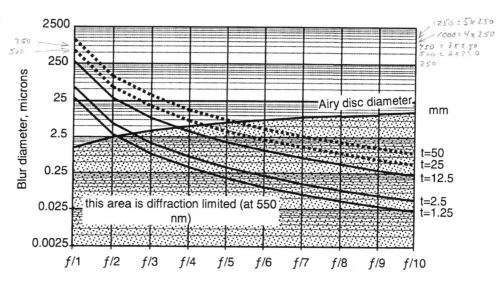

Figure 5.29
Spherical Aberration
of a Plane Glass Plate,
Refractive Index =
1.517 (BK7)

light, the spherical aberration in the focused image increases to 1.25 mm. The image blur is much larger than the Airy disk, which means that the glass plate significantly degrades the performance. We would have to stop down the aperture of the lens to about $f/5$ to reduce the size of the image blur to the Airy disk. At the point where the diffraction blur and the aberrations of the glass plate are equal, the glass plate will not have a detrimental effect on the image quality.

Third-order spherical aberration, coma, astigmatism, and distortion of a plane parallel plate depend on the index of refraction of the plate, n, and the thickness, d, and they are proportional to

$$\frac{(n^2 - 1)\, d}{n^3}$$

What happens if a plane parallel plate is inserted in a converging cone of light at 45°? If the tilted plate is inserted in a rotationally symmetric optical system in a convergent beam, the optical system is no longer rotationally symmetrical. A result of this is the presence of field aberrations such as coma, astigmatism, and lateral color in the on-axis field position.

Let us again assume that we have a diffraction-limited, very fast $f/1$ lens and we need to split the beam into two beams, one reflected at 90° and the other beam transmitted through the tilted plate beamsplitter. The reflected beam is unchanged after reflection off the plate, but the transmitted beam has on-axis astigmatism introduced by the tilted plate. The thicker the plate, the more astigmatism is present. The blur diameter associated with third-order astigmatism as a function of the thickness of the tilted plate and the $f/\#$ of the lens is shown in Fig. 5.30.

Is there a way to correct this on-axis astigmatism? There are a few different viable methods of correction. Astigmatism is proportional to the square of the tilt angle, so it is therefore not dependent on the sign of the plate tilt. Thus, astigmatism cannot be compensated with a second plate tilted in the opposite direction from the first plate and in the same plane of tilt. This is shown in Fig. 5.31. However, if the second plate is tilted in a plane which is orthogonal to the plane of tilt of the first plate, the astigmatism can be corrected for the most part. The second method of correction involves the use of a weak spherical surface on the tilted plate or a weak wedge instead of the plane parallel plate. Although it is

Figure 5.30

Third-Order Astigmatism Blur Diameter, in Micron, as a Function of the Thickness of a 45° Tilted BK7 Plate and the $f/\#$ of the Lens

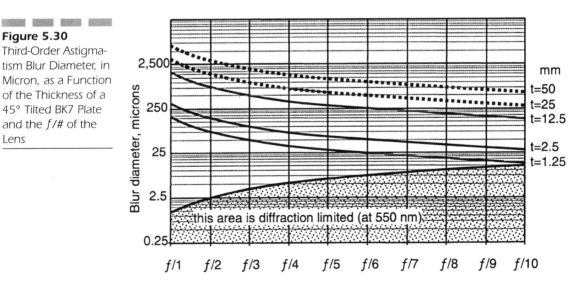

Figure 5.31
Correction of Astigmatism from a Tilted Plate in a Converging Beam

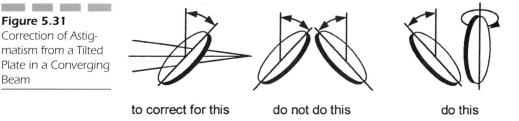

to correct for this   do not do this   do this

Figure 5.32
Tangential Coma Blur Diameter As a Function of the Thickness of a 45° Tilted Plate of Index 1.5 and the ƒ/# of the Lens

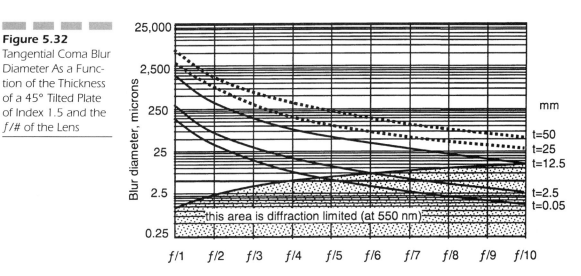

Figure 5.33
Lateral Color Blur Diameter in Microns as a Function of the Thickness of a 45° Tilted BK7 Plate and the ƒ/# of the Lens

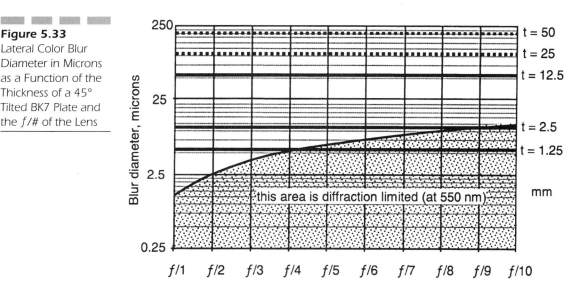

more difficult for fabrication, a good correction of the astigmatism can be achieved with a decentered cylindrical surface on the tilted plate. You can also think of this component as a wedged cylinder. In this way, the astigmatism, as well as a smaller residual of coma, are reasonably well corrected.

The most severe aberration introduced in an optical system by a tilted plate is astigmatism. However, coma and lateral color are also significant in the case of a fast lens even when the plate thickness is less than 1 mm. The residual tangential coma blur in a system with a 45° tilted plate is shown as a function of plate thickness and the $f/\#$ of the lens in Fig. 5.32. Lateral color blur in a system with a 45° tilted plate is shown as a function of plate thickness and the $f/\#$ of the lens in Fig. 5.33. Note that the lateral color is independent of the $f/$number. This is because the lateral color is only dependent on the height of the chief rays in the different wavelengths.

Glass Selection (Including Plastics)

Material Properties Overview

Every optical system works in its own particular wavelength region determined by the spectral characteristics of the light source, the spectral sensitivity of the sensor, as well as any other factors or components which alter the net sensitivity of the system. If an optical system is a visual system, the optical materials must be transmissive between approximately 425 and 675 nm, as determined by the photopic spectral response curve of the human eye. The photopic eye sensitivity is shown in Fig. 6.1. Optical glasses are the most commonly used materials in optical systems. However, there are some optical plastics with good transmission in the visible spectrum that can be injection molded. In high-volume production, this technology is significantly cheaper than classical glass manufacturing methods. Operating temperature range is very important when choosing optical materials. Optical materials change their index of refraction with temperature, and they also expand differently, changing the lens shape and optical power. Optical plastics have approximately one order of magnitude higher coefficient of thermal expansion than glasses.

If the temperature in an optical system rises to a few hundred degrees Celsius, plastic materials cannot be used because the plastic will melt. Most optical glasses can withstand temperatures of a few

Figure 6.1
Photopic Spectral Eye
Sensitivity Curve

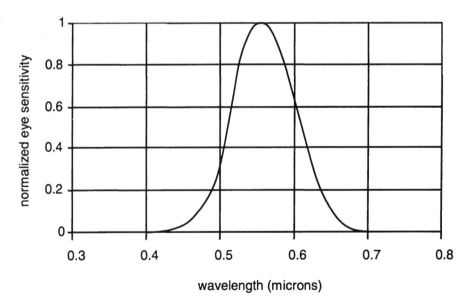

hundred degrees Celsius without changing their shape. In illumination systems close to the light source, the temperature can go up to 900°C. In this case, glass optics will melt too. Fused quartz or fused silica is often used in these systems because it can operate at temperatures close to 1000°C.

Manufacturers of optical components usually include in their catalogues information about the standard optical materials they use. An example of the general information on optical materials for the visible, near ultraviolet (UV) and near infrared (IR) spectral regions as provided by Melles-Griot is shown in Fig. 6.2.

The Glass Map and Partial Dispersion

The refractive index of all optical materials changes as a function of wavelength. The refractive index increases as the wavelength decreases. This means that optical systems with refractive components have chromatic aberrations. In fact, the performance of an optical system is often limited by chromatic aberrations rather than monochromatic aberrations.

In the time of Sir Isaac Newton, it was believed that it was not possible to correct chromatic aberrations by combining different types of glasses. Newton thought that the chromatic aberrations of all lenses were proportional to their powers, with the same constant of proportionality for all glasses. This is the reason why Newton built a reflecting telescope. In the eighteenth century, however, it was found that, with the proper choice of glasses and powers, it was possible to design an *achromatic doublet,* which was chromatically corrected for two wavelengths.

Let us consider two thin lenses made from two different glass types and cemented together as shown in Fig. 6.3. We want to find the condition for this doublet to be an achromatic doublet, chromatically corrected for the red C line wavelength 656.27 nm and for the blue F line 486.13 nm. Generally, the *crown* materials are less dispersive than the flints, and in an achromatic doublet we combine the less dispersive crown as the positive element and the more highly dispersive *flint* as the negative element. The central wavelength is usually chosen as the *d* line, which is 587.56 nm. If the power of the first lens is P_1 and the second lens P_2, then the total power of the doublet is

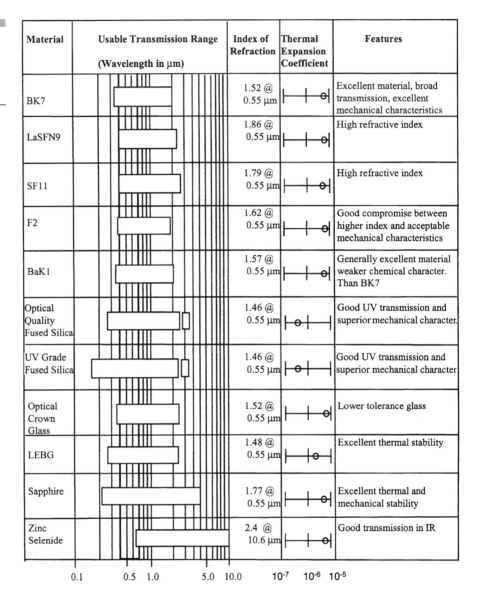

Material	Usable Transmission Range (Wavelength in μm)	Index of Refraction	Thermal Expansion Coefficient	Features
BK7		1.52 @ 0.55 μm		Excellent material, broad transmission, excellent mechanical characteristics
LaSFN9		1.86 @ 0.55 μm		High refractive index
SF11		1.79 @ 0.55 μm		High refractive index
F2		1.62 @ 0.55 μm		Good compromise between higher index and acceptable mechanical characteristics
BaK1		1.57 @ 0.55 μm		Generally excellent material weaker chemical character. Than BK7
Optical Quality Fused Silica		1.46 @ 0.55 μm		Good UV transmission and superior mechanical character.
UV Grade Fused Silica		1.46 @ 0.55 μm		Good UV transmission and superior mechanical character
Optical Crown Glass		1.52 @ 0.55 μm		Lower tolerance glass
LEBG		1.48 @ 0.55 μm		Excellent thermal stability
Sapphire		1.77 @ 0.55 μm		Excellent thermal and mechanical stability
Zinc Selenide		2.4 @ 10.6 μm		Good transmission in IR

0.1   0.5 1.0   5.0 10.0 10^{-7} 10^{-6} 10^{-5}

Figure 6.3
Focusing of White
Light with an Achro-
matic Doublet from
BK7 and SF2 Glasses

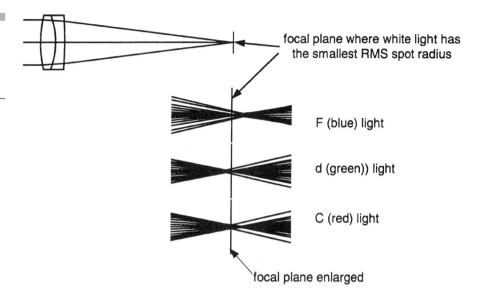

focal plane where white light has
the smallest RMS spot radius

F (blue) light

d (green)) light

C (red) light

focal plane enlarged

$$P = P_1 + P_2$$

An achromatic doublet will have the same power for the C line wavelength and the F line wavelength if

$$(P_1 + P_2)_C = (P_1 + P_2)_F$$

or

$$P_1 \frac{n_{1C} - n_{1F}}{n_{1d} - 1} = -P_2 \frac{n_{2C} - n_{2F}}{n_{2d} - 1}$$

where P_1 and P_2 are the powers of two thin lenses at 587.56 nm. The value $(n_F - n_C)$ is called the *main dispersion*. The ratio

$$V = \frac{n_d - 1}{n_F - n_C}$$

is called the V number or the Abbe number. The condition for a doublet to be an achromatic doublet becomes

$$\frac{P_1}{V_1} = -\frac{P_2}{V_2}$$

From this relation we can obtain the focal lengths of two components of an achromatic doublet as

$$f_1 = f \frac{V_1 - V_2}{V_1}$$

$$f_2 = -f \frac{V_1 - V_2}{V_2}$$

where f is the focal length of the doublet. The net result of this is to derive the powers of the less dispersive crown element and the more dispersive flint element, so that the combined doublet focal length in the red and blue wavelengths are the same. When this condition is reached, the central wavelength (green) is defocused slightly toward the lens.

In the second half of the eighteenth century, Ernst Abbe worked closely with Otto Schott on testing different types of optical glass and this encouraged the development of new glass types. It was found that the most suitable way to characterize optical glass was the specification of the index of refraction for the d line, n_d, and the Abbe number, V, which determines the glass dispersion. Manufacturers of optical glasses provide a glass map or an n_d/V_d diagram in which the Abbe number is plotted as the abscissa and the index, n_d, as the ordinate. The glass map from the Schott glass catalogue is shown in Fig. 6.4. Schott is the largest manufacturer of glass in the world, but there are other manufacturers, including Hoya, Ohara, Pilkington, Corning, and Sovirel.

The n_d/V_d diagram subdivides the various types of glasses into groups, each having a specific designation such as BK with BK7, BK1, and others. These designations are generally related to the fundamental materials used in the manufacture of the specific group such as LAFN31 which is a lanthanum flint glass. There is also a more general division of glasses into "crown" and "flint" glasses. The crown glasses are the ones with $n_d > 1.60$, $V_d > 50$ or $n_d < 1.60$, $V_d > 55$; the other glasses are flints. The available refractive indices range from 1.45 to 2 and the V number from 80 to 20.

Mathematically, the dependence of refractive index on the wavelength of light can be expressed in a few different ways, but none of the expressions is highly accurate over the entire glass transmission range. These relationships are empirically derived from measured data. The Sellmeier dispersion formula is

$$n^2 - 1 = \sum_i \frac{1}{\lambda^2 - \lambda_i^2} c_i \lambda^2$$

and the formula from the 1967 Schott catalogue is

Figure 6.4
Schott Glass Map
(Abbe Diagram)

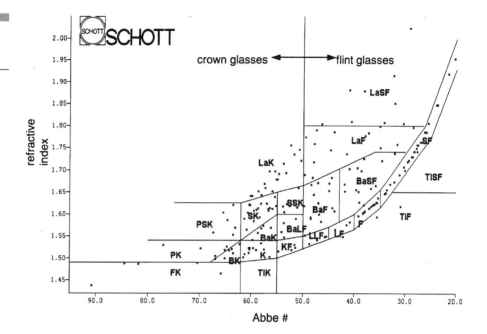

$$n^2 = A_0 + A_1\lambda^2 + \frac{A_2}{\lambda^2} + \frac{A_3}{\lambda^4} + \frac{A_4}{\lambda^6} + \frac{A_5}{\lambda^8}$$

These are the two most commonly used formulas. Beside these two formulas, the Hartmann and Conrady formulas are also offered in some of the lens design programs. The six constants that characterize glass dispersion vary considerably between glasses, and thus the general shapes of all dispersion curves are different.

In addition to the main dispersion $(n_F - n_C)$, which is the difference between the index of refraction for the blue and for the red line, the "partial dispersion" is also commonly used. Partial dispersion in the blue is the difference in index of refraction between 435.83 and 486.13 nm, and the red partial dispersion is the difference in index of refraction between 653.27 and 852.11 nm.

Perhaps even more important is the "relative partial dispersion," which is the ratio of the partial dispersion and the main dispersion. Generally, the relative partial dispersion is

$$P_{x,y} = \frac{n_x - n_y}{n_F - n_C}$$

A glass map with the relative partial dispersion as a function of the Abbe number from the Schott catalogue is given in Fig. 6.5.

The derived formulas for the design of an achromatic doublet provide a chromatic correction for F and C wavelengths. However, dependent on the choice of glasses, there will be a residual mismatch of dispersions, resulting in a larger or smaller "secondary spectrum." This secondary spectrum is the difference in image position between the central wavelength (green or yellow) and the now common blue and red image position. In order to eliminate the secondary spectrum, we should find a pair of glasses with different V values but the same relative partial dispersion. Abbe showed that the majority of glasses, the so-called normal glasses, exhibit an approximately linear relationship between the relative partial dispersion and the Abbe number. Thus

$$P_{x,y} \approx a_{x,y} + b_{x,y}\, V_d = (P_{x,y})_{\text{normal}}$$

which can be clearly seen in the relative partial dispersion map in Fig. 6.5. The reduction of the secondary spectrum requires the use of at least one glass type which does not lie on the $(P_{x,y})_{\text{normal}}$ line. The glasses that lie away from the line of normal glasses are often expensive and may be

Figure 6.5
Relative Partial Dispersion

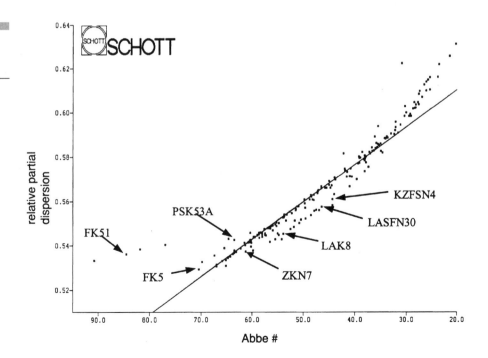

difficult to manufacture. Some of them are shown in the relative dispersion chart in Fig. 6.5. KZFSN4 is seven times more expensive than BK7, LaK8 is eight times, PSK53A is 11 times, and LaSFN30 24 times.

Parametric Examples of Glass Selection

In this section we will show how secondary spectrum can be corrected and to which level, with the right choice of glasses. It will also be shown how the spherical aberration and the secondary spectrum of a doublet are dependent on the $f/\#$. These data are shown as parametric analyses.

The first parametric study is shown for the case of an $f/10$ achromatic doublet using different glass combinations. Four doublets will be compared, the first using two normal glasses, and then using anomalous dispersion glasses for one or both of the elements. The first doublet is designed with two normal glasses: BK7 and SF2. It is an $f/10$ lens with a 100-mm focal length. The ray aberration curves are shown in Fig. 6.6. An

Figure 6.6
Secondary Spectrum Correction As a Function of Glass Selection

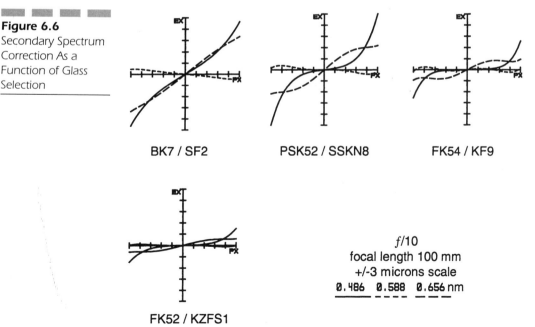

BK7 / SF2   PSK52 / SSKN8   FK54 / KF9

FK52 / KZFS1

$f/10$
focal length 100 mm
+/-3 microns scale
0.486   0.588   0.656 nm

explanation of these curves will be given in Chap. 10. The difference in the Abbe numbers between the two glasses should be sufficiently large, so that the shape or the power of each individual component is reasonable. Note that as the Abbe number difference between the two glasses of an achromatic doublet decreases, the relative powers of the positive and negative elements get stronger. This yields greater spherical aberration, as will be seen. For the $f/10$ lens the spherical aberration for the central wavelength is very small, the rms spot diameter is less than 1 μm. However, both the red and the blue foci are away from the green focus, which means that the secondary spectrum aberration is not corrected.

The second achromatic doublet is designed with PSK52 and SSKN8 glasses. SSKN8 is a normal glass, but PSK52 has anomalous dispersion. The green spot diameter is the same as in the case of the first doublet, but the polychromatic spot diameter is slightly smaller because the secondary spectrum is lower. Note here that we have an increase of spherical aberration which changes with wavelength. The blue has positive spherical aberration and the red has negative spherical aberration. This change in spherical aberration with wavelength is called *spherochromatism*.

The third case is a doublet with FK54 and KF9 glasses. Although KF9 is a normal glass, FK54 has an extremely high anomalous dispersion, resulting in a much better secondary spectrum correction than in the previous case. FK54 is over 30 times the cost of standard BK7 glass!

The fourth case is a doublet with FK52 and KZFSN4. Both glasses have anomalous dispersions. The polychromatic spot diameter is less than 1 μm, and the secondary spectrum is completely corrected. Furthermore, the spherochromatism is extremely well corrected as well.

The second parametric study is the analysis of spherical aberration and secondary spectrum correction as a function of the lens $f/\#$ for a chosen set of glasses.

The first pair of glasses are FK52 and KZFS1, as shown in Fig. 6.7. As demonstrated for an $f/10$ lens, the secondary spectrum can be very well corrected, since both glasses have anomalous dispersion. The Abbe numbers for these two glasses are very different; however, the relative partial dispersion is not. In the case of the $f/2$ lens, the spherical aberration is dominant, with the spot diameter close to 200 μm. As the $f/\#$ increases to $f/5$, the spot diameter decreases dramatically to about 3 μm. Chromatic aberration is more pronounced in this case. As the $f/\#$ increases to $f/20$, both the spherical and the chromatic aberrations are extremely well corrected, and the spot diameter is less than 1 μm.

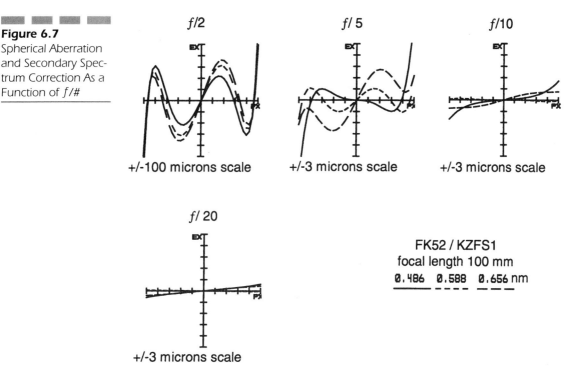

The second pair of glasses to be considered is LASFN31 and SFL6, as shown in Fig. 6.8. Although these glasses are high-index glasses and have anomalous dispersion, their relative partial dispersion is quite different and the secondary spectrum correction is poor. The difference in the Abbe numbers is not large. This makes it difficult to correct for the spherical aberration too, since we need to correct both aberrations simultaneously, trying to find the optimum balance between the two. In the case of the $f/2$ lens, the spot diameter is around 400 μm. As the $f/\#$ increases, the spot diameter decreases to about 2 μm at $f/20$, largely because of the uncorrected secondary spectrum.

The third parametric study is done for the case of an $f/4$ achromatic doublet using different glass combinations but allowing one surface to be aspheric for nearly complete correction of spherical aberration. This allows us to better see the change in residual spherical aberration with wavelength or spherochromatism. Four doublets are compared, with the same glass combination as the first parametric study. Ray aberration curves are shown in Fig. 6.9. In all four cases, the spherical aberration for the central wavelength is almost perfectly corrected with the asphere on

Figure 6.8
Spherical Aberration
and Secondary Spec-
trum Correction As a
Function of $f/\#$

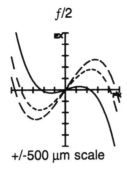

$f/2$

+/-500 µm scale

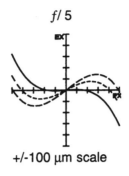

$f/5$

+/-100 µm scale

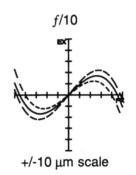

$f/10$

+/-10 µm scale

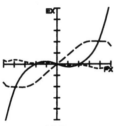

$f/20$

+/-3 µm scale

LASFN31 / SFL6
focal length 100 mm
0.486 0.588 0.656 nm

Figure 6.9
Secondary Spectrum
Correction As a Func-
tion of Glass Selec-
tion with One
Aspheric Surface

BK7 / SF2

PSK52 / SSKN8

FK54 / KF9

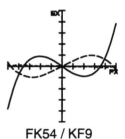

FK52 / KZFSN4

$f/4$
focal length 100 mm
+/-10 µm scale

0.486 0.588 0.656 nm

the front surface of the doublet. The spot diameter is determined by the chromatic blur of the blue and the red wavelength, and the difference between four cases of glass combination is small. Note that the shape of the ray aberration curves for the blue and the red color are similar in all cases.

How to Select Glass

Let us imagine that we have to design a wide field-of-view objective that operates in the visible spectrum. We will most likely start with five to seven elements and select a starting configuration based on our prior work, a patent, or we may elect to derive the design from basic principles. In most cases, it is sufficient to optimize a lens for only three wavelengths if the system is a visual system. However, in the case of a large field of view, it is better to work with five wavelengths properly weighted and a larger number of field points because of the potential problems with lateral color correction. At some point in the design, we will start to change the glass types either manually or by varying the glass characteristics automatically in the optical design program, and allowing them to move across the glass map until they settle in the locations which provide the lowest merit function. We will notice that the glasses tend to go from FK, PSK, across SK, LaK, LaSF, to SF. Very rarely a chosen glass will be KF, LLF, LF, or F. Even the glasses, such as BaLF, BaF, or BaSF, are not so often chosen in the optimization process. However, there are some types of glasses that are in the central region of the glass map, such as KzFS and TiF, that are often chosen in the glass optimization process. The reason for this is that, unlike BaF or BaLF glasses, which are the normal glasses, these glasses have anomalous relative partial dispersions and color correction, particularly secondary spectrum, is much easier with these glasses.

Generally, glasses in the central region of the glass map (BaLF, BaF, BaSF glasses) are not frequently used because they are normal glasses and secondary spectrum is not easily corrected with them. The other reason is that their *V* number has a medium value, which means that primary color is not easily corrected either. The exception is KzFS glasses, which have anomalous dispersion. Unfortunately, as will be discussed later, KzFS glasses are not preferred Schott glasses.

Now comes a very important step in the optical design—we have to check many parameters for each glass, including its availability, price, transmission properties, thermal properties, staining, etc., and make sure that the glass choice is the optimum one. Here the most important parameters an optical designer has to consider in the process of glass selection will be described. This will also be covered in Chap. 17 from a manufacturer's perspective.

AVAILABILITY   Glasses are divided into three groups: preferred, standard, and inquiry glasses. Preferred glasses are always available. Note that just because a glass is "preferred" does not mean that it is of good optical characteristics or low in cost, nor does it mean that the glass is easy to work in the shop; it only refers to availability. Quick delivery of standard glasses is generally possible as these glasses are generally in inventory. Inquiry glasses are available only on request, and they are normally not in stock. An optical designer should make every effort to design the system with preferred glasses. The optical design software program Zemax contains an option to use only preferred glasses from the Schott catalogue when a system is optimized with the "hammer" optimization and "substitute" glasses are used. This will be illustrated in the case studies in Chap. 21.

TRANSMITTANCE   Most optical glasses transmit light well in the visible and the near IR wavelength spectrum. However, in the near UV, the light is more or less absorbed by most glasses. If an optical system has to transmit UV light, the most commonly used materials are fused silica and fused quartz. Some optical glasses, such as a few SF glasses, have a reduced transmittance in the deep blue wavelength spectrum, and they have a yellowish appearance. Glass absorption as a function of wavelength is given in glass catalogues for 5- and 25-mm-thick glass plates.

INFLUENCE OF STRESS ON THE REFRACTIVE INDEX   Optically isotropic glasses become anisotropic through mechanically and thermally induced stress. This means that the s and p polarization components of light undergo refraction with different indices of refraction. High index alkali lead silicate glasses (dense flints) display a relatively large absolute refractive index change with a small stress birefringence. On the other hand, borosilicate glasses (crowns) exhibit a small absolute change in refractive index with a relatively large stress birefringence.

If the optical system has to transmit polarized light and has to maintain the state of polarization throughout the system or part of it, the choice of materials is very important. For example, when there is a prism in such a system with a relatively large mass, and there is a source of heat in its proximity, there can be a gradient of temperature inside the prism. It will introduce stress birefringence, and the polarization axis will be rotated inside the prism. In this case a better choice for the prism material should be one of the SF glasses rather than crown glasses.

CHEMICAL PROPERTIES   Optical glasses acquire their properties through their chemical composition, melting process, and finishing methods. In order to achieve the desired optical properties, optical glasses often exhibit reduced resistance to environmental and chemical influence. There is no single test method sufficient to describe the chemical behavior of all optical glasses. Four characteristics of glass resistance to environmental and chemical influence are given for each type of glass. In the Schott catalogue, glasses are sorted in four groups depending on their climatic resistance, which is the resistance to the influence of water vapor in the air. Water vapor in the air, especially under high relative humidity and high temperature, can cause a change in the glass surface in the form of a cloudy film that generally cannot be wiped off. Glasses are sorted in six groups depending on their stain resistance, which is a resistance to the influence of lightly acidic water without vaporization and possible changes in the glass surface. When the glass is in contact with an acidic aqueous medium, not only can stains appear on the glass surface, but the glass can also be decomposed. Optical glasses are divided into eight groups according to their resistance to acids. The last division of glasses is in four groups according to their resistance to alkalis.

THERMAL PROPERTIES   Optical glasses have a positive coefficient of thermal expansion, which means that glasses expand with an increase in temperature. The expansion coefficient, α, lies between 4 e-6 and 16 e-6/K for optical glasses. There are a few things that one should consider when designing an optical system to work in the given temperature range:

■ Thermal expansion or contraction of glass should not be in conflict with the expansion or contraction of the lens housing.

■ The optical system may have to be athermalized, which means that the optical characteristics of the system are unchanged with a change in the lens shape and index of refraction with temperature change.

▪ Change in temperature can cause temperature gradient in glass, and this can result in temperature-induced stress birefringence.

Most optical design programs have the capability of system optimization simultaneously at several different temperatures. The programs take into account both the expansion of glass elements and changing of their shape, expansion of the housing and the spacers between lenses, as well as the change of index of refraction of the glass materials.

Plastic Optical Materials

In a high-volume production environment, optical components or optical systems require low-cost materials and low-cost fabrication techniques. Plastic optics are used frequently today primarily for this reason. Plastic optical materials also have lighter weight, higher impact resistance, and offer more configuration possibilities than glass materials. Configuration flexibility is one of the greatest advantages of plastic optics. Aspheric lenses and elaborate shapes can be molded, for example, lenses with integral mounting brackets, spacers, and mounting features for easy alignment.

There are some issues, however, that must be considered when using plastic as an optical material. The principal disadvantage of plastic is its relatively low heat tolerance. Plastic melts at a much lower temperature than glass. It is less resistant to surface abrasion and chemicals. Adhesion of coatings on plastic is generally lower than on glass because of the limitation on the temperature at which the coatings are deposited, due to a low melting temperature of plastic. Further, the durability of coatings on plastic lenses is less robust than on glass. In addition, coatings on plastic often craze over time. The use of ion-assisted deposition of plastic coatings offers harder and more durable coatings on plastic.

The choice of optical plastic materials is very limited, which means that there is not a lot of freedom in the optical design process. A very important limitation is the high thermal coefficient of expansion and a relatively large change in refractive index with temperature. The refractive index of plastic materials *decreases* with temperature (it increases in glasses), and the change is roughly 50 times greater than in glass. The thermal expansion coefficient of plastic is approximately 10 times higher than that of glass. High-quality optical systems can be designed with a combination of glass and plastic lenses. In a combination with glass components in the system, plastic lenses can reduce the price and complexity of the optical system tremendously. When the optical power is mainly distributed over the glass components in the system, with one or two weak-powered plastic aspheric correctors, optical aberrations, especially distortion in wide field-of-view systems, can be very efficiently removed. Weak-powered plastic elements are used to minimize the effect on focus with temperature change.

Plastic optics can be injection molded, compression molded, or fabricated from cast plastic blocks. Fabrication of plastic elements by machining and polishing from cast plastic blocks is economical in the case of large optical elements, where the molding process has severe limitations. Compression molding offers a high degree of accuracy and control of optical parameters. However, injection molding is the most economical process. It offers moderate optical performance, which is acceptable in a lot of applications. Manufacturing of molds is an expensive process, but it pays off in high-volume production. During the system development phase, plastic optics can be very successfully diamond turned for prototyping, since the cost of diamond turning is lower than the cost of the manufacturing of molds. With today's high-quality diamond turning, the scattering effect from the turning grooves is most often under control, and if you have a good vendor, this should not be of concern for visible applications. Sometimes "postpolishing" is required to remove the turning mark residuals.

During the design of systems with plastic elements, the optical designer has to control the shape of the lenses more carefully than for glass elements. The shape (or bending) of the lens should be optimized for a good flow of the plastic material inside the molds. The thickness of the lens should be quite small, and the parting line, which is the line of contact of two molds, should go through the lens material. It is also important to eliminate inflection points on the lens surfaces in the case of compression molding. This limits the available lens shapes, and requires more parameters to be controlled in the optimization process. Additionally, the lens shape and the refractive index change with temperature have to be monitored, or the system has to be optimized for a given temperature range.

A few of the most commonly used plastic materials are acrylic (poly-methyl methacrylate), polystyrene, polycarbonate, and COC (cycloolefin copolymer):

Acrylic. The most common and important optical plastic material. It has a good clarity and a very good transmission in the visible spectrum, a high Abbe number (55.3), and very good mechanical stability. Acrylic is easy to machine and polish, and it is a good material for injection molding.

Polystyrene. Also a good plastic, cheaper than acrylic, but it has a slightly higher absorption in the deep blue spectrum. Its index of refraction (1.59) is higher than that of acrylic but the Abbe number is lower (30.9). It has a lower resistance to ultraviolet radiation and scratches than acrylic. Acrylic and polystyrene make a viable achromatic pair.

Polycarbonate. More expensive than acrylic, but it has very high impact strength and a very good performance over a broad temperature range. Polycarbonate is often used for plastic eyeglasses. A common form of polycarbonate in eyeglasses is CR39.

COC. A relatively new material in the optics industry, it has many characteristics similar to acrylic. However, its water absorption is much lower and it has a higher heat distortion temperature. COC is also brittle. A new brand name for COC is Zeonex. Comparative properties of optical plastics are shown in Table 6.1.

TABLE 6.1

Optical and Physical Properties of Optical Plastics

Property	Acrylic	Polystyrene	Polycarbonate	COC
Index @588 nm	1.49	1.59	1.586	1.533
Abbe#	55.3	30.87	29.9	56.2
$dn/dT \times 10^{-5}/°C$	-8.5	-12	-10	-9
Linear expansion coefficient/°C	6.5×10^{-5}	6.3×10^{-5}	6.8×10^{-5}	6.5×10^{-5}
Transmission (%)	92	88	90	91
Birefringence	Low	High/low	High/low	Low
Tensile strength (lb/in^2)	10,000	6000	9000	8700
HDT at 264 $lb/in^2(°C)$	92	82	142	120–180
Impact strength (ft-lb/in)	0.3	0.4	>5	0.45
Density (g/cm^3)	1.2	1.05	1.2	1.02
Water absorption (%)	0.3	0.02	0.15	0.01
Advantages	High stiffness, hardness, chemical resistance, and low cost	High index and low cost	Excellent impact resistance and high HDT	High stiffness, high HDT, low water absorption
Disadvantages	Brittle and heat resistant	UV absorption, birefringence, and low-impact strength	High birefringence, low Abbe #, and poor scratch resistance	Brittle

Spherical and Aspheric Surfaces

Definition of an Aspheric Surface

A *spherical* surface is defined by only one parameter, the radius or curvature of the surface. If the surface is refractive, with different indices of refraction before and after the surface, then the power of the surface is defined by the surface radius and the indices of refraction of the two media. *Radius* and *curvature* are reciprocal to one another.

Figure 7.1*a* shows a plano convex lens element with a spherical radius, imaging an axial point from infinity. The spherical aberration is quite evident. The high angle of incidence of the upper limiting ray of approximately 45° to the surface normal causes this ray to refract very strongly and ultimately to cross the axis significantly closer to the lens than rays closer to the optical axis. A spherical surface has the property that the rate of change of the surface slope is exactly the same everywhere on the surface, and thus the aberration is inevitable. Let us consider reducing the slope of the surface toward the outer periphery of the surface in order to flatten the shape in the region surrounding the outer rays. If we make the surface shape gradually flatter as we proceed outward from the optical axis, we can differentially reduce the refracting ray angle so that the net effect is to bring all of the rays to a common focus position, as shown in Fig. 7.1*b*. Figure 7.1*c* compares the spherical surface, which is steeper at its edge, with the aspheric surface, which is flatter at its edge. While correction of spherical aberration is not the only application of aspheric surfaces, it is one of the major application areas.

Aspheric surfaces cannot be defined with only one curvature over the entire surface because its localized curvature changes across the surface. An aspheric surface is usually defined by an analytical formula, but sometimes it is given in the form of a sag table for coordinate points across the surface. The sag of a surface is shown in Fig. 7.1. The most common form of an aspheric surface is a rotationally symmetric surface with the sag defined as

$$z = \frac{c\,r^2}{1+\sqrt{1-(1+k)c^2r^2}} \;+\; \Sigma a_i r^{2i}$$

where c is the base curvature at the vertex, k is a conic constant, r is the radial coordinate measured perpendicularly from the optical axis, and $a_i r^{2i}$ are the higher-order aspheric terms.

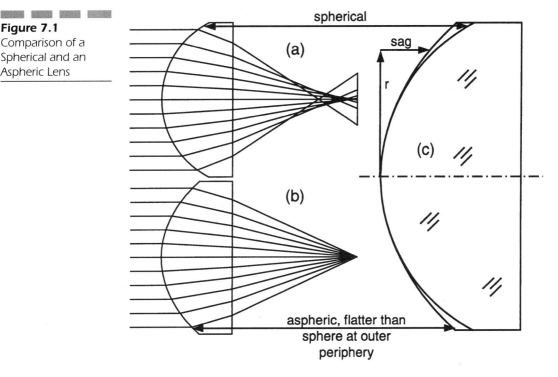

Figure 7.1
Comparison of a
Spherical and an
Aspheric Lens

When an aspheric surface is not rotationally symmetric, it is given either as a *biconic* surface with two basic curvatures and two conic constants in two orthogonal directions or as an *anamorphic asphere*, which has additional higher-order terms in two orthogonal directions.

Another form of aspheric surface is a *toroid* or *toric*. A toroid has, in effect, the shape of a doughnut. If a doughnut were sitting on the table, we all would agree that it had a basic outer diameter. If we now cut the doughnut vertically into two halves and we look at the cut, we see a circle whose diameter is less than the diameter by perhaps a factor of 5 or thereabouts. These two radii define a toroid, the overall outer radius of the doughnut and the smaller cross-sectional radius. If you work with torics in lens design, it is extremely important that you understand fully and completely the definition used by the computer program you are using, so take the time to study the manual in depth in this regard. In addition, if need be, set up a sample to assure that your understanding is correct. While a doughnut is one form of toroid, a football shape is another, and it is imperative to understand which one your equation is representing, especially if it is to be manufactured.

Conic Surfaces

In the case where the higher-order aspheric terms are zero, the aspheric surface takes the form of a rotationally symmetric conic cross section with the sag defined as

$$z = \frac{c\,r^2}{1+\sqrt{1-(1+k)c^2r^2}}$$

where c is the base curvature at the vertex, k is a conic constant, and r is the radial coordinate of the point on the surface. In Table 7.1, it is shown how a conic surface takes on the following surface types as a function of the conic constant, k, in the sag equation.

Figure 7.2 shows five surfaces having different conic constants but the same curvature. Most of us are generally familiar with the surface shapes described. One surface we do not come across often is the oblate ellipsoid, sometimes called the oblate spheroid. This can be thought of as the shape of the Earth as it rotates about its axis. Due to centrifugal force, the diameter is greater at the equator than in the polar direction. The oblate ellipsoid has its foci orthogonal to the optical axis.

Conic surfaces, either reflective or refractive, are free of spherical aberration for one particular set of conjugate points. Let us look into a set of different conic surfaces. A spherical surface forms an aberration-free image if the object is at the center of curvature of the surface. An ellipsoid forms an aberration-free image for a pair of real image conjugates

TABLE 7.1

Conic Section Types

Conic Constant k	Surface Type
0	Sphere
$k < -1$	Hyperboloid
$k = -1$	Paraboloid
$-1 < k < 0$	Ellipsoid
$k > 0$	Oblate ellipsoid

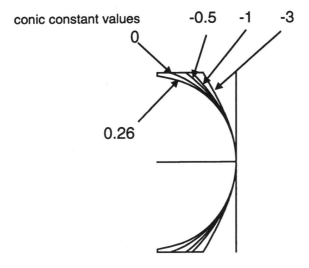

Figure 7.2
Conic Surfaces with
the Same Curvature
and Different Conic
Constants

on the same side of the surface and a hyperboloid for conjugates on two different sides of the surface. A parabolic mirror forms a perfect image of a point for an axial object at infinity. This is the reason why parabolic mirrors (sometimes combined with hyperbolic mirrors) are widely used in astronomical optics.

When the object point is moved axially from the position of the aberration-free conjugate, a certain amount of spherical aberration is introduced. If the point is moved laterally, other aberrations, such as coma, astigmatism, and field curvature, contribute to image blurring.

Application of Aspheric Surfaces in Reflective and Refractive Systems

Aspheric surfaces are widely used and often essential in reflective systems due to the small number of surfaces and typically large apertures. While a complex lens may consist of 18 spherical radii in order to minimize the aberrations, a reflective system can only have two surfaces in most cases. A simple spherical reflecting telescope suffers from spherical aberration and coma. A spherical mirror is shown in Fig. 7.3a. A point object at infinity is focused by the spherical mirror at a distance from the mirror equal to the one-half of the mirror radius, and this distance is the focal length. Third-order spherical aberration results, and the

Figure 7.3
Conic Surfaces for
Reflecting Systems

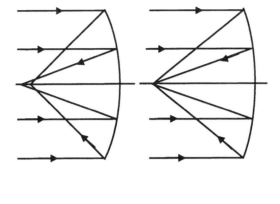

(a) spherical
reflector

(b) parabolic
reflector

wavefront error or OPD is proportional to the (aperture)^4, as described
in Chap. 5.

A parabolic reflecting telescope is shown in Fig. 7.3 b. This is a classic
example of how spherical aberration can be corrected. An infinitely distant axial point is imaged to a perfect aberration-free image point.
Unfortunately, the image quality degrades quickly when the object is
moved off axis. Coma is the aberration that restricts the field of view of
the parabolic telescope to a very small field.

A very common form of reflecting telescope is the two-mirror
Cassegrain telescope, with both mirrors being conic surfaces. The *classical
Cassegrain* telescope has a paraboloidal primary mirror and a hyperboloidal secondary mirror, as shown in Fig. 7.4. f1 is the location of the
image which would be formed by the large primary mirror, and ƒ2 is the
location of the image of the entire system. Note that the secondary mirror reimages f1 to f2. A similar configuration, called a *Ritchey-Chrétien*
Cassegrain telescope, has both primary and the secondary hyperbolic
mirrors. The classical Cassegrain performance is limited by off-axis
coma while the Ritchey-Chrétien Cassegrain is, in effect, a coma-free
Cassegrain. Its limiting aberration is astigmatism.

Another well-known type of telescope is a Schmidt telescope, which
is shown in Fig. 7.5. It consists of a spherical mirror with an aspheric
corrector plate located at the center of curvature of the mirror. Third-
order spherical aberration results in a wavefront aberration function
proportional to the (aperture)^4. If the aperture stop is located at the

center of curvature of the mirror, there is symmetry for all field positions. Apart from the aperture stop being obliquely viewed by the oblique bundle of off-axis rays, the oblique rays are focused in the same manner as the axial bundle. The chief ray is normally incident onto the mirror everywhere in the field of view. The image is formed on a spherical surface, with the image radius equal to one-half the mirror radius.

Without any aberration correction, the rays that are closer to the aperture edge are focused closer to the mirror than the paraxial rays. The wavefront distortion, which is proportional to the fourth power of the aperture radius, can be corrected with a wavefront distortion of the opposite sign introduced by the aspheric corrector plate placed in the aperture stop as in Fig. 7.5*b* to provide effective "parabolization" of the spherical mirror. A fourth-order aspheric deviation from the flat base surface of the glass corrector introduces a negative fourth-order

Figure 7.4
Cassegrain Telescope

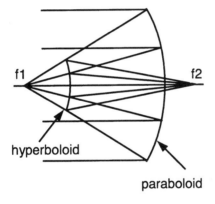

f1   f2

hyperboloid

paraboloid

Figure 7.5
Schmidt Telescope

mirror
radius

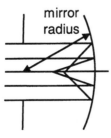
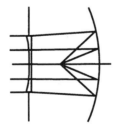
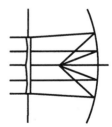

(a) spherical mirror   (b) classical solution   (c) weak spherical
power added to
the corrector

wavefront distortion. The aspheric refractive corrector reduces the spherical aberration. However, some chromatic aberration is introduced by the wedged shape at the outer periphery of the corrector, close to the aperture edge. In order to minimize this chromatic aberration, a very weak positive power is added to the corrector, such that the corrector has zero power at 0.7 of its aperture, as shown in Fig. 7.5c. This shape of the corrector is not only the optimum shape for the correction of spherochromatism (the variation of spherical aberration with wavelength), it is also best suitable for manufacturing.

The majority of optical systems are based on the use of spherical components because they are easier to manufacture. However, there are cases where aspheric optical components have a significant advantage over spherical ones. In astronomical optics, reflective aspheric components are widely used. Today, with the significant development of new plastic materials, low-cost molded aspheric refractive optics are finding their place in the large consumer market. Precision diamond grinding and compression molding of glass aspheric lenses is also becoming more common.

Refractive aspheric lenses are widely used in many kinds of illumination systems, from the condensers in projection systems and microscopes, to street lamps and searchlights. Since in many cases these are not imaging systems, the manufacturing tolerances on these components are somewhat forgiving. In the case where the optical components are not exposed to the heat from the light source, aspheric optical elements can be injection molded. In projection systems, aspheric condensers are often molded glass lenses, mostly from B270 glass. B270 is a very common low-cost glass similar to BK7, which is used extensively in "float glass" for low-cost windows and mirrors. Glasses that are moldable have a lower temperature at which they become soft than the standard optical glasses. Heat generated by the light source, or by the absorption of light by the optical components themselves, is often a severe problem in optical systems and requires the use of heat-resistant materials such as fused silica or fused quartz.

Another field where aspherics find their place is in systems for focusing of laser beams, for example, in CD players or data storage systems, when coupling light from a laser diode into a fiber, or when collimating light from laser diode arrays. These applications require high-precision optical components, as well as optically stable components in a given temperature range. The small size of these components makes them easier to mold. Optical plastics are used wherever they are acceptable, because of the much lower manufacturing costs. However, accurate lens shape and the very good temperature stability of glass aspheric lenses make them a better solution for applications where high precision is required. Glass aspheric lenses are manufacturable in diameters smaller than 25 mm. At diameters greater than 25 mm, aspheric glass lenses become too expensive for high-volume manufacturing. Single-glass bi-aspheric lenses are used for focusing or collimation of $NA = 0.5$ laser beams with diffraction-limited performance.

An infinitely distant object imaged through a planohyperbolic lens is focused to an aberration-free spot. This feature is used in the case of a planohyperbolic fiber lens (the cross section of the fiber has a planohyperbolic shape) to collimate the fast axis of the laser diode arrays, as shown in Fig. 7.6. The distance of the laser diode from the lens is determined by the index of refraction of the lens.

Improvements in the quality of injection-molded plastic optics make the use of them possible in camera lenses and projection lenses. The optical design of these kinds of lenses is difficult because many parameters have to be considered such as

▪ The location of the component in the lens to minimize the beam size over the plastic component.

▪ The shape of the lens to keep the molding parting line inside the component.

▪ Athermalization of the lens.

In high-performance lenses, plastic components should be away from the aperture stop because it is very difficult to achieve diffraction-limited performance with injection-molded plastic components. However, they can be extremely useful in the correction of field aberrations such as astigmatism, field curvature, and also distortion.

Much of the discussion thus far with respect to aspheric surfaces and their benefits has related to the correction of spherical aberration. If an

Figure 7.6
A Planohyperbolic
Collimator

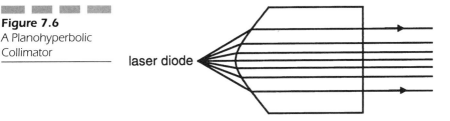

laser diode

aspheric surface is located at or near the aperture stop of a system, it will primarily affect or benefit spherical aberration, which is an axial aberration which, for the most part, carries across the field of view. As aspheric surfaces are located further from the stop, they can help to minimize some or all of the off-axis aberrations such as coma and astigmatism.

A good example of the application of an aspheric surface used for astigmatism correction is shown in Fig. 7.7. In Fig. 7.7a, we see a single-element lens with its aperture stop located far to the left of the lens. If the curved lens surface is spherical, the oblique rays create a footprint on the surface, which is larger in the plane of the figure than the orthogonal plane in/out of the figure. As we discussed in Chap. 5, this tends to refract and pull the rays in the plane of the figure inward from where they would otherwise focus. We now need to ask ourselves what would it take to push the focus position outward and compensate for this inward focus shift. The answer is to create a more negatively powered surface in the plane of the figure at the outer periphery of the lens. This is shown in proper scale in Fig. 7.7b and in an exaggerated form in Fig. 7.7c. This more negatively powered surface shape in the plane of the figure has virtually no effect in the orthogonal plane, hence the highly efficient correction of astigmatism by the aspheric surface.

Another common use of aspherics is in the thermal infrared where the cost of materials is extremely high. With the use of aspherics, the number of elements can be reduced to a minimum.

Guidelines in the Use of Aspheric Surfaces

The proper usage of aspheric surfaces is extremely important. This includes which surfaces to make aspheric and whether to use a conic section or, alternatively, a higher-order aspheric. The conic sections include paraboloids, hyperboloids, and ellipsoids, as discussed earlier in this chapter. The higher-order terms are surface departures from conic, which are proportional to r^4, r^6, r^8, r^{10}, and so on, where r is the radial distance from the optical axis. The simpler forms of reflective systems, such as the classical Cassegrain (paraboloidal primary, hyperboloidal secondary) and the Ritchey-Chrétien Cassegrain (two hyperboloids), were discussed earlier. The classical Cassegrain is limited by coma and field

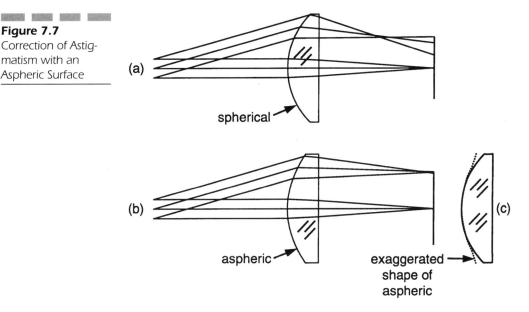

Figure 7.7
Correction of Astig-
matism with an
Aspheric Surface

curvature, and the Ritchey-Chrétien is, in effect, a coma-free Cassegrain which is limited by astigmatism and field curvature. Once the basic system is set up on your computer, varying the appropriate conic constants is all it takes to reach a viable solution.

But how do we decide which surface, or surfaces, in a lens system should be made aspheric, and how do we decide what form of aspheric to use? To answer this question, first consider Fig. 7.8, where we show: (1) the aspheric surface departures from the base spherical surface and (2) the departures from what we call the "nearest sphere" or "best-fit sphere" to the aspheric surface. The nomenclature is shown in the top of Fig. 7.8 with accurate data below.

If we now compute for our baseline spherical optical system a plot of the optical path difference, we should look for a form matching the basic profile or character of these data. For example, if the axial OPD plot resembles the form of the sag from the nearest sphere for the r^6 case, then varying the r^6 coefficient on a surface near the aperture stop will likely be beneficial. If we find a sharp increase or decrease in the OPD off axis at the edge of the pupil, then varying a higher-order term or two on a surface away from the stop will likely be beneficial. There are some basic guidelines, and these are listed here:

Figure 7.8
Aspheric Sags from
Vertex Sphere (Top)
and Best-Fit Sphere
(Bottom)

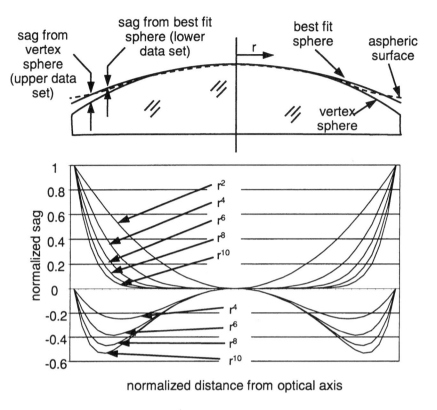

normalized distance from optical axis

1. Conic surfaces can be used for correcting third-order spherical aberration and other low-order aberrations.

2. If you have a nearly flat surface, then use an r^4 and higher-order terms rather than a conic.

3. If you have at least a somewhat curved surface, then you can use the conic along with higher-order terms if required.

4. It is generally best not to use both a conic and an r^4 surface, as they are mathematically quite similar. This is because the first term of the expansion of a conic is r^4. While they can both literally be used, the optimization process often tends to beat one against the other, yielding artificially large coefficients, and this may have an effect on the convergence of the optimization.

5. Use aspherics beginning with the lower-order terms and working upward as required. If you can stay with conics, this may make testing more manageable. You should be able to assess the need for adding terms based on the character of the OPD plot.

6. It is very dangerous to use a large number of aspheric surfaces, especially with higher-order terms. This is because they will beat against each other. This means that as one surface adopts a certain aspheric profile or contour, it may increase in its asphericity, with its effect cancelled by adjacent surfaces. For example, if the first of two closely located aspheric surfaces has significant surface departure from sphericity, the neighboring aspheric surface could very likely cancel this effect. While the lens may perform well on paper, we now need to manufacture two highly aspheric surfaces, a difficult and expensive task which may not be necessary.

7. If possible, optimize your design first using spherical surfaces, and then use the conic and/or aspheric coefficients in the final stages of optimization. This may help in keeping the asphericities to a more manageable level.

Specification of Aspheric Surfaces

It is important to specify an aspheric surface sufficiently enough to convey to the shop both what you want and what you need. The following items are most often included in specifying aspheric surfaces:

1. The surface to be aspheric is labeled aspheric on the component drawing.

2. You should include an equation of the surface shape along with the aspheric coefficients. A small sketch indicating the nomenclature and sign convention is recommended.

3. A table listing the sag as a function of the radial distance from the surface vertex normal to the optical axis, r, is imperative. You should list a sufficient number of data points to adequately sample the surface profile.

4. You should list how close the actual surface must come to the ideal design prescription. The form of this can be "surface to match nominal surface to within four visible fringes (or 0.001 mm) over clear aperture."

5. You may need to call out higher-frequency surface irregularities and/or surface finish. The higher-frequency irregularities can be called out by indicating the maximum slope departure from

nominal over the surface. Surface finish is normally called out by indicating the rms surface finish, in nanometers. This latter callout is generally used for diamond-turned surfaces, where surface roughness is sometimes a problem or where scattering and off-axis rejection is of major concern such as in space telescopes.

6. You should, if possible, indicate the form of testing to be used.

Do keep in mind that the more callouts you list and the more extensive the testing, the more costly your optics will be, and they will likely take longer to manufacture. Your callouts should indicate only what you need functionally for your system to work properly.

Design Forms

Introduction

In this chapter, we will discuss how we select the proper design form or configuration for both refractive and reflective image-forming systems. We will also consider fold mirrors and prisms since they have a significant influence on the system design configuration.

The proper system design form or "configuration" of an optical system is generally the key to a successful design effort. The term "configuration" here means the basic form of the system which includes not only the number of elements, but also the relative optical power and distribution of the elements within the lens system. For example, an achromatic doublet of two cemented elements, as shown in Fig. 8.1*a* is clearly different in form from a Cooke triplet, which consists of three separated elements, as in Fig. 8.1*b*, with two outer positive crown elements and a negative flint element at the center. The Cooke triplet can be used over wider fields of view than a doublet due largely to a reasonable degree of symmetry fore and aft of the central element, which is at or near the aperture stop. The doublet and triplet are very different configurations.

What if we were to add a single positively powered element immediately following a cemented doublet, as in Fig. 8.1*c*? Would the lens configuration be called a triplet? It certainly would not be a Cooke triplet as the symmetry is not present. This is a very different configuration or design form from a Cooke triplet. However, it is, of course, a three-element lens. The same is true for Fig. 8.1*d*, where we again have three elements, only here the third element is very near to the image and is serving to both flatten the field as well as correct astigmatism. All three of the three-element configurations are quite different in configuration or form, each with their relative advantages.

The selection of an optimum configuration prior to initiating a design effort provides the starting point from which the design optimization proceeds. While lens design software has improved significantly over the years, the programs are rarely capable of changing configurations, and never add or delete elements. Most of the time the program will reach an optimum or local minimum in the error function for the input configuration. For more information on the optimization process, see Chap. 9. The configuration selection is driven by many factors. Nearly every system specification can have an influence on the configuration. The major factors influencing the configuration selection are

Figure 8.1
Doublets and Triplets

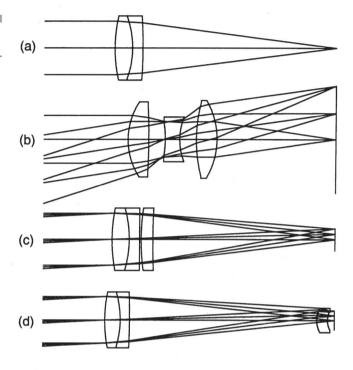

(a)

(b)

(c)

(d)

- Field of view
- Performance requirement
- $f/\#$
- Packaging requirements
- Spectral range

System Configurations for Refractive Systems

We will review a progression of configurations for lenses in order to illustrate just what differentiates one from another. The following configuration forms are shown in Fig. 8.2.

SINGLE-ELEMENT LENS (FIG. 8.2a)   A single-lens element has generally poor image quality and a very small field of view. Further, it suffers

Figure 8.2
Progression of
Configurations

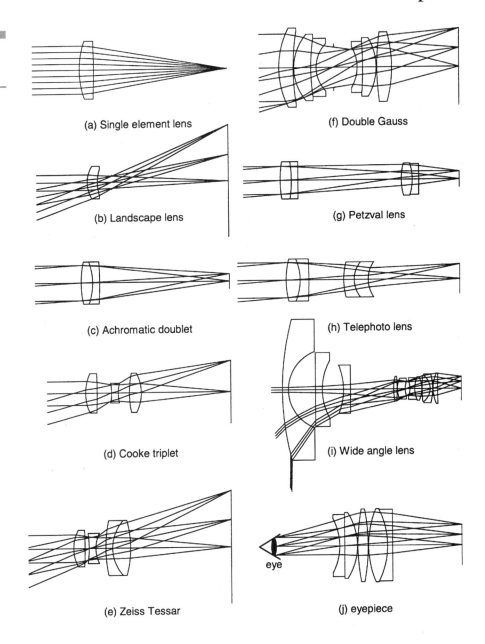

(a) Single element lens

(b) Landscape lens

(c) Achromatic doublet

(d) Cooke triplet

(e) Zeiss Tessar

(f) Double Gauss

(g) Petzval lens

(h) Telephoto lens

(i) Wide angle lens

eye

(j) eyepiece

from chromatic aberrations and it can only be used at a high $f/\#$. We often think of a single element as of "magnifying glass quality."

It is important to note that the performance, which may be poor for one application, may be just fine for another. For example, we have stated that a single element generally has poor image quality. This is, for the most part, true for most critical imaging applications such as camera lenses, machine vision optics, and other similar applications. However, if you are looking for a photon collector with little or no image quality requirements, then a single element may be quite adequate for the task. Another good example is the optics used for optical data storage and other microoptics applications. For data storage applications, a laser diode is imaged to a micron or submicron spot diameter. Since the laser is nearly monochromatic (there may be thermally induced shifts in wavelength), the field of view is nearly zero and the scale or size of the system is extremely small, we often find that a single aspheric element is sufficient for the task.

LANDSCAPE LENS (FIG. 8.2 *b*) While a landscape lens is also a single element, it has an aperture stop which is remote or separated from the lens itself. Further, the lens is bent or "curled" around the stop for symmetry reasons. This reduces the angles of incidence on the surfaces and thereby reduces off-axis aberrations. Earlier in Chaps. 3 and 5 we discussed in greater depth how minimizing angles of incidence within a lens system reduces aberrations. A landscape lens can have its aperture stop either aft of the element as shown or in front of the element. It can be shown that the aberrations are somewhat reduced if the stop is in front of the lens, and many early box cameras were constructed this way. The landscape lens has chromatic aberration as well as residuals of many of the other third-order aberrations.

We show in Fig. 8.3 two forms of landscape lenses, one with the stop aft of the lens and the other with the stop forward of the lens. It can be shown that the performance is slightly improved with the stop in front. Also shown are two photos of the front of an early Kodak Brownie camera with a flat window followed by its aperture stop and finally the lens, which is aft of the stop.

ACHROMATIC DOUBLET (FIG. 8.2*c*) The achromatic doublet is capable of bringing the red and blue wavelengths to a common focus, with the central green or yellow wavelength defocused slightly toward the lens. A typical achromatic doublet has a blur diameter approximately

Figure 8.3
Landscape Lens with
Stop Aft and Forward
of Lens

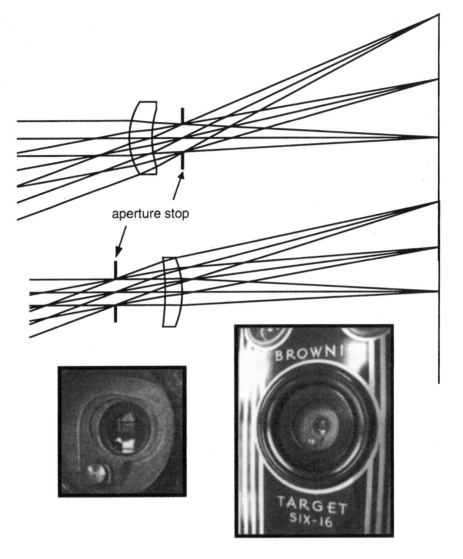

aperture stop

25 times smaller than an equivalent single-element lens (based on an $f/5$ lens in the visible spectral band).

A cemented doublet performs well only over a small field of view, and it cannot be used at low $f/\#$s due to higher orders of spherical aberration. In order to balance the inherent third-order spherical aberration of the cemented doublet, one can introduce a small airspace between the elements. This airspace will permit the balancing of fifth-order spherical aberration with the inherent third-order aberration for an improved overall level of performance. Further improvement can be realized by adding an additional element near the image, which can be used as a field flattener, and often it is possible to bend this element to balance and eliminate some or most of the astigmatism.

COOKE TRIPLET (FIG. 8.2*d*)   This three-element lens form takes advantage of symmetry in order to minimize the angles of incidence of the rays as they proceed through the lens over the field of view, and hence it is capable of an acceptable level of performance for many applications. The Cooke triplet was first designed in England by H. D. Taylor at the "Cooke and Sons" optical company. The Cooke triplet is the first configuration we have presented in this review that allows for the optimization and balancing of the seven primary, or third-order, aberrations, as well as the control of focal length. There are eight "useful" variables in a Cooke triplet, the six radii and the two airspaces. Element center thicknesses are generally not of significant use in aberration control and, for the most part, element thicknesses that yield reasonable manufacturing ease are best. We can thus control or optimize the following:

1. Spherical aberration
2. Coma
3. Astigmatism
4. Axial color
5. Lateral color
6. Distortion
7. Field curvature
8. Control of the focal length

It is important to note that just because we have the same number of useful variables as the number of primary aberrations (along with the focal length), this does *not* at all mean that the aberrations can be brought to zero or even close to zero. What it means is that for the *f*/number and focal length is selected, the aberrations can be reasonably well balanced against one another, especially the third-order aberrations. Thus, for example, an *f*/6 Cooke triplet covering a 10° full field of view over the visible spectrum will likely be capable of a reasonable level of performance. However, an *f*/1.4 Cooke triplet covering a 30° full field of view will probably provide fair to poor performance, at best. The low *f*/number will lead to significant spherical aberration residuals, and the wide field of view will lead to coma, astigmatism, and other off-axis aberrations.

ZEISS TESSAR (FIG. 8.2e) The Zeiss Tessar is derived from, and is an improvement upon, the Cooke triplet. Paul Rudolph of Zeiss Jena replaced the original single rear lens in the Cooke triplet with a doublet lens, resulting in a better lens performance, with higher resolution, excellent contrast, and very low levels of distortion.

There is a rule of thumb in lens design which says "clip it in the bud." What is meant here is that the best place to correct or eliminate aberrations is as close to where they are being introduced as possible. In the case of a Cooke triplet we see that the first positive lens element takes collimated light from infinity and bends or converges it into the second negative lens element. The second element takes the slightly converging light and diverges it into the third and positive element. And finally, the third positive element takes the slightly diverging light and bends the rays so as to create the required f/# of the lens. The amount of ray bending or redirection is greatest for the third element than for the first or second elements. Thus, the aberrations introduced by the third element will be greatest as well. This makes the third element an excellent candidate to convert to a doublet. As one of the more difficult aberrations to correct in a Cooke triplet is axial color (change in focal length with wavelength), making the third element a doublet (as in the Tessar) allows for a superior level of correction of this as well as other residual aberrations. Tessar designs can be effectively used at f/numbers down to f/4.5 or somewhat lower, depending on the relative mix and level of lens requirements. Further, as will become apparent, the Cooke triplet forms the basis of many more complex and high-performance configurations.

DOUBLE GAUSS (FIG. 8.2f) The double Gauss lens is yet a further extension toward improved performance at lower f/#s and wider fields of view. If we summarize what we have learned thus far, we can see more clearly the evolution of the double Gauss lens. The methodologies learned thus far include *splitting optical power* to minimize aberrations, using *negative-powered elements with smaller beam diameters* for field curvature correction as in the Cooke triplet, and using *symmetry fore and aft of the aperture stop* also for symmetry reasons, thereby minimizing the angles of incidence on the lens surfaces. As we will learn later, symmetry also allows for cancellation of several off-axis aberrations. The double Gauss lens uses at least two negatively powered elements near the stop and two or more positive elements on the outsides. Furthermore, there is a reasonably high level of symmetry surrounding the aperture stop.

The double Gauss lens is capable of good performance down to about $f/1.4$ and even lower. Indeed, there have been several $f/1.0$ double Gauss lenses in the 35-mm camera marketplace, including the famous *Nocitlux* designed by Walter Mandler of Ernst Leitz Canada. As with any lens, there are compromises, and at $f/1.0$ the lens is hardly diffraction limited; however, from an overall performance standpoint and light-gathering capability, the lens is a top performer. We should note that for the more demanding levels of performance (such as the Nocitlux), higher refractive index materials are often used as they can significantly reduce aberrations, as well as anomalous dispersion glasses for superior chromatic aberration correction. Glass selection is discussed in Chap. 6.

PETZVAL LENS (FIG. 8.2g) The Petzval lens represents a very different design philosophy. This lens is intended for smaller fields of view and only moderate $f/\#$s such as $f/3.5$ or slower. The design philosophy here is to use two separated doublets with the power task shared between the two. This yields lower secondary chromatic aberrations than a single doublet of the same net $f/\#$. This form of design is used for high-performance small field-of-view lenses as one might encounter in aerial reconnaissance for example. As noted, the Petzval lens is not well suited for wide-angle applications, as there is little opportunity for symmetry as with the Cooke triplet or the double Gauss.

TELEPHOTO LENS (FIG. 8.2h) The telephoto lens is a positive group of elements, followed by and separated from a negative group of elements. As we showed earlier in Fig. 8.1, this form of lens has a focal length longer than the physical length of the lens, hence the name "telephoto." The ratio of the physical length to the focal length is called the *telephoto ratio.* In the case of a fast lens (low f/number), with a low telephoto ratio below 0.6, for example, the lens configuration becomes quite complex. In the limit, if the light exits the second group collimated, we have a beam contractor or, in effect, a Galilean telescope, and the focal length is infinite. It is important to note that both the positive and the negative groups generally need to be separately achromatized in order to produce a complete lens with sufficiently low chromatic aberration.

It is interesting to think about the use of the words "telephoto lens." To a photographer a telephoto lens is generally a lens whose focal length is longer than a standard lens for a specific film format. For example, the standard focal length for a 35-mm camera lens is in the order of 50 to 55 mm. A lens with a 100- to 135-mm focal length or longer is generally considered to be a telephoto lens. This is because objects appear closer due to the smaller field of view covered by the longer focal length lens. An achromatic doublet with a focal length of 135 mm might, to a photographer, also be a telephoto lens because it brings the object closer. However, to a lens designer a telephoto must be of a form or configuration as shown using a positively powered front group and a negatively powered rear group so that the focal length is longer than the physical length of the lens.

WIDE-ANGLE LENS (FIG. 8.2i) A lens covering a substantially wider field of view than a normal lens (in photography, for example) is called a *wide-angle lens.* Thus, with a standard focal length of 50 to 55 mm in 35-mm photography, a wide-angle lens is generally considered to be 35 mm or less. In order to cover the wider fields of view, we often use a strong negatively powered front element or group of elements to bend the rays outward to cover the wider field angles. In order to still image from infinity, the light in the space between the main body of the lens and the front negatively powered group needs to be converging toward the object as seen in Fig. 8.2i.

We show the wide-angle lens as having a negatively powered three-element front group, with a multielement configuration for the prime lens group. What is happening here is that the prime lens group, itself, is covering a smaller field of view, with the field angle increase happening only at the negatively powered front elements.

EYEPIECE (FIG. 8.2j) An eyepiece is used to visually view and magnify the image from a microscope objective or a telescope objective, or alternatively to view a display such as in a head-mounted display system. The eyepiece is a very different configuration of lens system in that the aperture stop is not only quite remote from the main part of the lens, but it is in reality the pupil of the eye. Eyepieces are normally designed by tracing rays from the eye, which is the aperture stop, to the image plane, as shown in Fig. 8.2j.

As can be seen in Fig. 8.2j the rays at the extreme field of view are primarily using the outer periphery of the lens elements. Unfortunately, this often results in significant amounts of astigmatism, lateral color, coma, and distortion. These field aberrations can be quite significant and difficult if not impossible to correct for in extremely wide field-of-view eyepieces with 60° field of view or more. The careful use of higher-index glasses combined with one or more aspheric surfaces (if possible) can help to mitigate the problems to some extent. Eyepieces represent a very different configuration form than other lenses discussed thus far.

System Configurations for Reflective Systems

As with lenses, reflective or mirror systems can be of many varied configurations. There are, however, some inherent and very basic differences between refractive or lens systems and reflective or mirror systems. Lens systems are most often straight through and use the full clear aperture of the entrance pupil, as shown in the simple telephoto lens example in Fig. 8.4a. Mirrors have the fundamental challenge that they get in each others way as shown in the two-mirror Cassegrain system in Fig. 8.4b. The Cassegrain is the reflective analogy of the telephoto lens described earlier in this chapter. Note that the large objective lens shown as an achromatic doublet is the optical analogy of the large concave mirror

Figure 8.4
Telephoto and Cassegrain Configurations

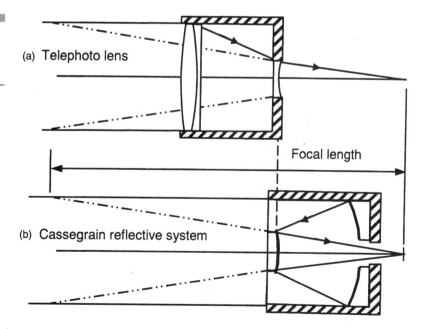

(a) Telephoto lens

Focal length

(b) Cassegrain reflective system

generally called the *primary mirror*. Further, the smaller lens in the refracting telephoto lens is the analogy of the small convex mirror in reflective systems, and it is generally called the *secondary mirror*. Note that both lens and mirror systems are in the form of the telephoto lens and the focal lengths are in fact equal in both systems.

The fundamental issue of the mirrors getting in each others' way in reflective systems leads to many significant differences between reflective and refractive systems.

We will review a progression of configurations for reflective systems in order to illustrate just what differentiates one from another. The following configuration forms are shown in Fig. 8.5:

PARABOLOID   A single parabolic mirror has zero spherical aberration on axis for an infinitely distant object. It is, however, limited by coma off axis. As shown, the image-forming light is often folded out to

Figure 8.5
Reflective
Configurations

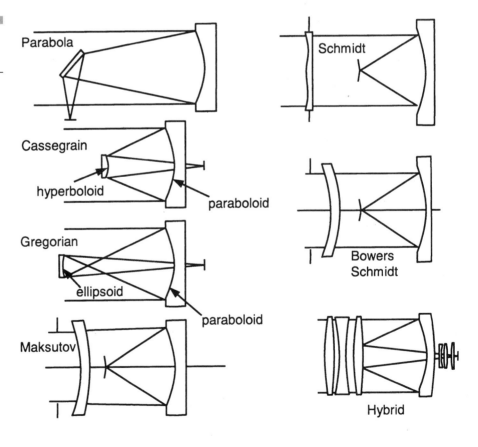

the side via a tilted flat-fold mirror sometimes called a diagonal mirror. In this case it is called a Newtonian telescope.

CASSEGRAIN   The Cassegrain form of reflective system is perhaps the most common. In its "classical" design form, a parabolic primary mirror and a hyperbolic secondary are used. With this prescription, coma is the limiting aberration and is the same as a single parabolic mirror of the same f/number. An improved level of performance is achieved by allowing the primary mirror to be hyperbolic along with a hyperbolic secondary. This solution is called the Ritchey-Chrétien form of Cassegrain. It is, in effect, a coma-free Cassegrain and is limited only by astigmatism and field curvature. The layout itself is virtually identical for the classical Cassegrain and the Ritchey-Chrétien forms. In most configurations, the Cassegrain has an inward or concave curving image field due to the Petzval contribution of the convex secondary mirror predominating over the concave primary mirror.

GREGORIAN   A concave parabolic primary mirror with a concave elliptical secondary mirror is the Gregorian form of telescope. In effect, the elliptical secondary mirror reimages the image formed by the primary mirror to its final position aft of the primary, as shown in Fig. 8.5. The Gregorian is not as common as it was some years ago, and the reason for this is that astronomers years ago did not believe that convex aspheric mirrors (as required by the Cassegrain) could be effectively fabricated and tested. With the advent of interferometry and other advances in optical metrology, testing convex aspheric surfaces became more viable and the more compact Cassegrain is now more widely used.

MAKSUTOV   The Maksutov uses a spherical primary mirror with a spherical weakly powered meniscus glass corrector plate to balance the spherical aberration of the mirror. This system has been popular with amateur astronomers for many years. A variation known as the *Maksutov-Cassegrain* locates the corrector closer to the primary, and an aluminized spot at the center of the convex surface acts like a Cassegrain secondary mirror, hence the name Maksutov-Cassegrain. This is the design form used in the well-known Questar telescope for many years.

SCHMIDT   The Schmidt system uses a thin aspheric corrector plate located at the center of curvature of a spherical primary mirror to effectively correct all orders of spherical aberration. The aperture stop is at the corrector plate, and is located at the center of curvature of the primary mirror. Due to this geometry, the chief rays at all field angles will be incident onto the primary perpendicular to its surface. To third order, the spherical aberration will be eliminated at all field angles. The image is formed on a spherical image whose radius is equal to the focal length of the primary mirror. Further, the Schmidt system works quite well at low f/numbers, even as low as f/1 or less.

The fact that most aberrations are zero is due to what is known as the "Schmidt principal," which has as its basis the aperture stop being located at the center of curvature of a spherical mirror. It can be shown that in addition to zero spherical aberration, there is no third-order coma, astigmatism, or distortion. The major residual aberration of tangential oblique spherical aberration is due to the ray obliquity on the corrector plate as well as the foreshortening of the entrance pupil at off-axis field angles. The Schmidt principal is an exceptionally powerful technique, which has been successfully applied in the design of many well-corrected optical systems. It is especially useful in wide-angle applications. A variation of the Schmidt, invented by Baker, uses three separated aspheric Schmidt plates for minimization of the off-axis residual aberrations.

There is a variation of the Schmidt system known as a *shortened Schmidt Cassegrain*. In this configuration, we move the corrector plate closer to the primary mirror to a location where the ray bundle diameter is similar to what it would be in a Cassegrain system. Then a convex mirror is mounted on the interior of the corrector plate and the system geometry now resembles a Cassegrain. The fact that the corrector plate and the aperture stop are no longer at the center of curvature of the spherical primary mirror are cause for coma and other off-axis aberrations; however, the overall performance is reasonably good for small fields of view. This form of system is extremely popular in contemporary amateur telescopes due to its robust performance combined with aggressive packaging.

BOWERS SCHMIDT   This configuration uses a thin corrector plate, which is a shell concentric about the aperture stop. The aperture stop is at the center of curvature of the primary mirror as with the Schmidt telescope. The net result is a level of performance that comes close to the Schmidt telescope but without aspheric surfaces. If an aspheric corrector plate is now located at the aperture stop, the aberration correction becomes incredible. This is because the shell corrector is concentric about the stop, and only the weak aspheric corrector suffers from obliquity effects.

HYBRID A hybrid system is a combination of several pure or classical solutions. What we have here is a combination of the following:

- A multielement corrector group of a diameter equal to the entrance pupil. This near zero power group of elements is typically three to five lens elements and can be of the same glass type with no chromatic aberrations because it is of zero net optical power. The primary purpose of the corrector group is to balance and cancel the spherical aberration of the spherical primary mirror.

- The bulk of the optical power is from the primary mirror as well as the secondary, which itself is an aluminized area on the aft surface of the corrector group.

- Finally, there is a field-correcting group just before the image plane. This group of elements can effectively flatten the field of view and correct the residual off-axis aberrations such as coma and astigmatism.

The beauty of working with the hybrid system configuration described here is that each functional attribute of the system is quite independent and can be easily understood. The front corrector group corrects the spherical aberration of the primary mirror. The fact that it is of near zero power means that a single glass will produce a result essentially free of chromatic aberration. And the rear field-correcting group can easily flatten the field curvature and simultaneously correct any residual off-axis aberrations, including coma and astigmatism.

UNOBSCURED APERTURE SYSTEMS Finally, there is a class of reflective optical systems generally known as "unobscured aperture systems." These include the three-mirror anastigmat (TMA) as well as other forms of all reflective nonrotationally symmetric optical systems. These systems eliminate the performance degrading and difficult to support secondary mirror of the Cassegrain. However, we sometimes require nonrotationally symmetric aspheric surfaces which are difficult to manufacture and test and which are generally costly. Figure 8.6 shows a typical form of TMA. This configuration is an afocal telescope and is per U.S. patent 5,173,801 by Cook. While the most obvious advantage of a TMA is in the hardware, especially as it relates to not needing to support a secondary mirror as in a Cassegrain form of system, the TMA can have significant advantages with respect to the suppression of unwanted diffracted light. Consider Fig. 8.7 where we show a system form developed for space applications. While there are variations on the basic theme, the essence of these systems is that they are used to image relatively close to a bright source as might be encountered in a space application when we are looking within several degrees of the sun. The sky is black, yet the intense solar radiation is just outside our field of view. In order to suppress the diffracted light, the aperture stop is reimaged to a location within the system where an aperture stop slightly smaller than that of the reimaged stop can be located so as to block the reimaged scattered light. This is known as a "Lyot stop" (pronounced "Leo," after the French astronomer Bernard Lyot).

Reflective Systems, Relative Merits

- *No chromatic aberrations.* There are no chromatic aberrations whatsoever in all-reflecting systems. According to ray tracing theory and the use of Snell's law, the refractive index of a mirror is -1.0 for all wavelengths. For this reason reflective optics can be extremely well suited for multispectral applications or situations where refractive materials are either expensive or unavailable.

- *Central obscuration.* Since mirrors get in each other's way, there is often a central obscuration associated with reflective optical systems such as in the Cassegrain configuration. This obscuration affects the net photon throughput, affects the image contrast or

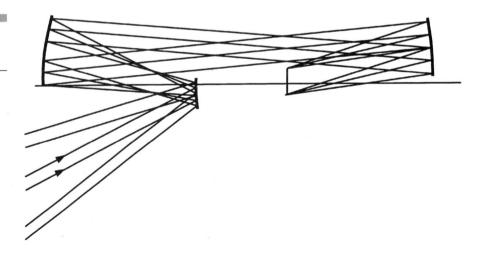

Figure 8.6
Three-Mirror
Anastigmat

Figure 8.7
Three-Mirror All-
Reflective System
Showing Lyot Stop
for Stray Light
Suppression

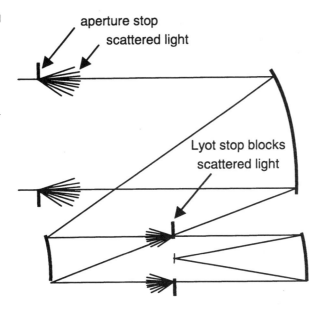

aperture stop

scattered light

Lyot stop blocks
scattered light

MTF (Chap. 15), and is difficult to mount and align. Needless to say, the support structure for this "secondary mirror" must be of a minimal obscuration to the incoming light, yet it must be strong and robust.

■ *Aspheric surfaces required.* Due to the limited number of surfaces that can be effectively used in a reflective optical system, there are rarely enough surfaces to allow for minimization of aberrations as with refractive lens systems. For example, a typical double Gauss lens system may consist of seven elements. If there were no cemented elements, we would have 14 radii with which to minimize ray bendings and hence minimize the residual aberrations. In our Cassegrain reflective system configuration, we have only two surfaces, which are far too few for aberration control, thus leading us to require the use of nonspherical surfaces. One has only to ask the question of how a reflective system with 14 mirrors would look to appreciate the difficulty of working effectively with more than two to three mirrors. It is important to realize that aspheric surfaces are not necessarily bad. There are many contemporary methodologies for producing aspheric surfaces on both mirror as well as transmissive lens surfaces. For mirrors, we have ultraprecision machining or diamond turning as the most common.

▪ *Small number of elements.* The small number of elements together with the baffling issues, and the fact that the mirrors get in each other's way, limits the field of view of reflective systems to be generally smaller than of the refractive systems.

▪ *Can be low weight.* Reflective systems can, in many cases, be made of aluminum, which is light in weight.

▪ *Inherently athermalized.* Reflective systems, if manufactured of a single material such as aluminum, are generally athermal. In other words, for a uniform temperature increase or decrease, the entire system expands or contracts by an amount dependent on the thermal coefficient of expansion. Since this is a uniform scaling of all system parameters, the image will still be in focus. If multiple materials and/or thermal gradients are used, then a careful assessment of the thermal properties of the imagery is critical. Zerodur is a glass material with almost zero coefficient of thermal expansion, and it is commonly used for large glass mirrors. The important message here is that reflective systems have the potential for being fully athermalized. To convince yourself that all reflective systems manufactured of the same material throughout are indeed athermal, consider the following explanation: Assume that our design is a Cassegrain reflective system where all components, including the mirrors and the support structure, are aluminum. If we heat or cool the system uniformly, then it will uniformly expand or contract according to the thermal coefficient of expansion of the material. This is, in effect, a scaling of the entire system. Imagine a drawing of our reflective system forming a perfectly focused image. Take this drawing to a copy machine and enlarge it by 20%. Now look at the drawing and ask yourself "is it still in focus?" Of course it is! Needless to say, if you have thermal gradients and/or different materials, you system may not be sufficiently athermal, and this may require active or passive athermalization.

▪ *Stray light susceptibility.* Reflective systems are often faced with problems associated with stray light. This stray light is often out-of-field light which directly or indirectly reaches the final sensor. The Cassegrain is an example of a system, which needs to be properly baffled to suppress unwanted stray light that may directly go through, missing both mirrors.

Refractive Systems, Relative Merits

■ *Straight through.* The system operates straight through without any central obscuration. This results in a potentially higher photon efficiency with none of the degradations associated with a central obscuration.

■ *Spherical surfaces, conventional manufacturing.* Since we can add lens elements and use the necessary techniques to minimize aberrations, we can most often use spherical surfaces and thus avoid expensive manufacturing methods often associated with aspheric surfaces.

■ *Can add a lot of components.* This makes it possible to design high-speed systems with large fields of view.

■ *Expensive materials and athermalization problems in the thermal infrared.* Refractive systems used in the thermal infrared (the MWIR which is the 3- to 5-µm spectral band and the LWIR which is the 8- to 14-µm spectral band) often require materials that are very expensive and have a high *dn/dt*. For example, germanium is extremely expensive, and furthermore it has a *dn/dt* = 0.000396/°C.

Mirrors and Prisms

Mirrors and prisms are the optical components used in optical systems to

■ Change the direction of light.

■ Fold an optical system for better packaging.

■ Provide a proper image orientation.

■ Combine or split the optical beams using beamsplitter coatings.

■ Disperse light with wedged prisms.

■ Provide the means for interpupillary distance change in binocular systems.

■ Expand or contract a laser's beam diameter, etc.

Flat mirrors fold the optical path in a system. Prisms also fold the optical path, except that the reflecting surfaces of the prisms behave like mirrors rigidly mounted with respect to each other. The optical designer has to be careful to leave enough space during the design of the optical system to place the mirrors and prisms where they are needed. Prisms have flat polished surfaces and have no optical power. If prisms are used for the proper image orientation and location, they use refraction at the input and output surfaces and reflection on the intermediate surfaces. Reflection is a total internal reflection if the incident cone of light is small enough and/or the magnitude of the prism angle is such that the condition for the total internal reflection is satisfied for all rays. Otherwise, surfaces are mirror coated.

Reflecting prisms are generally designed so that the entering and exit faces are parallel and perpendicular to the optical axis. This means that the prism can be represented as a plane parallel glass plate. The thickness of the glass plate is obtained by unfolding the prism around its reflecting surfaces. Unfolded prisms are shown in the form of a "tunnel diagram." The Penta prism with its tunnel diagram is shown in Fig. 8.8. We use a tunnel diagram so as to be able to use a single block of glass with no folding of the optical path during the design. The prism of refractive index, n, and the glass thickness, d (thickness of the unfolded prism), has its equivalent path length in air, also called the prism *apparent thickness*, and is equal to

$$d_{\text{apparent}} = \frac{d}{n}$$

If the prism is inserted in a convergent beam, the image will be shifted by

$$\Delta d = \frac{d(n-1)}{n} = d - \frac{d}{n} \quad (See \ Fig \ 8.9)$$

Prism thickness and the apparent prism thickness are shown in Fig. 8.9.

In the case of mirrors, the optical designer has to leave sufficient space in the optical path for the mirrors to be mounted. Prisms do not introduce aberrations only if they are located in a collimated beam. When the optical designer works on a system which contains, among other powered components, a prism of thickness d, he or she usually designs the system using a block of glass of thickness d because of the simplicity, smaller number of surfaces, and, consequently, faster ray tracing. In the case when the prism is located in a convergent or divergent beam, this block of glass introduces both monochromatic and chromatic aberrations, and

Figure 8.8
Penta Prism

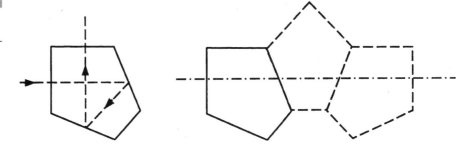

tunnel diagram of a penta prism

Figure 8.9
Prism Thickness and
the Apparent Prism
Thickness

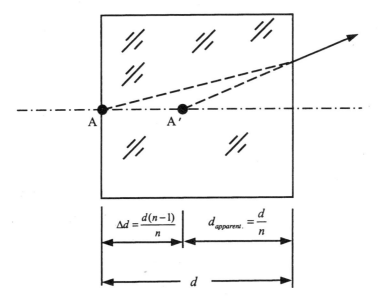

$$\Delta d = \frac{d(n-1)}{n} \qquad d_{apparent.} = \frac{d}{n}$$

$$d$$

has to be present during the optimization of the system. The whole process is iterative. The optical designer starts with the rough size of the prism, and optimizes the system with the prism equivalent block of glass. At some point in the design, the designer checks the diameter of the input and the output beam, as well as the ray angles through the prism, and makes the necessary adjustments to the prism size and positioning.

In a real system, the length of the prism along the optical axis is often significantly shorter than the glass equivalent plate, since the optical path in the prism is folded a few times. A good practice is to enter the real prism surfaces in the optical prescription when the design of the system is nearly finished and check the ray footprints on each prism surface as well as the optical performance of the whole system. It is also convenient to export a computer file with all the optical components in the system in a format that is suitable for the mechanical designer to design the mechanics around the optical components.

When a ray is reflected from a flat mirror, the incident ray, the normal to the mirror at the point of incidence, and the reflected ray, all lie in a single plane, which is called the *plane of incidence*.

Let us examine the orientation of the image after reflection from a flat mirror. If the observer looks directly at the object *AB* shown in Fig. 8.10, the point *A* appears to be the highest point of the object. If you look at the same object reflected from the mirror, it appears that the image is located behind the mirror in *A′B′*. However, this time it appears that point *A′* is at the bottom of the image. One reflection changes the orientation of the image in the plane of incidence, which means in our example that the image is upside down. If the object is a two-dimensional object, there is no change in the image orientation in the plane perpendicular to the plane of incidence. In our example, this means that the left side of the object after reflection is seen as coming from the left side. Let us imagine that object *AB* has no symmetry. It could be, for example, letter the R. When viewed directly, the letter is oriented properly, and we can read it. After reflection in the mirror, the letter R is upside down, and even if we rotate it around the direction of image propagation, we can never orient it so that it will be readable. In the case when the image is readable, which is shown in Fig. 8.11*a*, we call it a "right-handed image." If the image is as shown in Fig. 8.11*b*, where letter R is backwards, regardless of the orientation of the image, it is called a "left-handed image."

After multiple reflections off flat mirrors, or reflections inside a prism, a general rule says that an even number of reflections gives a right-handed image and an odd number of reflections gives a left-handed image. However, whether the number of reflections is even or odd does not tell us anything about the image orientation.

Note

Reflecting prisms are used to

- Erect the image in telescopes, which means that the top to bottom as well as the left to right are inverted, and the image is right-handed.

- Invert the image in one plane, either top to bottom or left to right. The image is left-handed.

- Deviate the optical axis, with inversion, erection, or no change in the image orientation.

Figure 8.10
Image Orientation after a Single Reflection from a Flat Mirror

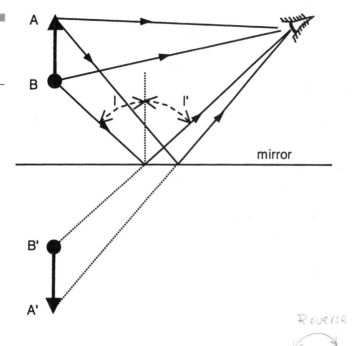

Figure 8.11
Right- and Left-Handed Images

(a) Right-handed image   (b) Left-handed image

■ Displace the optical axis, with inversion, erection, or no change in the image orientation.

■ Keep the unchanged image orientation, rotating the prism around the optical axis (these are prisms with no axis deviation or displacement) while the input image rotates around the center of the field of view (or around the input optical axis).

As an example of an inverting prism, we will look at the *Pechan prism*. A Pechan prism is shown in Fig. 8.12. It consists of two prisms with a small air gap between them. There are a total of five reflections, which means that a right-handed image entering the prism is changed to a left-handed image after the prism. All five reflections have a common plane of incidence, and the image is inverted in that plane. In the direction which is normal to the common plane of incidence, there is no change in the image orientation.

Tracing of an image through a Pechan prism is shown in Fig. 8.12. The object chosen for tracing is a circle-arrow object. After refraction through the surface, *AB*, the image is totally internally reflected from surface *BD*. This TIR limits the field of view of the Pechan prism, which is about 10° for medium-index glasses. After surface *BD*, the rays reach the surface, *AC*, which has to be mirror coated. Small arrows on each surface show the orientation of the circle-arrow object as it falls on a given surface. The image is then refracted by the two surfaces with the air gap between them, and then totally internally reflected off surface *EF*. The next surface, *GH*, also has to be mirror coated for the rays to be reflected, and after the final TIR on surface *EH*, the image exits the prism as an inverted image with no deviation. Although the Pechan prism consists of two prisms, which means that the alignment and mounting of the prism is rather complicated, it is very commonly used to invert the image. Most of the systems that require image rotation or derotation, have a Pechan prism that rotates half the rotation angle of the scanning entrance mirror. It can be used in a convergent beam, and the clearance for the prism rotation is the smallest of all inverting prisms with no deviation.

Figure 8.12
Pechan Prism

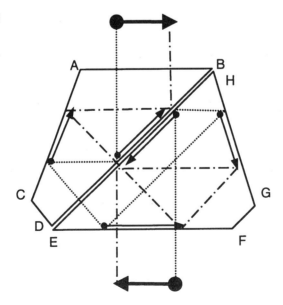

Figure 8.13
Roof Pechan Prism

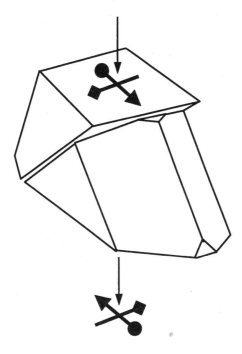

If surface *GH* of the Pechan prism is converted into a roof, where two sides of the roof are normal to each other, the prism is called a roof Pechan prism (Fig. 8.13). It has six reflections, so that the image is right-handed. The added reflection on the roof is in the plane which is normal to all other planes of reflection in the prism. This means that the image is right side up. Binoculars with a straight axis, which are Newton-type telescopes, use this type of prism for image erection. The prism is located between the objective and the image plane formed in the focus of the eyepiece. The light beam coming from one point in the field is split on the roof in two beam segments, each undergoing reflections on both sides of the roof. These two segments join after the roof, and form one image. However, if there is an error in the 90° roof angle, it may introduce the image blur or a double image. The tolerance on a roof angle is, in most cases, only a few arc-seconds.

An optical designer has to be careful when designing the prisms, or using standard prisms in an optical system, because the prisms have to work properly for a specified field of view, or a given cone angle. Mistakes can result in the appearance of ghost images. For example, in the case of the Pechan prism, there are three surfaces where the beam is totally internally reflected. If the incident cone angle is too large for a chosen index of refraction of the prism, there will be one part of the field, which will not be properly reflected through the prism as the rest of the image. It will only be refracted and pass directly through the prism. The other case when the ghost images can appear is when the prism is not large enough and some skew rays undergo one extra reflection off the side surfaces. This can happen in the right-angle prism. The ghost reflections may be eliminated making the prism larger, and cutting notches in the prism.

Let us now look at a right-angle telescope, determine what kind of prism can be used to deviate the optical axis by 90°, and also erect the image so that the viewer sees a noninverted right-handed image through the eyepiece. A Newtonian telescope creates an inverted right-handed image. If we put a screen in the focal plane of the objective, the image is both inverted and reverted. The objective creates an image which is right-handed, with an altered orientation both left to right and top to bottom. The eyepiece acts like a magnifier and it does not change the image orientation.

In order to determine the image orientation at any location along the optical system, it is convenient to write on a small piece of paper an object with no symmetry. This can be a letter R, and this piece of paper should be moved through the space, simulating reflections off the prism mirror surfaces. You should always be located such that the image is moving toward your eyes. Do not forget to rotate the image left to right and top to bottom in the case of the objective lens or a relay lens, and do not rotate it through the eyepiece.

Our problem is sketched in Fig. 8.14. The object is shown as a circle at the bottom, an arrow at the top, and a square on the left. After the objective, the image changes its orientation in both directions. Since the eyepiece does not change the image orientation, corrections in the image orientation in the horizontal and the vertical plane have to be done with the prism. If we would have only a plane mirror, the arrow and the circle would be properly oriented after the reflection. This is shown with the dotted lines drawn parallel to the optical axis. However, the square, which was turned to the right side after the objective, would stay on the right side after the reflection off the mirror. This means that we have to find a way to invert the image left to right, keeping only one reflection in the vertical plane. This can be accomplished using the right-angle prism and adding the roof on the hypotenuse surface. This prism is called the *Amici prism,* and it is shown in Fig. 8.15.

Figure 8.14
Right-Angle Telescope

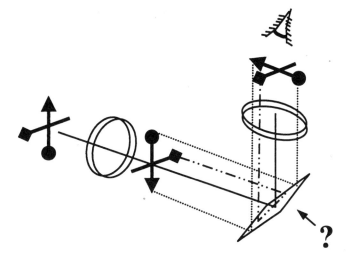

Figure 8.14
Right-Angle Telescope

There are many standard prism types. Useful information about prisms, their function, and the important dimensional relationships, can be found in the MIL-HDBK-141 (1962).

Design of Visual Systems

Visual optics includes the wide variety of optical systems creating imagery to be viewed directly by the human eye. This includes telescopes, microscopes, binoculars, riflescopes, camera viewfinders, head-mounted displays, magnifiers, and others. Common to all of these systems is that the eye is looking into some form of viewing optics such as an eyepiece.

Basic Parameters of the Human Eye

The basic optical parameters of the human eye are listed in Table 8.1. Note that most of the data are listed as "approximately" due to the natural variation from person to person.

The eye of an average young person can accommodate or focus to a distance in the order of 250 mm. Due to geometry, as a person looks at a progressively closer object, the two eyes must progressively increase their

Figure 8.15
Amici Prism

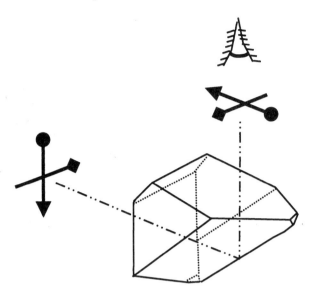

angle of convergence. In the limit, for an interpupillary distance or IPD of 63.5 mm and an object at a 250-mm distance, the full convergence angle is 14.5°. If we work with a monocular system such as a riflescope, then convergence is not an issue; however, for a binocular, or even more for biocular systems, it can be quite important. A *biocular* system is a large field-of-view "eyepiece" about 100 mm in diameter, used for viewing a screen or display with both eyes. Figure 8.16 shows a binocular and a biocular system. The design of biocular systems is extremely difficult, and complex multielement designs often are required. One of the difficulties to the designer is that each eye is looking through the extreme edges of the biocular system, and this is the region of the pupil which generally is the most difficult to correct for aberrations.

The user of a pair of binoculars will manually focus the binoculars. Since there is generally no convergence in the optical paths, the user will most often focus the binoculars at infinity, especially when looking at distant objects. This means that collimated light will exit the eyepiece and enter into the eye. If we were to design a riflescope, convergence is not an issue, so the user will focus the device to the most comfortable distance. This is often in the range of 2.5 to 6 m, although some research has shown that the resting state of the eye in a dark condition is more like 1 m. It is important to keep in mind that if you require the user to accommodate to some close distance, then you should consider converging the two optical paths (diverging them into the eyes).

TABLE 8.1

Typical Human Eye
Optical Parameters

Parameter	Value for Human Eye
Entrance pupil diameter (mm)	≈2.5 to ≈7
Focal length (mm)	≈16.9
f/number	≈2.4 to ≈6.8
Distance from cornea to point of rotation (mm)	≈13.5
Radius of cornea (mm)	≈8
Interpupillary distance (IPD) (mm)	≈63.5 average, 46 to 80 range
Accurate seeing area (degrees)	≈1
Normal viewing angles (horizontal) (degrees)	≈±5 to ±30
Total visual limit (horizontal) (degrees)	≈±108

You can appreciate the situation if you consider providing diverging light into the two eyes from a virtual object 250 mm in front of the user. While the user can easily accommodate to a 250-mm distance, if the two optical paths were parallel to each other, then there would be potentially significant eye strain, since when accommodating to 250 mm, a person will naturally converge his or her eyes by about 14.5° as noted previously.

There are a number of reasonably reliable eye models, and Figs. 8.17 and 8.18 show one of the more common models, the Lotmar eye model, for eye pupils of 7 and 3 mm in diameter, respectively. The data are shown for fields of view of 0, 22.5, and 45° off axis. You can see that there is a residual of spherical aberration which is, of course, more prominent with the larger pupil diameter. For the 7-mm-diameter pupil the spherical aberration equates to an rms blur diameter of about 8 min of arc on the retina, and this reduces to about 1.3 min of arc for the 3-mm-diameter pupil. The eye was permitted to refocus for each pupil diameter. Note the significant off-axis coma and astigmatism residuals.

It has been our experience that the residual eye aberrations are somewhat different for different eye models. This leads to the following question: Should the design of a visual system include the effects of the eye? In other words, should your optical design attempt to cancel the eye's residual aberrations based on one of the eye models? In all likelihood you could do a reasonably good job of accomplishing this from a lens

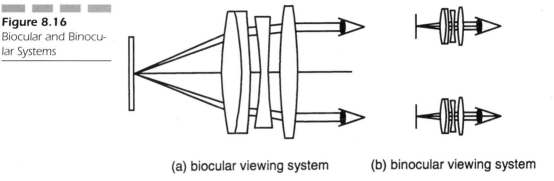

(a) biocular viewing system

(b) binocular viewing system

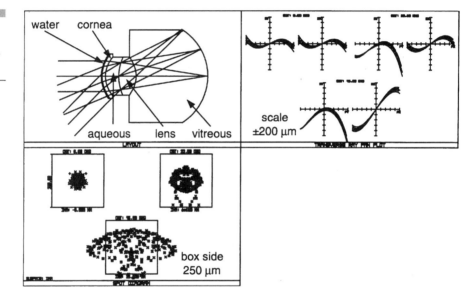

design standpoint; however, the disadvantage is that the eye's aberrations vary from person to person, and even the eye models available give somewhat different aberration residuals. It is generally accepted that it is most prudent to design visual optical systems assuming a perfect eye.

Pupil-Forming Systems

Most (but not all) visual systems are known as "pupil-forming systems." The term can be best illustrated by considering Fig. 8.19, where we show

Figure 8.18
Lotmar Eye Model
with 3-mm-Diameter
Pupil

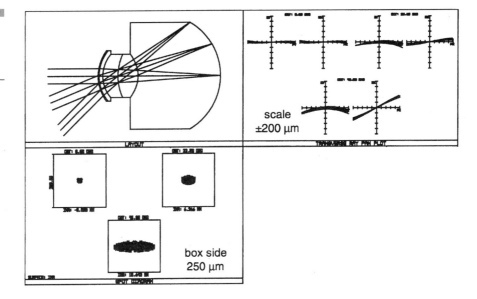

a magnifying glass in Fig. 8.19a and a simple telescope in Fig. 8.19b. The magnifying glass is *not* a pupil-forming system. What this means is that there is no well-defined exit pupil from the magnifying glass which needs to be mated or matched to the entrance pupil of the eye. On the other hand, the exit pupil of the telescope (Fig. 8.19b) is well defined, and in order to see the imagery, the entrance pupil of the eye *must* be lined up with the exit pupil of the telescope.

Another way to understand pupil-forming systems is to imagine placing a white card in place of the eye in both Figs. 8.19a and b. Further, assume a bright object which is being viewed. In the case of the magnifying glass, light from the object being viewed will fill a large area of diameter $d1$, which is effectively the diameter of the lens. In the case of the telescope, there will be a bright disk of diameter $d2$ at the exit pupil of the telescope. If the telescope were similar to a pair of 7×50 binoculars, this bright disk at $d2$ will be 7 mm in diameter. The important aspect of pupil-forming systems is that special attention must be given to assure that the light exiting the optical system does enter the pupil of the eye.

There is a very interesting subtlety, and this can be understood by looking at Fig. 8.20. In Fig. 8.20a the eye is looking straight ahead to the left. The center of the field of view will be at or near the fovea, which is the highest acuity of the retina, and the user will see in his or her

Figure 8.19
Pupil-Forming Optical
Systems

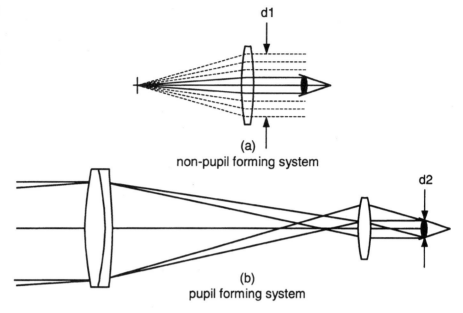

(a)
non-pupil forming system

(b)
pupil forming system

peripheral vision the field stop of the eyepiece. In effect, the imagery will fall within a well-defined circle, and there is black outside of the circle. If the person now looks upward to the left by rotating his or her eye about the center of rotation of the eye, some of the light from the edge of the field will not enter the eye's pupil. In situations where we have a wide apparent field-of-view telescope with a 7-mm or smaller exit pupil diameter, when the person looks upward toward the edge of the field stop, the light may even disappear completely! It is possible to move your eye closer to the eyepiece so that the exit pupil of the telescope is located at the point of rotation of the eye. Now the person can see clearly the field stop in the eyepiece when looking toward the field stop; however, when looking straight ahead, the field stop and the outer periphery of the field of view may completely disappear from view! This is a very striking, as well as a weird, effect, and if you ever have the opportunity to see it, it is worthwhile to do so.

Requirements for Visual Optical Systems

When designing visual systems, a number of parameters must be considered which are unique to these systems, and these are listed here:

Figure 8.20
Telescope Showing
Effect of Eye Rotation

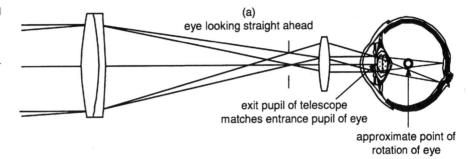

(a)
eye looking straight ahead

exit pupil of telescope
matches entrance pupil of eye

approximate point of
rotation of eye

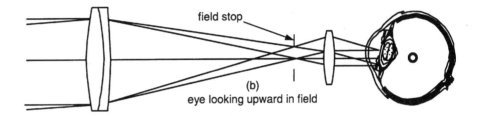

field stop

(b)
eye looking upward in field

▦ For many hand-held systems such as binoculars, the user will most often bring the object of greatest interest to the center of the field of view. For this reason, off-axis performance is generally more forgiving for these systems.

▦ The *eye relief* is the clearance or distance from the last optical element in the viewing optics to the front of the cornea of the eye. For eyeglass users the generally accepted minimum eye relief is in the order of 25 mm. Larger values are, of course, helpful, but the diameter of the eyepiece grows with the eye relief.

▦ It is clear from the transverse ray aberration curves for the Lotmar eye model that the resolution of the eye degrades quite severely as we move away from the center of the field of view. The visual acuity of the eye is maximum at the fovea, which is very close to the center of the field of view of the eye. As an object moves away from the fovea, the visual acuity decreases dramatically, and at ±20° the visual acuity is only about 10% of that at the fovea. This is well illustrated in Fig. 18.21 where we show a series of letters of different sizes. The increase in letter size is inversely proportional to the decrease in visual acuity as we move from the fovea. Thus, if you look at the small spot at the center of the figure, all of the letters should be approximately equally resolved.

Although it is true that the visual acuity drops quickly away from the fovea, visual systems should have relatively good imagery at the edge of the field. How good depends on the application of the system. For example, consider a binocular. A viewer can rotate his or her eyes to look toward the edge of the field, in which case the field periphery is the sharpest area the eye sees. This could lead us to a conclusion that the edge of the field should be very well corrected for aberrations. However, we all know that any object in the field can be brought to the center by simply rotating the whole binocular. So the quality of the imagery at the edge of the field in a binocular does not have to be as good as in the center, but it should not be so bad that the image blurring and coloring is immediately noticed when the eyes are pointed to the field edge.

Figure 8.21
Illustration of Visual Acuity As a Function of Distance from the Fovea

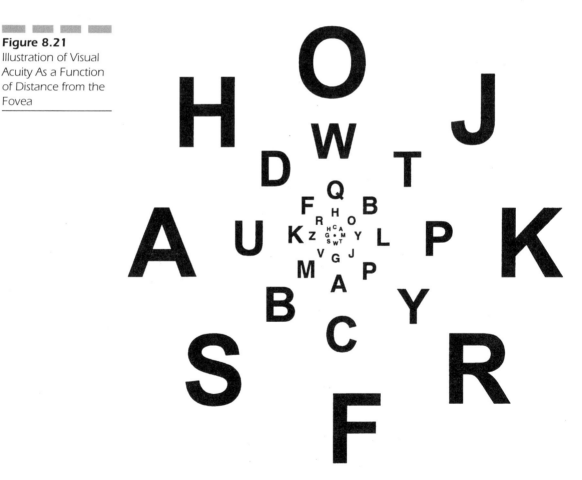

■ The most general and accepted metric that we often hear is that *the eye resolves 1 min of arc*. What this means is that the eye can resolve the capital letter E when each of the dark or bright bars forming the horizontal portions of the letter subtend 1 min of arc, as shown in Fig. 8.22. This means that the eye resolves 2 min per line pair, or 0.5 line pair per minute of arc.

■ With respect to visual systems design, providing an image blur diameter from 1 min of arc to perhaps 3 min of arc is generally considered an acceptable level of performance in the center of the field of view. At the edge of the field, 20 to 40 arc min of image blur maximum may be acceptable.

Table 8.2 shows a list of the typical tolerances associated with optics for the eye, as developed by Mouroulis. These data are for binocular viewing with two eyes. Note that items that the eye normally does not do are specified with tighter tolerances. This includes dipvergence (one eye looking upward and the other eye downward) and divergence for example.

There is a rule of thumb that says that the eye can resolve a contrast of 5%. Thus, it is not uncommon to determine where the MTF drops to 0.05 and to conclude that the eye will resolve this spatial frequency. There is perhaps a better way, and this is to use the so-called aerial image modulation (AIM) curve. This is the relationship between the contrast and the number of line pairs per millimeter that the eye can resolve. Walker shows these data, and they are summarized in Fig. 8.23. The AIM curve is a relationship between the modulation required to resolve

Figure 8.22
Eye Resolution

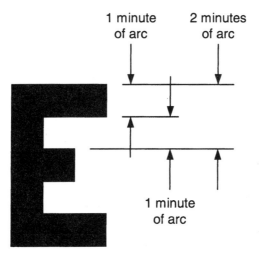

1 minute
of arc

2 minutes
of arc

1 minute
of arc

TABLE 8.2

Typical Tolerances
for Visual Systems

Parameter	Typical Specification
Divergence (degrees)	0
Convergence (degrees)	1.5
Dipvergence (min)	8
Magnification difference (%)	±0.5 to ±1
Brightness difference (%)	±10
Best focus if fixed	−1 diopter[*]
Accommodation	−3 diopter to infinity is OK
	−1 to −1.5 diopter is best
Axial chromatic aberration	−1 to −1.5 diopters
Lateral color (min)	2
Image quality	Integral of MTF from 0 to 20 line pairs/degree

[*]A *diopter* is the reciprocal focal length, in meters. As used in the table, −1 diopter means that the eye will be focusing to a distance of 1 m in front of the user.

a given target and the spatial frequency of the target. Using the Lotmar eye model with a 7-mm pupil diameter, we see that if the eye were diffraction limited in Fig. 8.23a, we would resolve 1.6 min/line pair, or 0.8 min/line; however, due to the spherical aberration, we can only resolve in the order of 3 min/line pair, or 1.5 min/line (Fig. 8.23b). In the case of a 3-mm pupil diameter the diffraction-limited eye would resolve about 1.6 mm/line pair (Fig. 8.23c) and the Lotmar model predicts about 1.7 mm/line pair (Fig. 8.23d). It is important to realize that these data are only as accurate as the Lotmar eye model and the referenced AIM threshold data. While they may not be precisely accurate, they do give us a good indication of the resolution of the eye.

Distortion is an important criterion in visual systems. As we learned in Chap. 5, distortion is more of a mapping error rather than an image-degrading aberration. Its primary effect is to the cosmetic appearance of the image. Distortion is a change in magnification with field of view. Generally, positive or negative distortion in the order of 2 to 2.5% is small enough to be almost imperceptible. Large amounts of distortion can be annoying in any visual system. Telescopes with a large apparent field of view often have up to 10% of distortion at the edge of the field.

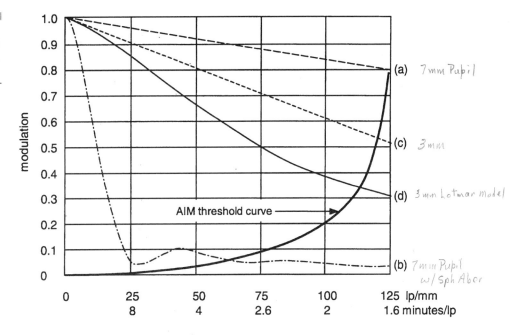

Figure 8.23
Use of AIM Threshold Curve to Predict Visual Resolution

Summary on Design of Visual Systems

It is best in the design of visual optical systems to assume a perfect eye model. Most of the computer design programs have the ability to model a so-called perfect lens or paraxial lens, which is used following the system being designed to emulate the eye (or any form of focusing lens). What these so-called perfect lenses do is to simply compute the angular aberrations in the optical system and multiply them by the focal length of the perfect lens which is input by the user. So if you are interested in the actual image blur on the retina, then use a perfect lens of focal length 16.9 mm, which is the approximate focal length of the eye.

If you are designing a visual system which must be fixed focus, it is probably best to design and produce your system to provide light which is either collimated to the user or appears to come from a distance of from 3 to 6 m. Some data suggest a value closer to 1 m. If you can allow the user to refocus, this is generally preferred.

The image quality, or blur diameter, should be in the order of 1 min of arc to 3 min of arc. The specific application will have a lot to do with how good the imagery needs to be. You should design your system for an exit pupil diameter of at least 7 mm (10 to 12 mm is better), and then evaluate the performance both at the 7-mm-diameter pupil as well as with reduced pupil diameters. Some detailed specifications include the eye being decentered to various positions within the exit pupil of the optical system. If the system is to be used in bright conditions, you may even evaluate your design in the 2.5- to 3-mm pupil diameter region.

Our final note on the design of visual systems is to keep in mind that the eye is quite forgiving. Persons who get a new eyeglass prescription often notice color fringing or lateral color when looking toward the outer periphery of their field of regard (30 to 40° from straight ahead, for example). After several days to several weeks, this color fringing often seems to disappear. The eye, along with the rest of the human visual system, is, in effect, a very powerful computer with impressive image-processing capabilities. This does not mean that you can ignore the aberrations and image quality of visual systems, what it does mean, however, is that, in many situations, the optics for visual systems can be more forgiving than you might think.

The Optical Design Process

The optical design process includes a myriad of tasks that the designer must perform and consider in the process of optimizing the performance of an imaging optical system. While we often think primarily of the robustness of the optimization algorithm, reduction of aberrations, and the like, there is much more to do. The designer must be at what we sometimes call "mental and technical equilibrium with the task at hand." This means that he or she needs to be fully confident that all of the following are understood and under control:

- All first-order parameters and specifications such as magnification, focal length, f/number, full field of view, spectral band and relative weightings, and others.

- Assure that the optical performance is being met, including image quality, distortion, vignetting, and others.

- Assure that the packaging and other physical requirements, including the thermal environment, is being taken into account.

- Assure that the design is manufacturable at a reasonable cost based on a fabrication, assembly, and alignment tolerance analysis and performance error budget.

- Consider all possible problems such as polarization effects, including birefringence, coating feasibility, ghost images and stray light, and any other possible problems.

Once every one of these items has been addressed and is at least recognized and understood, we start with the sketch of the system. First, the system is divided into subsystems if possible, and the first-order parameters are determined for each subsystem. For example, if we are to design a telescope with a given magnification, the entrance pupil diameter should be chosen such that the exit pupil size matches the eye pupil. A focal length of the objective and the eyepiece should be chosen such that the eyepiece can have a sufficiently large eye relief. Now, when the specs for each subsystem are defined, it is time to use the computer-aided design algorithms and associated software to optimize the system, which will be discussed in the rest of this chapter. Each subsystem can be designed and optimized individually, and the modules joined together or, more often, some subsystems are optimized separately and some as an integral part of the whole system.

What Do We Do When We Optimize a Lens System?

Present-day computer hardware and software have significantly changed the process of lens design. A simple lens with several elements has nearly an infinite number of possible solutions. Each surface can take on an infinite number of specific radii, ranging from steeply curved concave, through flat, and on to steeply curved convex. There are a near infinite number of possible design permutations for even the simplest lenses. How does one optimize the performance with so many possible permutations? Computers have made what was once a tedious and time-consuming task at least manageable.

The essence of most lens design computer programs is as follows:

- First, the designer has to enter in the program the starting optical system. Then, each variable is changed a small amount, called an *increment*, and the effect to performance is then computed. For example, the first thickness may be changed by 0.05 mm as its increment. Once this increment in thickness is made, the overall performance, including image quality as well as physical constraints, are computed. The results are stored, and the second thickness is now changed by 0.05 mm and so on for all variables that the user has designated. Variables include radii, airspaces, element thicknesses, glass refractive index, and Abbe number. If you are using aspheric or diffractive surfaces, then the appropriate coefficients are also variables.

- The measure of performance as used here is a quantitative characterization of the optical performance combined with a measure of how well the system meets its first-order constraints set by the user such as focal length, packaging constraints, center and edge thickness violations, and others. The result of the computation is a single number called an *error function* or *merit function*. The lower the number, the better the performance. One typical error function criteria is the rms blur radius, which, in effect, is the radius of a circle containing 68% of the energy. Other criteria include optical path difference, and even MTF, as described in Chap. 15.

- The result is a series of derivatives relating the change in performance (P) versus the change in the first variable (V_1), the change in performance (P) versus the change in the second variable (V_2), and so on. This takes on the following form:

$$\frac{\partial P}{\partial V_1}, \frac{\partial P}{\partial V_2}, \frac{\partial P}{\partial V_3}, \frac{\partial P}{\partial V_4} ...$$

■ This set of partial derivatives tells in which direction each parameter has to change to reduce the value of the sum of the squares of the performance residuals. This process of simultaneous parameter changes is repeated until an optimum solution is reached.

A lens system consists of a nearly infinite number of possible solutions in a highly multidimensional space, and it is the job of the designer to determine the optimum solution.

Designers have used the following analogy to describe just how a lens design program works:

■ Assume that you cannot see and you are placed in a three-dimensional terrain with randomly changing hills and valleys. Your goal is to locate the lowest elevation or altitude, which in our analogy equates to the lowest error function or merit function. The lower the error function, the better the image quality, with the "goodness" of performance being inversely proportional to the elevation.

■ You are given a stick about 2 m long, and you first stand in place and turn around tapping the stick on the ground trying to find which direction to walk so as to go down in elevation.

■ Once you determine the azimuth resulting in the greatest drop in elevation, you step forward in that direction by 2 m.

■ You now repeat this process until in every direction the elevation goes up or is level, in which case you have located the lowest elevation.

■ But what if just over a nearby hill is an even lower valley than you are now in? How can you find this region of solution? You could use a longer stick, or you could step forward a distance several times as long as the length of your stick. If you knew that the derivative or slope downward is linear or at least will continue to proceed downward, this may be a viable approach. This is clearly a nontrivial mathematical problem for which many complex and innovative algorithms have been derived over the years. But the problem is so nontrivial as well as nonlinear that software algorithms to locate the so-called global minimum in the error function are still elusive. Needless to say, the true global minimum in the error function may be quite different or distant from the current location in our *n*-dimensional terrain.

Figure 9.1 shows a two-dimensional representation of solution space as discussed previously. The *ordinate* is the error function or merit function, which is a measure of image quality, and the *abscissa* is, in effect, solution space. We may initiate a design on the left and the initial optimization brings the error function to the first minimum called a *local minimum* in the error function. We then change glasses and/or make other changes to the design and ultimately are able to move the design to the next lower local minimum. Finally, we add additional elements and make other changes and we may be able to reach the local minimum on the right. But how do we know that we are at, or even close to, a global minimum? Here lies the challenge as well as the excitement of lens design!

It is important here to note that reaching global minimum in the error function is not necessarily the end goal for a design. Factors including tolerance sensitivity, packaging, viability of materials, number of elements, and many other factors influence the overall assessment or "goodness" of a design. Learning how to optimize a lens system is, of

Figure 9.1

Illustration of Solution Space in Lens Design

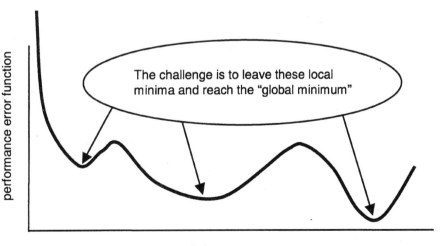

course, quite critical to the overall effort, and learning how to reach a viable local or near-global minimum in the error function is very important to the overall success of a project.

How Does the Designer Approach the Optical Design Task?

The following are the basic steps generally followed by an experienced optical designer in performing a given design task. Needless to say, due to the inherent complexity of optical design, the processes often becomes far more involved and time consuming. Figure 9.2 outlines these basic steps:

1. The first step in the design process is to *acquire and review all of the specifications*. This includes all optical specifications including focal

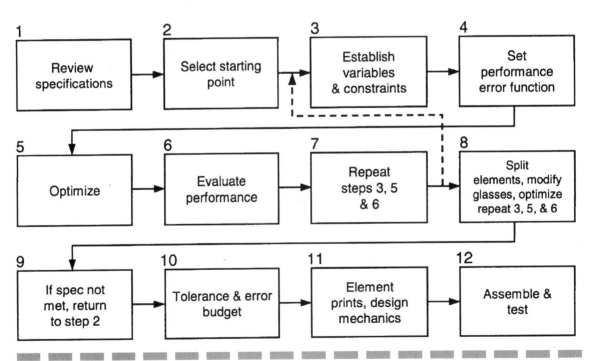

Figure 9.2 Lens Design and Optimization Procedure

length, f/number, full field of view, packaging constraints, performance goal, environmental requirements, and others.

2. Then we *select a representative viable starting point*. The starting point should, wherever possible, be a configuration which is inherently capable of meeting the specifications for the design. For example, if the specifications are for an f/10 monochromatic lens covering a very small field of view and having an entrance pupil diameter of 5 mm, then the lens may very well be a single element. However, if the requirements call for an f/1.2 lens over a wide spectral band covering a 40° full field of view, then the solution may very well be a very complex six- to seven-element double Gauss lens form. If we were to use a single element for this latter starting point, there would be no hope for a viable solution. Finding a good starting point is very important in obtaining a viable solution. The following are viable sources for starting points:

You can use a *patent* as a starting point. There are many sources for lens patents including Warren Smith's excellent book *Modern Lens Design*. There is also a CD-ROM called "LensView," which contains over 20,000 designs from patents. These are all searchable by a host of key parameters. While the authors of this book are not patent attorneys, we can say with confidence that you may legally enter design data from a patent into your computer and work with it in any way that you would like to. If your resulting design is sold on the market, and if the design infringes on the patent you used (or any other for that matter), you could be cited for patent infringement. It is interesting to note that the purpose of our patent system in this country is to promote inventions and innovation. This is done by offering an inventor a 17-year exclusive right to his or her invention in exchange for *teaching* in the patent how to implement the invention. Thus, you are, in effect, invited to use the design data and work with it with the goal of coming up with a better design, which you can then go out and patent. By this philosophy, inventors are constantly challenged to improve upon an invention, which, in effect, advances technology, which is what the patent process is all about. Needless to say, we urge you to be careful in your use of patents.

You could use a so-called hybrid design. We mean a hybrid to be the combining of two or more otherwise viable design approaches so as to yield a new system configuration. For example, a moderate field-of-view Tessar lens design form can be combined with one or more strongly negatively powered elements in the front to create an extremely wide-angle lens. In effect, the Tessar is now used over a field of view similar to its designed field, and the negative element or elements bend or "horse" the rays around to cover the wider field of view. An *original design* can, of course, be a viable starting point. As your experience continues to mature, you will eventually become comfortable with "starting from scratch." With today's computer-aided design software, this works most of the time with simple systems such as doublets and triplets; however, with more complex systems, you may have problems and will likely be better off resorting to a patent or other source for a starting point.

3. Once you have entered your starting point into the software package you are using, it is time to *establish the variables and constraints*. The system variables include the following: radii, thicknesses, airspaces, surface tilts and decenters, glass characteristics (refractive index and Abbe number), and aspheric and/or other surface variables, including aspheric coefficients. The constraints include items such as focal length, f/number, packaging-related parameters (length, diameter, etc.), specific airspaces, specific ray angles, and virtually any other system requirement. Wavelength and field weights are also required to be input. It is important to note that it is not imperative (nor is it advisable) to vary every conceivable variable in a lens, especially early in the design phase. For example, your initial design optimization should probably be done using the glasses from the starting point, in other words do not vary glass characteristics initially. This will come later once the design begins to take shape and becomes viable. You may also want to restrict the radii or thicknesses you vary as well, at least initially. For example, if adjacent elements have a very small airspace in the starting design, this may be for a good reason, and you should probably leave them fixed. Also, element thicknesses are very often not of great value as variables, at least initially, in a design task, so it is usually best to keep element thicknesses set to values which will be viable for the manufacturer.

4. You now will *set the performance error function and enter the constraints*. Most programs allow the user to define a fully "canned" or automatically generated error function, which, as discussed earlier, may be the rms blur radius weighted over the input wavelengths and the fields of view. In the Zemax program the user selects the number of rings and arms for which rays will be traced into the entrance pupil (rays are traced at the respective intersection points of the designated number of rings and arms). Chapter 21 shows a detailed example of how we work with the error function.

5. It is now time to *initiate the optimization*. The optimization will run anywhere from a few seconds for simple systems to many hours, depending on just how complex your system is and how many rays, fields of view, wavelengths, and other criteria are in the system. Today, a state-of-the-art PC optimizing a six- to seven-element double Gauss lens with five fields of view will take in the order of 5 to 10 s per optimization cycle. Once the computer has done as much as it can and reaches a local minimum in the error function, it stops and you are automatically exited from the optimization routine.

6. You now *evaluate the performance* using whatever criteria were specified for the lens. This may include MTF, encircled energy, rms spot radius, distortion, and others.

7. You now *repeat steps 3 and 5 until the desired performance is met*. Step 3 was to establish the variables and constraints, and step 5 was to run the optimization, and these steps are repeated as many times as necessary to meet the performance goals. You will often reach a solution that simply does not meet your performance requirements. This is very common during the design evolution, so do not be surprised, depressed, or embarrassed if it happens to you...it happens to the best of us. When it does happen, you may need to *add or split the optical power of one or more of the lens elements* and/or to *modify glass characteristics*. As we have discussed previously, splitting optical power is extremely valuable in minimizing the aberrations of a lens.

8. There is a really simple way of splitting an element in two, and while it is not "technically robust," it does work most of the time. What you do is insert two surfaces in the middle of the current element, the first of which will be air and the second is the material of your initial element. The thickness of each "new" element is one-half of the initial element and the airspace should be small, like 0.1, for example. Now simply enter twice the radius of the original element for both *s*1 and *s*2 of the new elements. You will end up with two elements whose net power sum is nearly the same as your initial element. You can now proceed and vary their radii, the airspace, and, as required, the thicknesses.

9. If you still cannot reach a viable design, then at this point you will need to *return to step 2 and select a new starting point*.

10. Your final task in the design process is to *perform a tolerance analysis and performance error budget*. We will be discussing tolerancing in more depth in Chap. 16. In reality, you should be monitoring your tolerance sensitivities throughout the design process so that if the tolerances appear too tight, you can take action early in the design phase and perhaps select a less sensitive design form.

11. Finally, you will need to *generate optical element prints, contact a viable lens manufacturer, and have your elements produced.* You will also need to work with a qualified mechanical designer who will *design the cell or housing* as well as any required interfaces. It is important to note that while we list the mechanical design as taking place at this point after the lens design is complete, it is extremely important to work with your mechanical designer throughout the lens design process so as to reach an optimum for both the optics as well as the mechanics. Similarly, you should establish a dialog with the optical shop prior to completing the design so as to have time to modify parameters which the shop feels needs attention such as element thicknesses, glass types, and other parameters.

12. Once the components are in house, you will need to *have the lens assembled and tested.* Assembly should be done to a level of precision and cleanliness commensurate with the overall performance goals. Similarly, testing should be to a criterion which matches or can be correlated with your system specifications and requirements. We discuss testing in Chap. 15.

Sample Lens Design Problem

There was a very interesting sample lens design problem presented at the 1980 International Lens Design Conference. The optimized design for an $f/2.0$, 100-mm focal length, 30° full field-of-view double Gauss lens similar to a 35-mm camera lens was sent out to the lens design community. One of the tasks was to redesign the lens to be $f/5$ covering a 55° full field with 50% vignetting permitted. Figure 9.3 shows the original starting design, as well as the design after changing the f/number and field of view, without any optimization.

Sixteen designers submitted their results, and they spent from 2 to 80 h working on the problem. We will present here three representative solutions in Fig. 9.4. The design in Fig. 9.4*a* is what we often call *a happy lens.* What we mean is that the lens is quite well behaved with no steep bending or severe angles of incidence. The rays seem to "meander" nicely

Figure 9.3
Starting Design for
Sample Lens Design
Problem

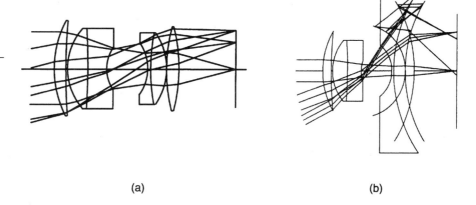

(a)   (b)

through the lens. It is a comfortable design. We show to the right of the layout a plot of the MTF. MTF will be discussed in detail in Chap. 10. For the purpose of this discussion, consider the MTF to be contrast plotted in the ordinate as a function of the number of line pairs per millimeter in the abscissa. The different curves represent different positions in the field of view and different orientations of the resolution patterns. The higher the curves, the better the contrast and the overall performance. The MTF is reasonable for most of the field positions. As will be discussed in Chap. 21, a good rule of thumb for the MTF of a 35-mm camera lens is an MTF of 0.3 at 50 line pairs/mm and 0.5 at 30 line pairs/mm.

The design in Fig. 9.4*b* has a serious problem; the rays entering the last element are at near-grazing angles of incidence. Notice that the exit pupil at full field is to the right of the lens (since the ray cone is descending toward the axis to the right), and at 70% of the field the exit pupil is to the left of the lens (since the ray cone is ascending to the right and therefore appears to have crossed the axis to the left of the lens). This is a direct result of the steep angles of incidence of rays entering the last element. The variation in exit pupil location described here would not itself be an issue unless this lens were used in conjunction with another optical system following it to the right; however, it does indicate clearly the presence of the severe ray bending which will inevitably lead to tight manufacturing and assembly tolerances. Further, the last element has a near-zero edge thickness which would need to be increased. The lens is large, bulky, and heavy. And finally, the MTF of this design is the lowest of the three designs presented.

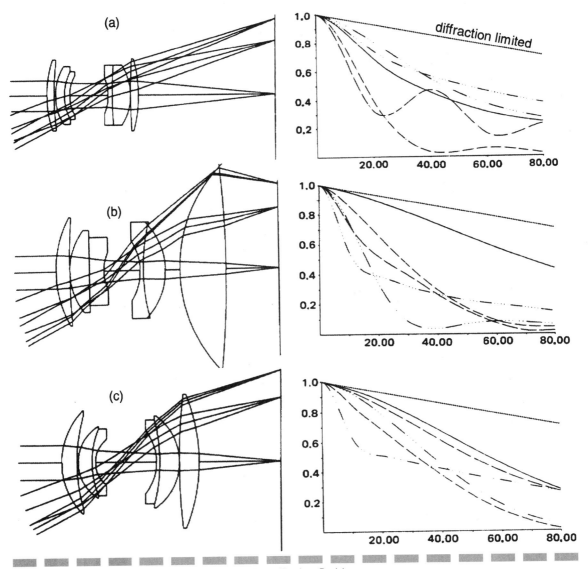

Figure 9.4 Representative Solutions to Sample Lens Design Problem

Finally, the design in Fig. 9.4*c* is somewhat of a compromise of the two prior designs in that it is somewhat spread out from the design in Fig. 9.4*a* but does not have the problems of the design in Fig. 9.4*b*. The MTF of the design in Fig. 9.4*c* is the best of the three designs.

Comparing the three designs is very instructive as it shows the extreme variability of results to the same problem by three designers.

The question to ask yourself is what would you do if you subcontracted the design for such a lens, and after a week or two the designer brought you a stack of paper 200-mm thick with the results of the design in Fig. 9.4b. And what if he or she said "wow, what a difficult design! But I have this fabulous solution for you!" Prior to reading this book, you might have been inclined to congratulate the designer on a job well done, only to have problems later on during manufacturing and assembly. Now, however, you know that there may be alternate solutions offering superior performance with looser tolerances and improved packaging. Remember that even a simple lens has a near infinite number of possible solutions in a multidimensional space.

10

Computer Performance Evaluation

What Is Meant by Performance Evaluation

The performance characteristics of an imaging optical system can be represented in many ways. Often the final optical performance specification is in terms of the modulation transfer function (MTF), encircled energy, rms blur diameter, or other image quality criteria. These criteria relate in different ways to the image quality of the system. Image quality can be thought of as *resolution* or how close two objects can approach each other while still being resolved or distinguished from one another. Image quality can also be thought of as image sharpness, crispness, or contrast.

As discussed earlier, imagery is never perfect. It is limited by geometrical aberrations, diffraction, the effects of manufacturing and assembly errors, and other factors. The characterization of image quality by the methods described in the following sections will help you to assess just how your system performs with respect to its imagery.

It is important to realize that the image quality or resolution of the entire system is not totally dependent on the optics, but may include the sensor, electronics, display device, and/or other system components making up the system. For example, if the eye is the sensor, it can accommodate for both defocus and field curvature, whereas a flat sensor such as a CCD cannot. In this chapter, we will be discussing only the optics contribution to image quality.

What Is Resolution?

When we think about the image quality of an image-forming optical system such as a camera lens, the first parameter that often comes to mind is *resolution* or *resolving power.* Classically, the ability of an optical system to separate two closely spaced point sources at the nominal object distance is generally considered to be the resolution.

Consider a perfect optical system which has an entrance pupil diameter, D, and focuses to an image with a given $f/\#$. Two point sources closely spaced will be imaged through the optical system, each of them forming a diffraction pattern. If two perfect diffraction patterns as in Fig. 10.1 are separated by the radius of the Airy disk (the radius of the

Figure 10.1
Two Resolvable
Images of Closely
Spaced Point Sources

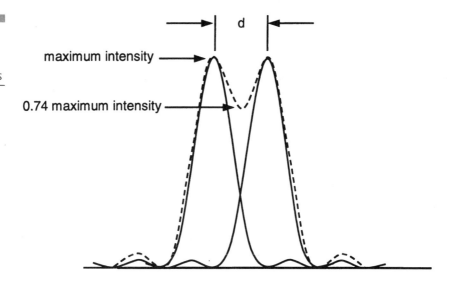

Figure 10.1
Two Resolvable
Images of Closely
Spaced Point Sources

first dark ring in the diffraction pattern), then the intensity midway between the two peaks in the pattern drops to 0.74 of the maximum intensity, and the two point images are said to be *resolvable*. This is the Rayleigh criterion for resolution. This, of course, assumes that the ultimate media or sensor is not the limiting factor. The separation, *d*, of the two points in the image plane is

$$d = 1.22 \, \lambda \, f/\#$$

in units of wavelength. In object space, in radians, this becomes (λ and pupil diameter in ssme units)

$$\alpha = \frac{1.22 \, \lambda}{\text{entrance pupil diameter}}$$

or very frequently used as a rule of thumb, the resolution, in arc seconds, for the visible spectral range is

$$\alpha = \frac{136}{\text{entrance pupil diameter}}$$

where the entrance pupil diameter is given in millimeters. This is interesting and certainly of value in understanding the limiting resolution of the optical system with given first-order optical parameters, but it really does not help us to understand the performance of a specific optical system design. As will be shown in this chapter, there is far more to the characterization of optical performance than the theoretical resolution.

Ray Trace Curves

Most of the methods used in computing image quality, such as the modulation transfer function, spot diagrams, encircled energy, and the like, are functionally robust and represent different, yet similar representations of the net performance of the optical system as designed. However, there are two disadvantages with these metrics. First, they can sometimes take too much time to compute. This, however, is less and less of a problem as PCs have become faster and faster. Second, the real problem is that while these metrics do help to show the overall net resulting image quality, they do not provide a detailed indication to the designer of the specific aberrations present in the design over the field of view and over the spectral bandwidth. While some information can at times be derived, more often the user really cannot tell what aberrations are present and at what magnitudes. These data are important to the designer as an aid in correcting the residual aberrations.

The solution is to generate what are called *transverse ray aberration curves* or simply *ray trace curves*. With these graphical data, a reasonably experienced designer can immediately tell just how much spherical aberration, coma, astigmatism, field curvature, axial color, lateral color, and field curvature are present. In addition, in many cases the user can also tell what orders of these aberrations are present. Finally, with this knowledge, the designer can often make a reliable judgment as to what to do next regarding further optimization of the lens. In spite of some fabulous advances in performance simulation and modeling, transverse ray aberration curves are still invaluable to the serious designer.

We show in Fig. 10.2 the basic formation of the ray trace curve. This perspective figure (Fig. 10.2) shows a lens exit pupil with the lens imaging to an off-axis image position. First, consider tracing the *chief ray* to the image. The height on the image of the chief ray is our reference point, and is generally taken to be the image height. Now let us trace a ray through the top of the exit pupil. This ray, which is called the *upper marginal ray,* hits the image higher than the chief ray for the aberration shown, which is coma. Now let us trace a ray through the bottom of the exit pupil. This ray, which is called the *lower marginal ray,* also hits the image higher than the chief ray, and in fact for classical third-order coma it hits the image the same distance above the chief ray as the ray from the top of the pupil. In other words, both of the rays from the top and the bottom of the exit pupil hit the image vertically displaced by the same amount.

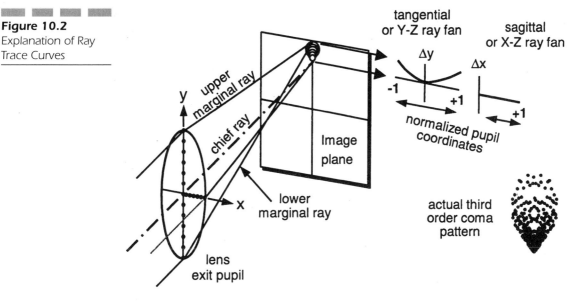

Figure 10.2
Explanation of Ray
Trace Curves

We will now proceed to establish a set of coordinate axes for our ray trace curves. In the first set of coordinates (on the left in Fig. 10.2), the abscissa is the normalized exit pupil radius in the y direction, and the ordinate is the distance above or below the chief ray on the image that our ray intersects the image plane (Δy). Thus, both the upper and lower marginal rays form the end points on the curve. We now proceed to trace rays through each of the black dots from $y = +1$ to $y = -1$, with the intersection points relative to the chief ray plotted on the curve. For third-order coma the result will be a quadratic or parabolic curve since third-order coma is quadratic with aperture.

Now we establish a second set of coordinate axes, as shown on the right in Fig. 10.2. Here we have the normalized X coordinate in the exit pupil in the abscissa, and the displacement of the ray in the x direction (Δx) as the ordinate. As it turns out, for third-order coma there is no x departure at all for these rays in the exit pupil. We will show why this is the case shortly. For now, you will see that for third-order coma a quadratic curve in the "tangential" ray fan and zero departure for the x rays in the "sagittal' ray fan are the results. If any lens designer who is "worth his or her salt" sees this form of ray aberration curves with a quadratic in the tangential ray fan and virtually zero in the sagittal curve, then he or she should conclude instantly that the lens has third-order coma.

Since these ray trace curves are so fundamentally important to the optical designer's work, a more in-depth discussion is in order. As you will see, there are here, as with many other areas of optical design, subtleties that could easily be misleading if not fully understood. Consider our coma pattern where the ray trace curves suggest zero x departure of the rays hitting the image, which implies or suggests zero x width to the image blur. Yet we all know that coma does have width in the x direction. Just what is going on, and why are the data misleading?

Figure 10.3 will explain the situation. Here we trace rays around the periphery of the exit pupil from positions 1 through 8. From our prior discussion, we know that the chief ray is our reference, and that rays 1 and 5 from the top and bottom of the exit pupil both hit the image high, above the chief ray. If you follow the numbers in Fig. 10.3, you will see how *one* rotation around the exit pupil results in *two* rotations around an ellipse in the image, and since positions 1 and 5 are both high, then positions 3 and 7 which are 180° opposed will be at the bottom of the elliptical pattern. Neither of these rays will have any x departure at all! Thus, the ray trace curve for rays traced in the x direction was

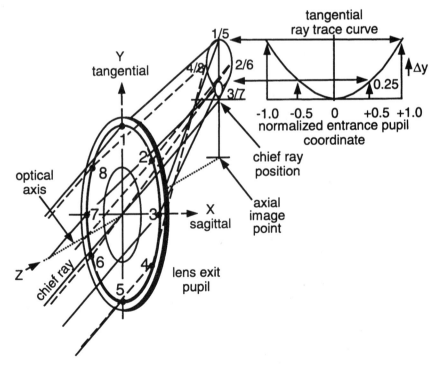

Figure 10.3
Formation of Comatic Image Blur

a horizontal line in Fig. 10.2. So where is the *x* spreading of the coma pattern coming from? The answer is from rays at positions 2, 4, 6, and 8, which are called *skew rays*. Since the rays making up the ray trace curves contain only the *y* (or *tangential rays*) and the *x* (or *sagittal rays*), the designer sees no indication or evidence whatsoever of the *x* spreading of the imagery. This is a real subtlety, and it is a fine example why one should never be totally dependent on only one form of image evaluation or analysis. By looking only at the ray trace curves, one could easily conclude that such a system had virtually zero *x* spreading of the off-axis imagery, and this could make its performance ideal for some system applications. For the most part use of the ray trace curves are wonderfully helpful and revealing; however, do be aware of subtleties as pointed out earlier.

A further illustration of the ray trace curves, Fig. 10.4 shows how spherical aberration is formed and how the ray trace curves are derived. In the top of Fig. 10.4 the image is located at paraxial focus and it should be clear how each of the rays entering the lens from the left results in a corresponding intercept on the image plane and how this is plotted as the ray trace curve. Ray 1 strikes the image lowest and results

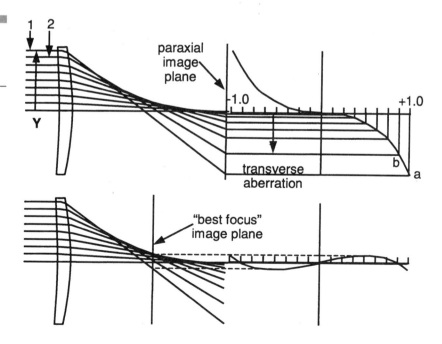

Figure 10.4
Formation of Ray
Trace Curves for
Spherical Aberration

in point *a* in the plot. Ray 2 is the next ray lower down entering the entrance pupil and it results in point *b* in the plot, and so on. Since third-order spherical aberration is cubic with aperture, the resulting curve is cubic. Note the symmetry above and below the optical axis. Now consider what happens if we relocate the image plane to the "best focus" position. Following the same logic in generating the ray trace curves, we see a much lower departure of the ray intercept points making up the curve. This is true and quite real, and it tells us that the image blur diameter when we refocus the image will be significantly reduced from that at paraxial focus. As an exercise, what will the ray trace curve be for a perfect image where the image plane is intentionally defocused toward the lens? The answer is a straight line sloped upward to the right. So let's use this as an aid in further understanding ray trace curves. Since defocus yields a sloped but otherwise straight line, we can easily determine what any ray trace curves for any lens will look like as we go through focus by simply drawing or imagining a sloped straight line as a new coordinate axis. This is an invaluable tool as you can now immediately assess the relative improvement after refocusing a given lens. And since field curvature is a quadratic change in focus with field of view, you can with a little practice assess immediately the benefits of curving your sensor if this is possible.

We show in Fig. 10.5 ray trace curves for various typical aberrations and combinations of aberrations:

Figure 10.5*a* is pure defocus. As noted earlier, defocus will produce equal sloped straight lines in the sagittal and tangential ray fans. Recall that the tangential ray fan is in the *y-z* direction and typically oriented parallel to the field-of-view direction. The sagittal ray fan is orthogonal to the tangential ray fan, and typically, the sagittal fan is fully symmetrical which is why we sometimes show only one-half of the fan.

Figure 10.5*b* shows straight lines at different slopes. This is a combination of astigmatism (which is the difference between the slopes of the two curves) and defocus.

Figure 10.5*c* shows that if we best focus for the residual astigmatism off axis as we might do with a curved image surface, we find the result here where an equal and opposite ray fan slope results in the tangential and sagittal directions.

Figure 10.5*d* is negative or undercorrected third-order spherical aberration, which is, of course, a cubic with aperture.

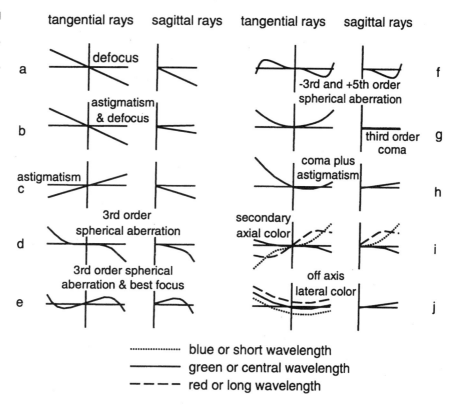

Figure 10.5
Typical Transverse Ray
Aberration Curves

The data in Fig. 10.5e are the same third-order spherical aberration as in Fig. 10.5d, only we have refocused the image to a more optimum focus position to minimize the residual blur diameter.

Figure 10.5f shows negative third-order spherical aberration, which is being balanced by positive fifth-order spherical aberration.

The data in Fig 10.5g are for pure third-order coma, which, as we know from before, is quadratic with aperture. We also know from before that the sagittal curve indicates zero image blurring in the sagittal direction. This may be misleading and is due to the nature of coma formation and the fact that the ray aberration curves show only the rays along two lines in the pupil plane.

The data in Fig. 10.5h are for a combination of third-order coma and astigmatism.

Figure 10.5i shows a combination of some negative third-order spherical aberration at the central wavelength as well as secondary axial color and spherochromatism. The secondary axial color is the focus difference between the central wavelength and the common red and blue focus, which together are focused beyond the central wavelength. The spherochromatism is the change in spherical aberration with wavelength.

Finally, Fig. 10.5*j* shows an off-axis ray trace curve with primary lateral color or color fringing along with a small amount of coma and astigmatism.

It should be quite apparent that the ray trace curves for each of the aberrations has its own distinctive form, and this is what makes them so useful. The aberrations, in effect, add algebraically, so it is easy to tell almost immediately what aberrations are present in a given design at the different field positions.

Spot Diagrams

Spot diagrams are the geometrical image blur formed by the lens when imaging a point object. This is a more functionally useful form of output; however, it is sometimes difficult to distinguish the specific aberrations present. Figure 10.6 shows the spot diagrams for a Cooke triplet form of lens and shows both the transverse ray aberration curves as well as spot diagrams for the same field positions. Generally, the rms spot radius or diameter is output with the spot diagrams. The rms spot diameter is the diameter of a circle containing approximately 68% of the energy. This metric can be of great value, especially when working with pixelated sensors where one often wants the image of a point object to fall within a pixel.

Note that in optical design software we often come across the terms "spot radius," "spot diameter," and "spot size." While the designer is most interested in spot diameter, the software generally outputs spot radius. The use of the words "spot size" is fine for relative comparison (for example, "the spot size has increased by a factor of 2"); however, the term can cause undo confusion when tied to a specific value. For example, "the spot size is 50 μm" does not really tell us whether this is the radius or diameter of the image blur, nor does it tell us whether this is for 100% of the energy or some other value such as the rms. Be careful in interpreting these forms of data from the software you are using.

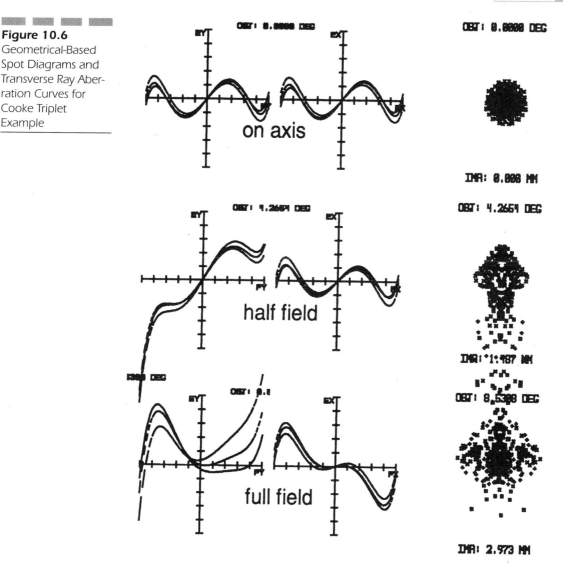

Optical Path Difference

As we know from Chap. 4, if the peak-to-valley optical path difference (P-V OPD) is less than or equal to one-fourth of the wavelength of light, the image quality will be almost indistinguishable from perfect. It is known as diffraction limited. Just like transverse ray aberration curves, we can plot the optical path difference, and Fig. 10.7 shows a typical OPD

Figure 10.7
Optical Path Difference for Cooke Triplet Example

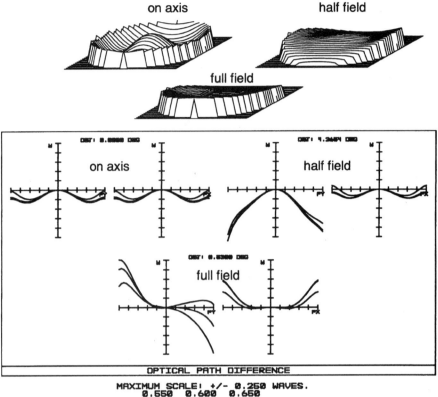

plot for our sample Cooke triplet. In addition to the plotted data, we show also perspective views of the three-dimensional wavefront departure from perfect. Note how the curve and the perspective view correlate so well for the on-axis field position with the bump in the center as well as the turned up edge of the wavefront clearly evident in both data.

It is a little difficult without extensive experience to quickly determine the residual aberrations from an OPD plot, so you will generally want to compute the more standard ray trace curves for this purpose.

Encircled Energy

Encircled energy is energy percentage plotted as a function of image diameter. One good example of how we might use encircled energy is to specify an imaging optical system using a CCD sensor. Let us assume

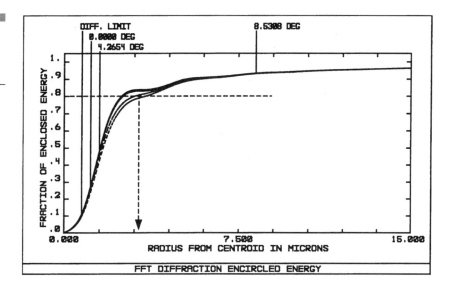

Figure 10.8
Encircled Energy for
Cooke Triplet
Example

that the pixel pitch of our sensor is 7.5 μm. A good reliable and simple specification is that 80% of the energy from a point object shall fall within a diameter of 7.5 μm. Figure 10.8 shows an encircled energy plot for our sample Cooke triplet. Eighty percent of the energy is contained within a diameter of approximately 6 μm, which is a good match to the sensor. It also leaves some margin for manufacturing tolerances.

MTF

MTF is perhaps the most comprehensive of all optical system performance criteria, especially for image forming systems. Figure 10.9 is a representation of what is happening. We begin with a periodic object or target, which is varying sinusoidally in its intensity. This target is imaged by the lens under test, and we plot the resulting intensity pattern at the image. Due to aberrations, diffraction, assembly and alignment errors, and other factors, the imagery will be somewhat degraded and the brights will not be as bright and the darks will not be as dark as the original pattern.

Let us define some terms:

$$\text{Modulation} = \frac{I_{max} - I_{min}}{I_{max} + I_{min}} \qquad \text{MTF} = \frac{\text{modulation in image}}{\text{modulation in object}}$$

Figure 10.9

Illustration of the
Meaning of the
Modulation Transfer
Function

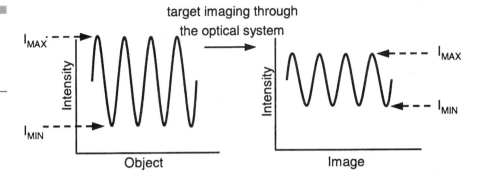

The modulation is simply the maximum intensity minus the minimum intensity divided by the maximum plus the minimum. The MTF is the ratio of the modulation in the image to the modulation in the object as a function of spatial frequency, which is generally in the form of line pairs per millimeter. Thus, the modulation transfer function represents *the transfer of modulation from the object to the image by the lens as a function of spatial frequency.*

There is another term, *contrast* (sometimes called *contrast ratio*), which is given by

$$\text{Contrast} = \frac{I(\text{max})}{I(\text{min})}$$

Figure 10.10 shows several typical MTF curves. We show the MTF of a perfect optical system, a perfect system with a central obscuration (such as a Cassegrain telescope), and a typical real system. The MTF of the perfect obscured system has more diffraction due to its obstruction, and thus a lower MTF. The cutoff frequency, which is where the MTF goes to zero, is

$$\nu_{\text{cutoff}} = \frac{1}{\lambda(f/\#)}$$

The example shown is an $f/2$ lens in the visible (0.55-μm wavelength), and the cutoff frequency is approximately 882 line pairs/mm. A good rule of thumb to remember is that an $f/2$ lens used at 0.5 μm has a cutoff frequency of 1000 line pairs/mm, and it is very easy to scale from here. For example, an $f/4$ lens has twice the Airy disk diameter and thus half the cutoff frequency of 500 line pairs/mm.

Recall that the MTF tells us how well the modulation in the object is transferred to the image by the lens. We show below the MTF curve in Fig. 10.10 a graphical representation of an object and of the resulting image at a low spatial frequency, a midspatial frequency, and at a high spatial frequency. Think of the low spatial frequency as being very large tree trunks with a bright sky between them, and the high spatial frequency as being tiny close tree branches with the same bright sky between them. The sky is the same for both and the darkness of the bark on the tree trunks and the branches is the same, hence the modulation of the objects is identical. However, the modulation of the image will be far lower at the higher spatial frequency of the branches since the MTF is lower.

The MTF is generally computed or measured for bars that are *radial* and for bars that are *tangential*. The radial bars are like the spokes on a bicycle, and the tangential bars are tangential to the edge of the bicycle tire as well as tangential to the edge of a circular field of view. The tangential and radial bars are orthogonal to each other. There is a subtlety of some

Figure 10.10
Typical MTF Curves

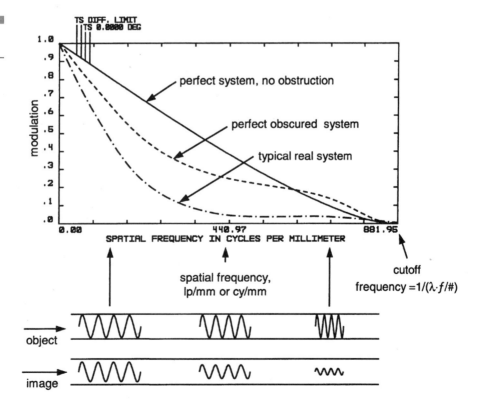

significance with respect to the preceding nomenclature and how we handle it in the various software packages. Consider a lens designed to cover a rectangular format with a $3 \times 4 \times 5$ aspect ratio. We said earlier that the radial bars are parallel to spokes on a wheel and tangential bars are orthogonal to the radial bars. This is fine and true, and the definition is valid. How-ever, given the pixelated nature of many of today's sensors such as CCD or CMOS detector arrays, or alternatively pixelated display devices as used in LCD and similar projection systems, it makes far more sense if the resolution bars and their associated nomenclature were consistent with the pixel rows and columns. What we mean is that it will be far more appropriate to refer to bars that are vertical and bars that are horizontal, rather than radial and tangential as shown in Fig. 10.10. Fortunately, most of the software packages use the terminology *sagittal* and *tangential*, and these refer to entrance and exit pupil coordinates. Tangential ray aberrations in the exit pupil are up and down in the plane of the paper in the y direction and sagittal ray aberrations are in and out of the plane of the paper in the x direction. We show in Fig. 10.11 the orientation of the bars using both nomenclatures. Regardless of which software package you use, it is imperative to understand what assumptions are being made with respect to target orientation.

The bottom line here is as follows:

■ Tangential aberrations are in the y direction and blur horizontal bars. Tangential bars are horizontal.

■ Sagittal aberrations are in the x direction and blur vertical bars. Sagittal bars are vertical.

■ Radial bars are bars parallel to spokes on a wheel.

Figure 10.11
Target Orientation
Conventions

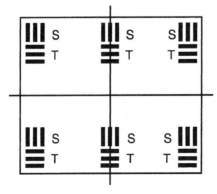
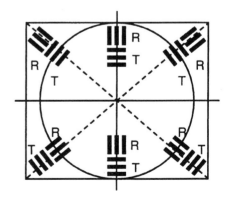

■ Tangential bars are tangent to the rim of a wheel.

A good example of a computed MTF and the resulting image quality is shown in Fig. 10.12. Here we plot the MTF of a double Gauss lens at four fields of view (on axis, 0.33 field, 0.67 field, and at the edge of the field). Note how the radial (vertical) bars at the edge of the field of view are more degraded than the tangential (horizontal) bars. The simulated 3-bar imagery is at 50 line pairs/mm, and the modulation at this spatial frequency is about 0.65 for the tangential bars and about 0.4 for the radial bars. The simulated imagery clearly shows this difference in the two target orientations. If you ever have a question with respect to this, we recommend that you read the manual for the software package you are using, phone the user support person, or, even better, run several examples.

Figure 10.13 shows the MTF of a perfect system as a function of obstruction ratio in a system such as a Cassegrain telescope. As the central obscuration increases, the amount of light that is diffracted away from the central maximum of the Airy disk increases, there is a corresponding reduction in MTF. As this is happening, the diameter of the central maximum actually decreases somewhat, resulting in a high-frequency MTF that is actually slightly above the unobscured perfect system MTF. This can sometimes be used to advantage if you are trying to separate two close point objects or stars as in astrometric work. Figure 10.14 shows point-spread functions for aberration-free systems with no central obscuration, 0.2, 0.4, 0.6, and 0.8 diameter central obstructions, respectively.

It can be shown that the MTF of an image-forming optical system can actually be less than zero. As an otherwise well-performing system is degraded in performance due to defocus, aberrations, and/or manufacturing errors, the MTF degrades. As the performance continues to degrade, eventually the MTF may drop below zero. In extreme cases, the MTF will oscillate above and below zero. Any time the MTF is less than zero, that constitutes a phase reversal, which is where the dark bars become bright and the bright bars become dark. This can be seen visually by covering one eye and looking at Fig. 10.15. Hold the page at a distance closer than you can comfortably accommodate or focus. Do not try to focus on the pattern. This might range from 25 to 75 mm or more, depending on your accommodation. Then move the page closer or further from your eye very slowly, and you should be able to see very clearly the phase reversal. What is especially interesting here is what you are observing is the phase reversal of the imagery formed by your own

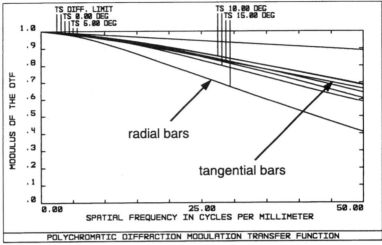

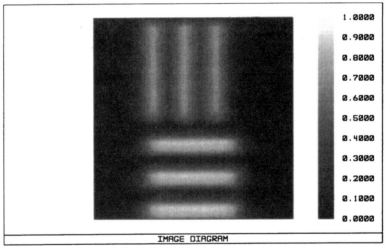

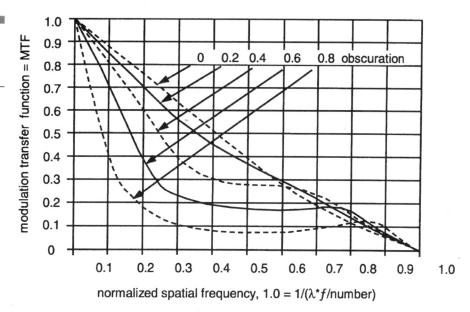

Figure 10.13
MTF of a Perfect System As a Function of Central Obscuration

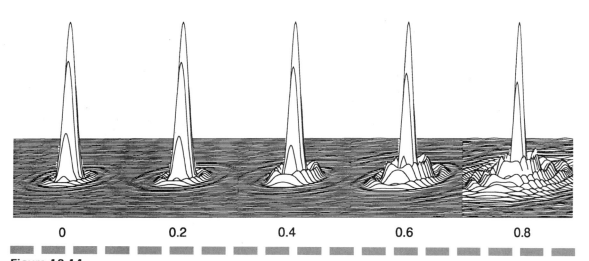

Figure 10.14
Point Spread Functions of an Aberration-Free System As a Function of Central Obscuration

cornea and eye as imaged onto your own retina...there is no other optics whatsoever! The phase reversal should be very clear and striking. If for some reason you cannot see it, Fig. 10.16 is a digital photograph of the effect. The camera lens was focused at infinity and located several inches from the figure to obtain the photo. Figure 10.17 shows the MTF of an otherwise perfect aberration-free $f/5$ system with approximately 1.5 waves of defocus. Note how at about 50 line pairs/mm the MTF is negative at about 0.1. In this spatial frequency range, there will be a phase reversal, as shown and discussed earlier.

Figure 10.15
Radial Bar Pattern for Demonstration of Spurious Resolution

Figure 10.16
Spurious Resolution
Using Defocused
Digital Camera

phase change
black to white

Figure 10.17
Spurious Resolution
Due to Approximately
1.5 Waves of
Defocus

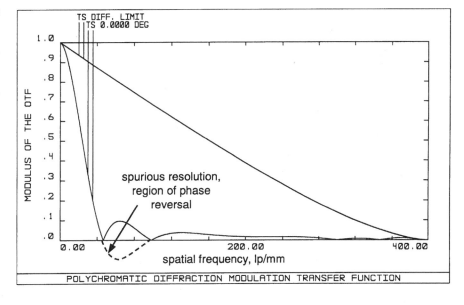

spurious resolution,
region of phase
reversal

POLYCHROMATIC DIFFRACTION MODULATION TRANSFER FUNCTION

Gaussian
Beam Imagery

Coherent light generated by lasers has properties different from light generated by other sources which we usually deal with in more conventional optical systems. If we look through a telescope at a distant object, the light intensity across the entrance pupil and aperture stop is uniform, and this is generally known as a "top-hat" intensity profile or distribution. A telescope objective, if it is free of aberrations, focuses a point object into an Airy disk pattern, with the diameter determined by the f/number or numerical aperture of the objective lens. In this case, a uniform top-hat distribution in pupil space transforms mathematically (by the optical system) to an Airy disk in image space. Laser beams emitted from rotationally symmetric resonators, such as HeNe or YAG lasers with a TEM00 output, have an intensity distribution across the beam which is in the form of a gaussian intensity profile, as shown in Fig. 11.1. A gaussian intensity distribution in pupil space will mathematically transform to a gaussian in image space if the beam is not truncated by the aperture of the optical system, which is, of course, different from the uniform pupil transformation. Note that all of the material in this chapter assumes an aberration-free optical system. It is important to include the effects of lens aberrations in the final assessment of image quality and spot size.

The optical design of systems through which laser beams propagate is, therefore, very different from the design of conventional nonlaser systems used in either the visible or some other wavelength region. First,

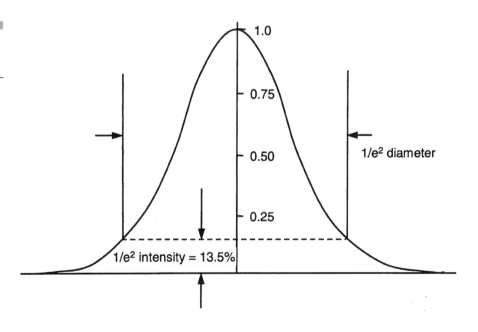

Figure 11.1
Gaussian Intensity
Distribution

color correction of the laser-based optical system is much easier because the wavelength band of the laser light is extremely narrow. For HeNe lasers fully monochromatic light at 0.63282 μm can be used. Some lasers emit multiple spectral lines, or wavelengths can be changed. In these cases the optics must be designed to cover the functional spectral bandwidth. Laser diodes sometimes have a wavelength shift with temperature, which also must be taken into account by the optics.

Laser systems are often corrected for a small field of view because the laser beam enters the system parallel to the optical axis. Even if the real usable field of view is zero, the optical design must be optimized over a small, yet finite field of view in order to accommodate assembly and alignment tolerances. If one were to design such a system at identically zero field of view, it is possible that the performance may seriously degrade within 1 or 2 mrads off axis. However, there are systems, such as laser scanning systems, where some components of the system have to be designed for a large field of view. There are also systems where it is required to focus a laser beam to a very small spot, which requires the design of diffraction-limited optics with very large numerical apertures. In some cases, the laser beams have high-power densities, and they can cause damage to the optical components. Transparency of the material from which the components are made can be degraded, and the material used for cementing of components and coating of the components can be damaged. The choice of optical materials is thus very important. In some cases, dust particles on the components in the system can also absorb enough energy to damage the surface of the component. Scattering from surface defects is a greater problem in laser systems than in visible and other incoherent systems.

The coherence length of a gaussian laser beam is large, and it appears as if the wavefront emerges from one point. If a gaussian beam is not truncated by the optical system, it emerges from the system as a gaussian beam. The narrow spectral line width of a laser beam and a well-defined wavefront permit very precise focusing and control of the beam.

Beam Waist and Beam Divergence

The beam emitted from a laser in TEM00, or fundamental mode, has a perfect plane wavefront at its beam waist position and a gaussian transverse irradiance profile that varies radially from the axis, which can be described by

$$I(r) = I_0 \exp\left(-2\,\frac{r^2}{r_0^2}\right)$$

or by the beam diameter

$$I(d) = I_0 \exp\left(-2\,\frac{d^2}{d_0^2}\right)$$

where I_0 is the axial irradiance of the beam, r and d are radius and diameter of a particular point in the beam, and r_0 and d_0 are the radius and diameter of the beam where irradiance is $(1/e^2)I_0$, or 13.5% of its maximum intensity value on axis. To define the propagation characteristics of a laser beam, the value, r_0, is accepted as a definition of the radial extent of the beam. Finite apertures in the optical system or inside the laser itself, along with diffraction, cause the beam to diverge or converge. There is no such thing as a perfectly collimated beam without any spreading. Spreading of a laser beam is defined by diffraction theory. A laser beam converges to a point where the beam is smallest, called the *beam waist,* and it then diverges from this point with a full angular beam divergence, θ, which is the same as the convergence angle. The beam divergence angle, θ, is the angle subtended by the $1/e^2$ diameter points in the far field, where the irradiance surface asymptotically approaches the full divergence angle, θ, shown in Fig. 11.2. At the location of the beam waist, the laser beam wavefront is flat, and it quickly acquires curvature on both sides of the waist. The beam waist diameter, d_0, depends on the full divergence angle, θ, as

$$d_0 = \frac{4\lambda}{\pi\theta}$$

where λ is the wavelength of the laser beam and θ is in radians. It is important to know that the product of the divergence and the beam waist diameter is constant. The beam diameter grows with the distance, z, from the beam waist according to

$$d(z) = d_0 \sqrt{1 + \left(\frac{4\lambda z}{\pi d_0^2}\right)^2}$$

The radius of the wavefront at the beam waist location is infinite. As we move away from the beam waist, it then passes through a minimum at some finite distance, after which it rises again, approaching the value of z as z approaches infinity. The radius of the laser beam wavefront can be expressed as

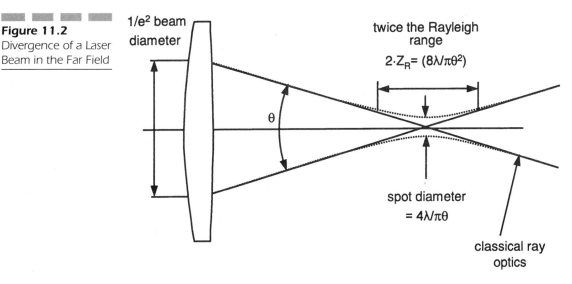

Figure 11.2
Divergence of a Laser
Beam in the Far Field

$$R(z) = z\left[1 + \left(\frac{\pi d_0^2}{4\lambda z}\right)^2\right]$$

The minimum value of the wavefront radius occurs at what is known as the *Rayleigh range*, where the beam diameter has the value $(\sqrt{2})d_0$. At the extremes of the Rayleigh range, the image diameter is thus 41% larger than at the center of the beam waist. The Rayleigh range is sometimes used as a depth of focus number; however, do keep in mind that this reasonably large increase in spot diameter may not be acceptable. The Rayleigh range as measured from the beam waist is

$$z_R = \frac{d_0}{\theta} = \frac{4\lambda}{\pi\theta^2} = \frac{\pi d_0^2}{4\lambda}$$

The wavelength of the laser radiation, the beam waist diameter, the beam divergence angle, and the Rayleigh range are four parameters that completely describe a gaussian beam.

Collimation of Laser Beams

There is no perfectly collimated beam, since the beam divergence and the beam waist are defined by diffraction theory. It can be shown that the minimum beam spread between two points at a distance, z, happens when the starting beam waist is equal to

$$d_0 = 2\left(\frac{\lambda z}{\pi}\right)^{1/2}$$

This relationship is equivalent to the one that gives the Rayleigh range, which tells us that the Rayleigh range is the distance inside which a gaussian beam has the minimum spreading. The diameter of the beam at the Rayleigh range is

$$d = \sqrt{2}\,d_0$$

If we use a beam expander at the output of a laser, we can increase the distance of the minimum spreading. If we additionally adjust the beam expander so that the beam waist is located in the middle of the starting Rayleigh range, then the beam will spread to $\sqrt{2}\,d_0$ over a distance twice as long.

Propagation of Gaussian Beams and Focusing into a Small Spot

In 1983 S.A. Self derived a simple algorithm for tracing a gaussian beam through an optical system. The formula that he used had a very similar form to the paraxial lens formula that gives the relationship between the object position, focal length of a lens, and the image position. For a gaussian beam, Self calculates the Rayleigh range and the beam waist transformation by each lens in the system:

$$\frac{1}{s + [1/(s-f)]z_R^2} + \frac{1}{s'} = \frac{1}{f}$$

where s is the distance from the lens to the waist on the object side and s' is the distance from the lens to the new waist position. When the incident beam has its waist located in the front focal plane of a positive lens, the emerging beam has its waist at the rear focal plane.

Optical design programs generally use the gaussian beam propagation algorithm derived by A.E. Siegman. These programs usually calculate the radial beam size (semidiameter), the narrowest radial waist, surface coordinates relative to the beam waist, the semidivergence angle, and the Rayleigh range. This assumes the TEM00 fundamental mode, that is, a gaussian irradiance distribution. Lasers may be able to produce a number of other stable irradiance distributions, or modes, but they are not as compact as a gaussian beam, and they all have regions or holes in the beam, that is, where the irradiance drops to zero within the irradiance distribution. The mixed mode beams, defined by the beam quality factor, M, can also be traced in the current generation of optical design programs, in which case the M factor scales the embedded TEM00 gaussian mode.

In many applications, the goal is to focus the laser beam down to a very small spot. If the optical system focusing the beam is diffraction limited, the spot diameter at $1/e^2$ of the peak irradiance is defined by

$$d = \frac{4\lambda f}{\pi d_0}$$

The depth of focus is proportional to the square of the $f/\#$ of the focusing lens. If we define the allowable increase in the diameter of the focused spot, the depth of focus can be calculated using the formula for the spot diameter as the function of the distance from the waist location

$$d(z) = d_0 \sqrt{1 + \left(\frac{4\lambda z}{\pi d_0^2}\right)^2}$$

Truncation of a Gaussian Beam

Let us assume that we have a diffraction-limited lens with a given aperture diameter. The intensity profile of the focused spot is dependent on the intensity distribution of the radiation filling the aperture of the lens. For a uniformly illuminated aperture, the diffraction pattern is the classical Airy disk. It has a central bright spot and progressively weaker rings, with the first dark ring (intensity falls to zero) at a diameter, d:

$$d = 2.44 \, \lambda f/\#$$

When the illumination in the aperture is not uniform, the intensity profile in the focused spot does not have zero-intensity points, and the measure of the spot diameter is usually accepted as the diameter at which the intensity drops to $1/e^2$ of the peak intensity. When a gaussian beam falls onto the aperture of a lens, it may be truncated by the lens aperture. Let us define the truncation ratio as

$$T = \frac{d_0}{D}$$

where d_0 is a gaussian beam diameter measured at $1/e^2$ of the peak intensity and D is the aperture diameter of the lens. In the case of the lens aperture being two times the gaussian beam waist diameter ($T = 0.5$), the beam is truncated only below the intensity level of 0.03%. We can say that the effect of truncation is negligible, and the gaussian beam after transformation by the lens remains gaussian. On the other hand, when T, the truncation ratio, becomes a large number, and only the narrow central portion of the gaussian beam is transmitted through the lens, this case corresponds to the uniform illumination of the lens aperture, and the transmitted beam has the intensity distribution similar to the Airy disk. The beam waist diameter, d_0, given in units of $\lambda f/\#$ for a few values of truncation ratio is given in Table 11.1. Two cases of the intensity distribution for $T = \infty$ and $T = 1$ are shown in Fig. 11.3.

In Fig. 11.4, we show how the truncation ratio affects the performance of an otherwise perfect system with a gaussian intensity profile incident onto the entrance pupil of an imaging optical system. The abscissa is the ratio of the physical aperture to the $1/e^2$ beam diameter. The ordinate on the right is the normalized spot radius (normalized to $1/e^2$ spot radius).

The easiest way to understand this somewhat complex data is to think of a constant $1/e^2$ beam diameter of, say, 25 mm. If the physical aperture diameter were 1 m, there would be virtually zero truncation of the beam, and a gaussian intensity distribution in the entrance pupil would

TABLE 11.1

Beam Waist Diameter in Units of λ $f/\#$ for Several Values of the Truncation Ratio

Truncation Ratio T	d_0 at $1/e^2$ Intensity (in units of $\lambda f/\#$)	d_0 (Intensity Goes to Zero)	Truncation Intensity Level (%)
≈∞	1.64	2.44	100
2	1.69	—	60
1	1.83	—	13.5
0.5	2.51	—	0.03

Figure 11.3
Truncation of a
Gaussian Beam and
the Intensity Distribu-
tion at the Image
Plane

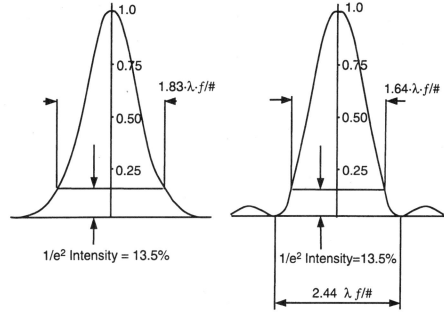

Gaussian intensity distribution   Airy disc intensity distribution

Figure 11.4
Truncated Gaussian
Beam Imagery. Aper-
ture Changes Relative
to a Constant $1/e^2$
Diameter

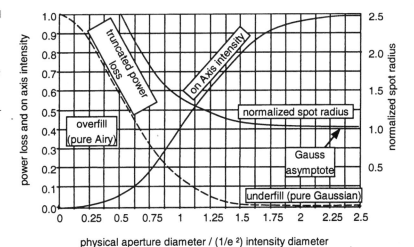

transform into a perfect gaussian image profile at the image. The normalized spot radius asymptotically approaches unity as the physical aperture diameter continues to increase. The far right data point in the abscissa in Fig. 11.4 is for the physical aperture being 2.5 times the $1/e^2$ beam diameter. If, on the other hand, the physical aperture diameter were in the order of 3 mm for our 25-mm $1/e^2$ beam diameter, the ratio of the physical aperture diameter to the $1/e^2$ beam diameter would be about 0.125 and the intensity profile would be nearly a top-hat, which means that we will acquire diffraction and the imagery will be close to a classical Airy disk pattern. It should be clear that if we truncate the beam at the $1/e^2$ beam diameter, the $1/e^2$ diameter of the image will be approximately 40% larger than an untruncated beam.

The ordinate on the left in Fig. 11.4 is for both the power loss and the on-axis intensity, with the abscissa having the same meaning as before. As the ratio of the physical aperture to the $1/e^2$ beam diameter increases, the on-axis intensity increases, and the loss in power due to truncation decreases.

In Fig. 11.5 we show a similar plot, only here we have the $1/e^2$ beam diameter changing relative to a constant physical aperture diameter. The interpretation and rationale is the same as for the prior data.

There are several important messages from Figs. 11.4 and 11.5. First, a small beam truncation will affect the $1/e^2$ beam diameter of the resulting image. Truncating the beam to the $1/e^2$ beam diameter will increase the $1/e^2$ spot diameter by approximately 40%. In order to increase the spot diameter by less than 10%, we require a physical aperture about 50% larger than the $1/e^2$ beam diameter. The other data which are of significance are the intensity and power loss associated with a given truncation factor. If our physical aperture were 50% larger than the $1/e^2$ beam diameter, our on-axis intensity would be 80% of that from an untruncated beam.

Application of Gaussian Beam Optics in Laser Systems

Gas lasers have their place in many applications in the large consumer market. HeNe lasers, which emit red light at λ = 632.82 nm, are used in bar code readers, the printing industry, machine vision, etc. They are

Figure 11.5
Truncated Gaussian
Beam Imagery. $1/e^2$
Diameter Changes
Relative to a Constant
Aperture Diameter

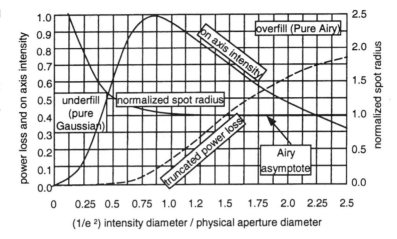

(1/e²) intensity diameter / physical aperture diameter

used wherever the packaging constraints are not too tight and the use of visible light is convenient. The beam out of the HeNe laser is TEM00 rotationally symmetric gaussian beam, which can be easily transformed with the properly designed optics to the required spot size.

Solid-state lasers can generate very powerful beams. They can emit TEM00 gaussian beams. There are a significant number of types of solid-state lasers that emit in the near IR spectral region. A YAG laser, which emits light at the wavelength of 1.064 μm, can be frequency doubled to produce green light at 532 nm. It can also be frequency tripled or quadrupled. YAG lasers are used in industrial applications where high power is needed, such as writing on metal, welding, cutting, hole drilling, etc. One disadvantage of solid-state lasers is that they are generally expensive. One field that makes an extensive use of lasers is medicine, especially in the area of diagnostics, cancer treatment, and eye surgery. The laser wavelengths used range from the UV below 200 nm to 10.6 μm in the far infrared. CO_2 gas lasers emit radiation at 10.6-μm wavelength, and they are widely used both in medical and industrial applications.

UV lasers are used in lithography to achieve the submicrometer imagery needed to make integrated circuit chips. The minimum spot size is determined by the wavelength and the numerical aperture of the focusing optics. The shorter the wavelength, the smaller the spot to which the laser beam can be focused. This pushes the microprocessor industry to use increasing shorter wavelengths. Currently, a common wavelength is 193 nm.

Laser diodes are undergoing an extremely rapid development. Advantages of laser diodes include their low cost and small physical size. They are also available in a wide range of wavelengths. Laser diode devices provide continuous output power or may be analog or digitally modulated. Pulsed laser diodes typically operate with pulses shorter than 100 ns. Laser diodes can be packaged in the form of a linear diode array or a two-dimensional array. Laser diodes may be temperature tuned by approximately 0.3 nm/°C. Applications of laser diodes range from telecommunications, data communications, sensing, thermal printing, laser-based therapeutic medical systems, satellite telecommunications to diode pumping of solid-state lasers.

The primary optics-related disadvantage of laser diodes is that they emit nonrotationally symmetric beams, which are more difficult to collimate or focus into a small spot. Most beams from laser diodes have a gaussian intensity profile along one axis, often called the *fast axis*, and a nongaussian intensity profile along the *slow axis* perpendicular to it. The beam diverges much faster in the fast axis than in the slow axis. Achieving collimation of the beams from laser diodes is not a trivial task because of the lack of the rotational symmetry in the beam. It requires anamorphic optics, in most cases treating the fast and the slow axes separately. A good way to collimate a beam from a laser diode is to use two crossed glass or gradient index fibers, perpendicular to the optical axis of the laser beam shown in Fig. 11.6, each fiber having a different focal length. This is an elegant solution because of the small packaging. Two perpendicular fibers have small focal lengths. The fiber that collimates the fast axis has a focal length of a few hundred micrometers, and is can be a gradient index fiber with a radial gradient index distribution, or a planohyperbolic homogeneous fiber. The fiber that collimates the slow axis has a longer focal length. In this way, the output of most laser diodes can be transformed into near rotationally symmetric beams. The beam can then be focused with a rotationally symmetric lens into a single-mode or a multimode fiber core that transmits the laser signal along the fiber axis.

A very convenient way of transmitting laser beams is through fibers. The laser beam from a laser diode can be coupled into a fiber with a coupling efficiency as high as 90% and transformed to a clean beam, easier to collimate at the output of the fiber. In telecommunications, the laser light-carrying signals are transmitted through hundreds of kilometers of fiber.

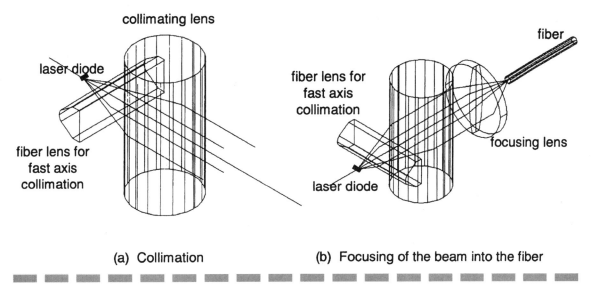

(a) Collimation   **(b)** Focusing of the beam into the fiber

Figure 11.6 Collimation and Focusing of the Beam from the Laser Diode

F-θ Lenses in Laser Scanners

Laser scanners represent one of the more common applications of gaussian beam optics, and a very special lens form is required in these laser scanning systems. Figure 11.7 shows a very generic laser scanning system. An HeNe or similar laser beam is first expanded by a beam expander, and the expanded laser beam is directed toward a multifaceted polygon scan mirror. As the mirror rotates at a constant rotational speed, angle θ changes linearly.

Most lenses follow the relationship that the image height $Y = f \tan \theta$. This is literally true for lenses with zero distortion. If such a lens were used in our laser scanner, then the velocity of the scanned laser spot would increase in proportion to the tangent of the scan angle. Since most laser scanners require a linear spot velocity, conventional lenses are not viable. Lenses which follow the relationship, $Y = f\,\theta$, will produce a linear scan velocity, and these lenses are called F-theta, or F-θ, lenses. In effect, they are lenses in which negative or barrel distortion is intentionally introduced in order to counteract the increased image velocity in conventional low-distortion optics. Fortunately, these forms of remote-aperture stop lenses (the stop is on the polygon facet) tend to

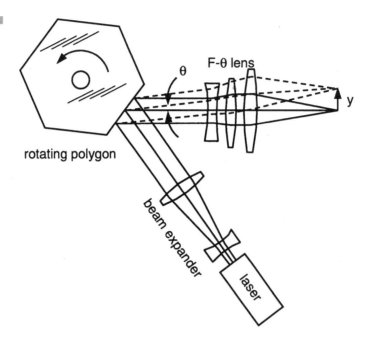

inherently have negative distortion. It is critical, however, to assure uniform spot size through scan, and this is sometimes challenging. In order to loosen the tolerance on pyramidal error of the polygon, the collimated beam from the laser is often focused with a cylindrical lens to a line on the polygon facet. This requires an anamorphic *F*-θ lens to focus the beam to a spot. This form of optical system is used in laser printers.

12

Basics of Thermal Infrared Imaging in the 3- to 5- and 8- to 12-μm Spectral Bands (Plus UV Optics)

The Basics of Thermal Infrared Imaging

Thermal infrared imaging is generally considered to be in the medium or midwave IR (MWIR) that extends from 3 to 5 μm and in the longwave IR (LWIR) that extends from 8 to 14 μm. In these wavelength bands, we are looking at thermal or heat sources rather than visible light. There are many different applications for thermal infrared imaging such as nondestructive testing whereby an IR camera can image a machine, such as a CNC lathe, to look for overheating of the bearings, or we can image houses looking for heat losses in the winter. In the medical arena, doctors can look for various abnormalities indicated by localized skin temperature variations. In nuclear power plants IR cameras are invaluable to quickly search for thermal leaks in the cooling system. Boarder control and security are other areas where IR imaging has become crucial. Figure 12.1 shows several industrial and commercial examples of thermal infrared imaging. The applications of thermal infrared imaging are continuing to develop at a rapid pace.

The human eye is sensitive to the spectral band from approximately 0.4 to 0.7 μm, and thus the eye cannot see this longer-wavelength thermal energy. It takes special detectors or sensors to record the energy, and, needless to say, the imaging optics must efficiently transmit these wavelengths. While there are many applications for the near infrared (NIR), which includes the regions from 0.85 to 1.6 μm, for telecommunications as well as 1.06 μm, which is the Nd: YAG wavelength often used in the applications where the higher power is needed, for the most part ordinary optical materials can be used. The near IR will thus not be considered in this chapter, but rather the MWIR and LWIR where special optical materials and other design considerations are mandatory.

We see in Figure 12.2 the spectral transmittance of a 1.8-km air path in the 3- to 5-μm MWIR and 8- to 14-μm LWIR spectral bands. Water and CO_2 absorption bands limit the use of wavelengths to these two bands within the atmosphere. We also see the radiant exitance for black bodies ranging in temperature from 100 to 1000 K. For reference, ambient temperature is about 300 K, and you can see that the peak radiant exitance at this temperature occurs at about 10 μm. For this reason, LWIR systems tend to have the highest sensitivity. However, LWIR detectors are more expensive and difficult to produce than their MWIR

Figure 12.1

Examples of Industrial
Applications of Thermal Infrared Imaging
(*Courtesy of FLIR
Systems, Boston*)

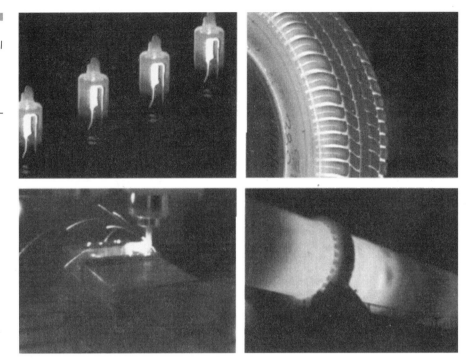

Figure 12.1

Examples of Industrial
Applications of Thermal Infrared Imaging
(*Courtesy of FLIR
Systems, Boston*)

counterparts. In addition, with today's image-processing hardware and algorithms, excellent MWIR imagery can be obtained.

In a simplified form shown by Riedl, the signal-to-noise ratio for an IR system is

$$S/N = (W_T \varepsilon_T - W_B \varepsilon_B)(\tau)\left(\frac{D^*}{\sqrt{\Delta f}}\right)\left[\frac{\tau d'}{4(f/\#)^2}\right]$$

where ε = emissivity

W = radiant exitance (W/cm^2)

W_T = target exitance (W/cm^2)

ε_T = target emissivity

WB = background exitance (W/cm^2)

ε_B = background emissivity

D^* = specific detector detectivity (cm · Hz$^{1/2}$ · W^{-1})

Δf = noise equivalent bandwidth (Hz)

τ = optical transmission

Figure 12.2
Top: Transmittance of
1.8-km Horizontal Air
Path at Sea Level;
Bottom: Radiant
Exitance of Blackbody
As Function of
Temperature

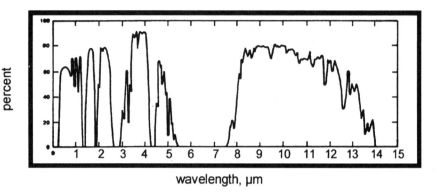

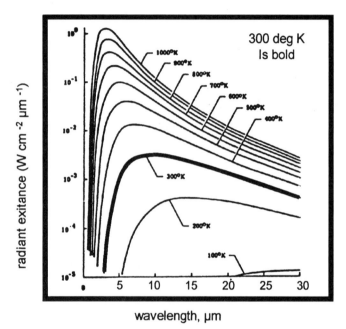

Figure 12.2
Top: Transmittance of
1.8-km Horizontal Air
Path at Sea Level;
Bottom: Radiant
Exitance of Blackbody
As Function of
Temperature

d' = detector size (cm) assuming square detector

$f/\#$ = f/number of optics

The first factor in the previous equation relates to the object we are imaging. This factor gives us the exitance difference between the primary object of interest or target and the area surrounding the object of interest or background. The second factor is the transmission of the atmosphere or other medium in which the system is immersed and the transmission through the optical elements. The third factor relates to the focal plane array and is the detectivity divided by the noise equivalent bandwidth. The fourth and final factor has the sensor size (the width of a pixel) and the optical transmittance in the numerator and the $(f/\#)^2$ in the denominator. This is where the optics becomes critical since the signal to noise is inversely proportional to the square of the $f/\#$. This drives many IR systems to extremely low $f/\#$s in order to reach the desired signal-to-noise ratio. In addition, with some of the new uncooled microbolometers, the $f/\#$ often needs to be $f/0.8$ or even lower. The reader is referred to the reference for further information on this important relationship.

With respect to the optics, most glasses simply do not transmit above about 2.5 μm. Certain special glasses do transmit up to 4.5 μm, and fused silica transmits up to about 4 μm. Infrared-transmitting materials are therefore essential, and as we will see there is only a limited selection available. These IR-transmitting materials are generally expensive and have other problems.

The Dewar, Cold Stop, and Cold Shield

Since we are looking at thermal or heat sources with a thermal-imaging system, for maximum system sensitivity most thermal imaging systems use cryogenically cooled detectors which operate at the liquid nitrogen temperature of 77 K or even lower. If these detectors, or *focal plane arrays* (FPAs), are allowed to "see" any thermal energy other than the energy contained within the scene being viewed, then the sensitivity is reduced. In addition, if the magnitude of this nonscene energy changes or modulates over the field of view, then we often see cosmetically undesirable image anomalies. In order to achieve maximum sensitivity and avoid image anomalies, the IR FPA is cryogenically cooled and mounted into a thermally insulated "bottle," or *dewar*, assembly.

Figure 12.3 shows a typical generic detector/dewar assembly intended for an IR-imaging application. Before we show how the dewar works, we need to see just how it interfaces with the rest of the optical system. The smaller figure on the upper right of Fig. 12.3 shows an entire scanning-imaging IR system. Light (actually infrared radiation) enters from the left into the larger lens generally called the *collecting optics*. After forming an intermediate image, the light is collimated by the second smaller lens. A further purpose of the second lens is to form an image of the larger collecting optics element, which is the system aperture stop, onto the scan mirror. After the light reflects from the scan mirror, it enters the region within the circle in Fig. 12.3. This area is shown enlarged and in more detail in the larger figure.

Collimated light from the scan mirror first enters the *focusing lens,* which is generally outside of the cryogenically cooled dewar and which focuses the light onto the FPA after passing through the dewar window. Note that for the example shown in Fig. 12.3 the detector array is a linear array extending in and out of the figure. The following attributes make up the dewar assembly:

▤ The dewar is an evacuated bottle very similar to a classical Thermos bottle.

▤ The entrance window must, of course, transmit infrared radiation.

▤ A *cold finger* butts up against the aft end of the FPA to keep it cryogenically cold. The cold finger itself is a high-specific-heat metal rod of iron or steel, which is wrapped with a coil through

Figure 12.3
Top: Entire ir Imaging System; Bottom: Typical Generic Detector Dewar Assembly

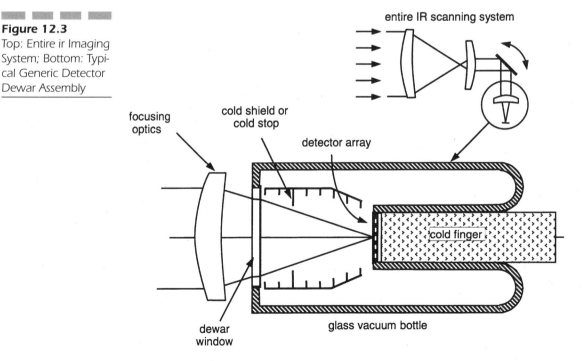

which liquid nitrogen is pumped (or other similar operation). This, in turn, cools the aft end of the FPA.

■ A baffle, called a *cold shield* or *cold stop,* is located as shown inside the dewar. We will define these later.

Now—an *extremely important point*—to evaluate what is happening with respect to the imaging light and potential stray light which may lead to undesirable image anomalies, you need to figuratively *put your eye at the detector and look out the front and ask yourself "what do you see?"* Let's do just that and put our eye figuratively at the center of the FPA where the light focuses.

■ Within the solid angle forming the imaging cone of light, we "see" *scene energy.* We are, in effect, looking into the exit pupil of the optics, and we see a solid angle of radiation coming from the nominal scene. We do not see the image, but rather the exit pupil. This is much like looking toward a round porthole window in a ship from some distance away.

■ At angles just outside the imaging cone, yet not quite hitting the cold shield baffle, we have radiation, which is not scene energy, nor is it cryogenically cold. This sliver of solid angle, which is a circular annulus, represents energy from the interior of the system, which literally reaches the FPA.

■ Outside this sliver of interior system energy, we see the cold shield or baffle within the dewar. Since this component is cryogenically cold and finished with a nonreflective coating, it emits little or no radiation.

If, in the previous example, the cold shield had been the same diameter as the light cone, then the detector would only have been able to see scene energy. In effect, we have in this example made the cold shield slightly larger in diameter than it needed to be, and we will show why later. For now, we simply conclude that there is a sliver of solid angle, which is outside the scene energy and inside the cold shield, which can record energy from the system interior.

Cold Stop Efficiency

An IR system is said to be *100% cold stop efficient* if the detector can see or record energy only from the scene. What this really means is that with 100% cold stop efficiency, the detector records energy from both the cone of light representing scene energy and from the cryogenically cold thermal baffle, known as a cold stop (remember that being cryogenically cold, there is virtually no energy emitted from the cold stop itself). We have used the example in the previous section of putting your eye figuratively at the detector and looking out toward the front of the system and asking yourself "what do you see?" If, for every pixel on your FPA, you can convince yourself that your eye sees only the solid angle representing the imaging light (scene energy) and also portions of the thermal baffles representing the cold stop, then the system is indeed 100% cold stop efficient.

Note in Fig. 12.3 we have shown a series of small stray-light baffles within the cold stop or cold shield area. Without these stray-light suppression features, there may be stray radiation paths which will cause unwanted radiation to reach the FPA.

In Fig. 12.4 we show on the left a system which is not 100% cold stop efficient and one which is 100% cold stop efficient on the right. Casually looking at these figures shows little difference. The lower sets of figures are enlargements of the areas within the dashed circles of the upper figures. Note on the left figure how the aperture stop is both on the front element of the system as well as on the rear surface of the two-element reimaging lens group. Furthermore, if we place our eye at the lower end of the FPA as shown and look toward the scene, we see the solid angle representing the scene as well as a solid angle above this region and below the cold shield which is not coming from the scene but rather from some portion of the system interior. This nonscene energy is, in effect, analogous to stray light in a conventional optical system operating in the visible portion of the spectrum. If this nonscene energy is "warm," then the sensitivity of the detector will be reduced from its nominal value; however, if this nonscene energy changes in magnitude over the FPA or through scan, then you will have image anomalies similar to ghost images in a conventional visible system. This is bad, especially when the image anomalies are confused with the real scene which can sometimes happen.

On the right-hand figure in Fig. 12.4, we see how the aperture stop at the front element is reimaged into the cold stop plane inside the dewar assembly. Here if we look out from anywhere on the FPA, we see only scene energy and no system interior energy whatsoever. This system is said to be 100% cold stop efficient. The *cold stop efficiency* is the ratio of the total solid angle reaching a given pixel which comes from the scene

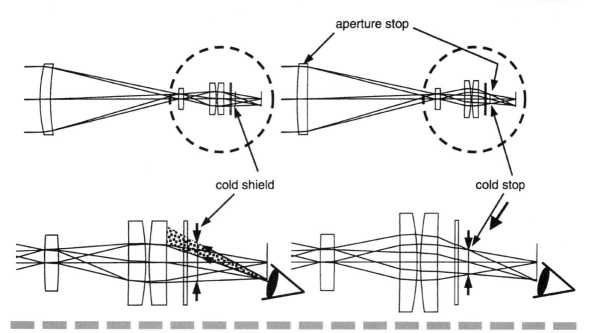

Figure 12.4
Left: System Which Is Not 100% Cold Stop Efficient; Right: System Which Is 100% Cold Stop Efficient

to the total solid angle reaching the same pixel from the entire opening in the thermal baffle or cold shield. For example, we might find that the solid angle reaching a given pixel from the scene is 90% of the total solid angle which can possibly reach the same pixel within the thermal baffle. This system has 90% cold stop efficiency. The photons or thermal energy which come from some interior portion of the system or its housing is unwanted energy in the form of stray light. Its presence lowers the net sensitivity of the system, but even more importantly if the magnitude of this nonscene energy changes or modulates over the field of view, or through scan if we have a scanning system, then we likely will see image anomalies on the video display.

Figure 12.5 shows a paraxial IR lens system that is designed to be 100% cold stop efficient. In order to make it clear just how the ray paths proceed through the system, we show three permutations of the design. Note that the design itself is identical in all three cases. Figure 12.5a shows three field angles, on axis and ±3°; Fig. 12.5b shows only the extreme fields of view; and Fig. 12.5c shows only the maximum field of view. Note that in all three cases the front objective lens is reimaged into the cold stop plane. Rays from various fields of view are shown all superimposed. It is clear how the ray cones all overlay on the front element and also on the cold stop surface. It should also be clear how, if you locate your eye anywhere along the FPA and look toward the front of the system, the only solid angle you can see is from the solid angle of image-forming energy. Anything outside these solid angles will view the aft side of the cold stop and/or the interior of the cryogenically cold dewar.

Scanning Methods

A typical imaging sensor, such as a CCD or a CMOS chip in a commercial camcorder or other video-imaging system, uses a full two-dimensional detector array or focal plane array (FPA). Very much analogous to film, these FPAs record the entire image essentially at once. In the thermal infrared these FPAs are called staring, mosaic, or two-dimensional (2-D) detectors.

Figure 12.5
Paraxial IR Lens with 100% Cold Stop Efficiency

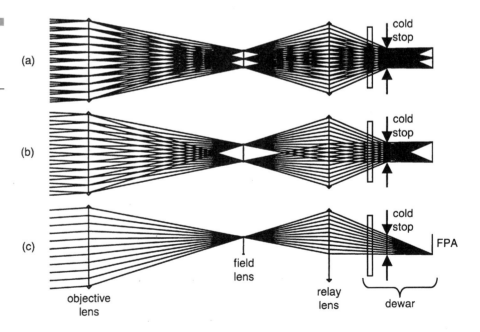

Infrared detectors are still quite costly and difficult to manufacture, and for this reason detector arrays which are much smaller in extent than a full two-dimensional array are often used with appropriate scanning to allow imaging coverage over the full desired two-dimensional field of view. If we use a small detector array, we can create a full two-dimensional field of view by following the steps outlined here and illustrated in Figure 12.6.

- Scan the field of view in the azimuth or horizontal direction over the full width of the field. This is the upper swath shown in Fig. 12.6.

- Then simultaneously increment the field down by the vertical extent of the array while reinitializing to the original azimuth position on the left.

- Now we scan again in azimuth.

- This procedure is repeated until we cover the full vertical or elevation field of view, after which the entire process is repeated again and again to create the full two-dimensional field of view.

This process is known as *serial scanning*. It requires two scan motions, one for the azimuth or horizontal scan and another for the elevation or vertical scan. A good analogy to serial scanning is mowing your lawn. If you have a large rectangular lawn area, you likely use a lawn mower with a width of 0.5 m or so. And you likely mow in one direction, increment by the width of the mower at the end, and mow back. This is known as a *bidirectional scan* in optical terms. Needless to say, you can mow a large area lawn with a small lawnmower, just like covering a large field of view with a small detector array. The previous methodology can be used with very small arrays with as few as one element or pixel or with larger arrays, with up to 25% of the vertical field of view or more.

Figure 12.6
Serial Scanning

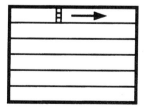

serial scan

Electronics called *scan-converting electronics* is used to combine or multi-plex the data streams and create a virtually seamless standard video signal such as composite video, NTSC, etc.

Operationally, serial scanning requires two scan motions as discussed previously, horizontal and vertical. This can be implemented using two scanning or moving mirrors, one for the scan in the horizontal direction and the other in the orthogonal vertical direction. While, theoretically, both functions could be accomplished with one mirror scanning in both directions, this is rarely done due to the high bandwidths and other difficulties of the two required motions. And do keep in mind that the scan motion in the scan direction should be linear in its angular velocity.

A second form of scanning is known as *parallel* or *pushbroom* scanning, as shown in Fig. 12.7. Here we use a long detector array covering the entire vertical field of view, and the scanning motion is in the azimuth or horizontal direction (or conversely). For this form of scanning only one scan mirror is required. Comparing this with a serial scan system shows the parallel or pushbroom system requires one scan mirror motion and uses an FPA with more elements. The serial scan system uses two scan mirrors and has more complex scan-converting electronics but a far simpler detector. It is clear that there are many trade-offs to be made in selecting the optimum configuration for a given application.

It is important to note that while we are describing scanning over the field of view in the azimuth and elevation directions, what really happens in the optical systems is quite different. What literally happens is that the optical system is, in reality, quite stationary in space during this scanning operation (except of course for the scan mirror itself). The motion of the scan mirror or mirrors scans or translates the image of object space over or across the image plane or FPA. So if your eyes could

Figure 12.7
Parallel or Push
broom Scanning

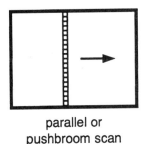

parallel or
pushbroom scan

see in the thermal infrared and you were able to view the focal plane area, what you would see is the image being translated or scanned across the FPA in azimuth and elevation (unless it is a parallel scan operation in which case the image is moved only in the one direction).

Staring or mosaic arrays are full two-dimensional focal plane arrays which sense the entire scene instantaneously with no mechanical scanning required. Figure 12.8 shows such an FPA. Here we are trading off a more complex sensor for simplified mechanics and scan-converting electronics. It is interesting to note that every commercial camcorder and digital camera uses staring arrays without scanning. The analogous form of sensors are, of course, far more costly in the thermal IR.

We will now show how we can implement a scanner in an IR system for a pushbroom scan. Consider Fig. 12.9 where we show what is, in effect, an astronomical telescope. The system takes collimated radiation of diameter D covering a field of view of $\pm\alpha$ and outputs collimated radiation of diameter d covering a full field of view of $\pm\theta$. For this system the magnification, M, can be stated as D/d or equivalently θ/α. In order to scan over the $\pm\alpha$ full field of view, we could do either of the following:

Figure 12.8
Staring Array (No
Scanning Required)

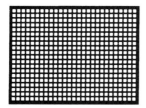

Figure 12.9
Scanning
Methodology

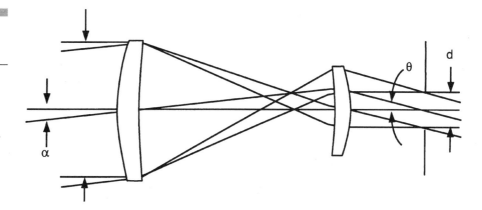

- Place a large flat mirror in front of the system and scan it by ±α/2.
- Place a small flat mirror aft of the second collimating lens at the exit pupil and scan it by ±θ/2 = ±α(D/d)/2 = ±αM/2, where M is the magnification.

Consider the following example: Assume D is 100 mm in diameter and the magnification is 10×. Thus, d = 10 mm. And assume that α = ±2°, which means that θ = ±20°. We can now accomplish our scanning with either of the following scenarios:

- We can use a mirror approximately 100 × 140 mm and scan it by ±1°.
- We can use a mirror approximately 10 × 14 mm and scan it by ±10°.

While both of these methods will work, there are many reasons why scanning the smaller mirror by the larger angle is preferred. The smaller mirror is much easier to manufacture. If we require a given surface flatness on the mirror, this requirement will hold for the larger mirror as well as the smaller mirror, and the smaller mirror will be far less costly. Also, the smaller mirror system will be far better from a packaging point of view.

We show in Fig. 12.10 an oscillating mirror for a pushbroom scanning system as described previously. The scanning mirror is located at the exit pupil of the telescope as we described earlier. In Fig. 12.11 we show how a polygon mirror can also be used for a pushbroom scanner. A polygon mirror is a very interesting device. Each facet is, in effect, a separate scan mirror. And each facet, as it scans, rotates not about the vertex of the facet but rather about the center of rotation of the entire polygon. This creates a different motion in space and must be properly modeled to assure proper system performance. Note in Fig. 12.11 that we show the mirror in its "neutral" position as the top mirror. Once again, let's put our eye at the FPA and look outward. We reflect off of the prime facet to the left and ultimately view object space. Everything is fine. Now look at the bottom polygon mirror, which is shaded. The polygon has rotated around in a counterclockwise direction in the order of 20°, and we show it in a position where the next facet behind the prime facet is just allowing a sliver of radiation to be reflected from it. With the polygon in this position, the prime imaging facet reflects the light down to the left while the adjacent facet reflects the sliver of light up to the left. In a situation as shown, the angle of the radiation from the adjacent facet is likely larger than that from the prime facet, and for this reason the radiation may or may not actually reach object space. If it does not, it will strike an interior structure of the system and may create an image anomaly. This reflection from the adjacent facet is known as *ghosting,* and if the ghost radiation makes it to object space it is called *external ghosting* while if it strikes an interior portion of the system it is *internal ghosting.* In order to prevent ghosting, the electronics is often shut off just prior to the adjacent facet enters into the imaging beam of radiation. Occasionally, one can build a valid case to allow a small fraction of the pupil to be ghost radiation, but it really does need to be small. The fraction of the time that the polygon is actually being used for imaging relative to the total amount of time it is running is called the *scan efficiency.* Scan efficiencies in the vicinity of 80% are not uncommon.

Figure 12.12 shows how we can create a serial scan motion. Recall that in a serial scanning system we scan in the azimuth direction while imaging onto a small detector array of only a few elements. We then increment in elevation and scan in azimuth once again. The process is repeated until we build up a full two-dimensional field of view. Such a system requires two scan-mirror motions, one for the azimuth scan and the other for the elevation direction. In Fig. 12.12 we show the upper mirror as the elevation mirror. For the azimuth scanning mirror we show two potentially viable mirror motions, one is incorrect and the other correct. As it turns out, if we rotate the mirror about the incorrect axis of rotation, we will be

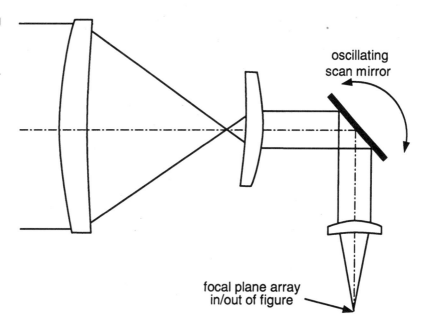

Figure 12.10
Oscillating Mirror for
Pushbroom Scanning

oscillating
scan mirror

focal plane array
in/out of figure

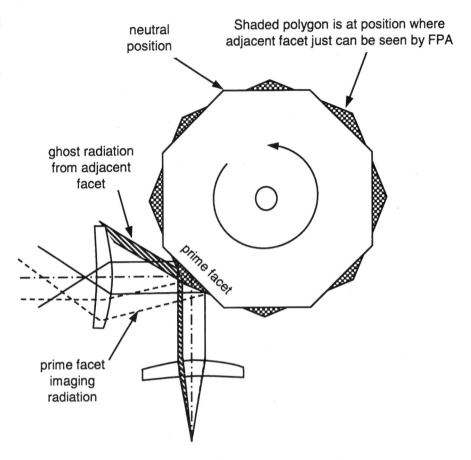

scanning an arc in the azimuth direction as opposed to a straight horizontal direction, which will result from the correct axis. It is extremely important that you fully understand just how the mirrors in your system work and how they scan in object space. How can we be sure that we do not have a problem? There are several ways. First, you really should have an accurate model of your scanning system in your computer lens design program. Take nothing for granted! You must spend the time to assure that your model is an accurate representation of the real world and show how the scan motion works. One interesting thought is to set up your system in reverse from the FPA out toward object space. If you do this, you need only trace a central ray along the optical axis rather than a full cone of rays, as this ray will, in effect, be the chief ray at the given field and/or scan position. Now rotate your mirror or mirrors and monitor at what azimuth and elevation the ray leaves the system for object space. This will tell you directly whether you are scanning a straight line or not. One final hint: take some very simple "drugstore" mirrors and make a crude setup in the lab to model the mirror motions. If in such a setup you exaggerate the mirror motions, you can generally get a valid indication of what is going on. Regardless of how you approach the situation, we cannot overemphasize the importance of modeling properly your system with respect to its scanning motions and the resulting imagery into object space.

IR Materials

While there are many glass types available for visible systems, there are only a very limited number of materials that can be effectively used in the MWIR and LWIR spectral bands. Table 12.1 shows the more common materials and their most important properties. Figure 12.13 shows a plot of the transmittance of the more common IR transmitting materials. It is important to note that these data include surface reflection losses, and often a significantly higher transmittance results after applying high-efficiency antireflection coatings.

Figure 12.12
Two Mirrors (Azimuth and Elevation) for Serial Scanning

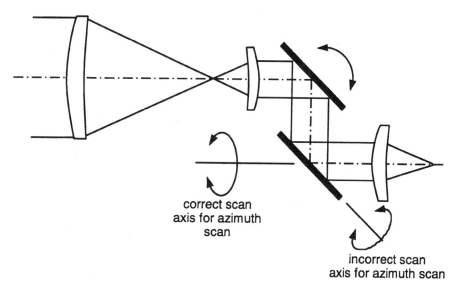

correct scan
axis for azimuth
scan

incorrect scan
axis for azimuth scan

Figure 12.13
Spectral Transmit-
tance of ir Materials,
Including Surface
Losses

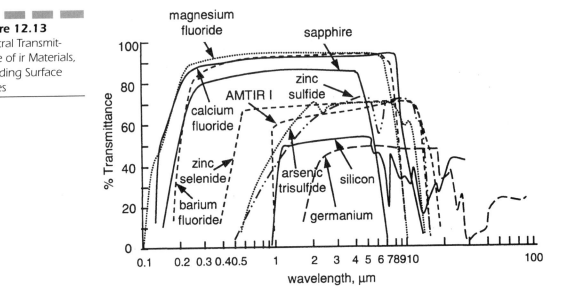

Figure 12.14 shows a "glass" map where we plot the refractive index in the ordinate versus the *V#* in the abscissa for common infrared-transmitting materials. Recall that the *V#* is inversely proportional to the material's dispersion, and note how for germanium is nearly 1000 in the LWIR (very low dispersion) versus about 100 in the MWIR. You can use this glass map in much the same way as you would for visible systems.

We will now discuss each of the common IR materials:

GERMANIUM   *Germanium* is perhaps the most common of infrared materials. It is used in both the LWIR where it is the crown or positive component of an achromatic doublet and in the MWIR where it is the flint or negative component of an achromatic doublet. This anomaly is due to the differences in its dispersion properties in the two spectral bands. In the MWIR germanium is approaching its lower absorption band and hence the refractive index is more rapidly changing, thus leading to a greater dispersion. This, in turn, makes it appropriate for the negatively powered element of an achromatic doublet.

With respect to germanium's optical properties, two parameters are of major significance. First, the refractive index of germanium is just over 4.0, which means that shallow curves (long radii) are feasible. As we will see later, along with the higher refractive index aberrations are easier to reduce, which is a significant benefit to the designer. Another parameter

TABLE 12.1

Properties of Common Optical Materials in the Thermal Infrared

Material	Refractive Index at 4 μm	Refractive Index at 10 μm	$dn/dt/°C$)	Comments
Germanium	4.0243	4.0032	0.000396	Expensive, large dn/dt
Silicon	3.4255	3.4179*	0.000150	Large dn/dt
Zinc sulfide, CVD	2.2520	2.2005	0.0000433	
Zinc selenide, CVD	2.4331	2.4065	0.000060	Expensive, very low absorption
AMTIR I (Ge/As/SE:33/12/55)	2.5141	2.4976	0.000072	
Magnesium fluoride	1.3526	†	0.000020	Low cost, no ctg required
Sapphire	1.6753	†	0.000010	Very hard, low emissivity at high temperature
Arsenic trisulfide	2.4112	2.3816	‡	
Calcium fluoride	1.4097	†	0.000011	
Barium fluoride	1.4580	§	−0.000016	

*Not recommended.
†Does not transmit.
‡Not available.
§Transmits up to 10 μm but drops abruptly.

of significance is the *dn/dt*, which is the change in refractive index with respect to temperature. For germanium the *dn/dt* is 0.000396/°C. This is a very large value, especially when compared with 0.00000360/°C for ordinary glasses such as BK7 glass. This can cause a large focus shift as a function of temperature. This defocus is generally so large that these systems often require some form of athermalization (focus compensation versus temperature).

Germanium is a crystalline material and, as such, can be grown in either polycrystalline form or monocrystalline form, which is also called single-crystal germanium. Depending on the manufacturing and refining processes, single-crystal germanium may be more costly

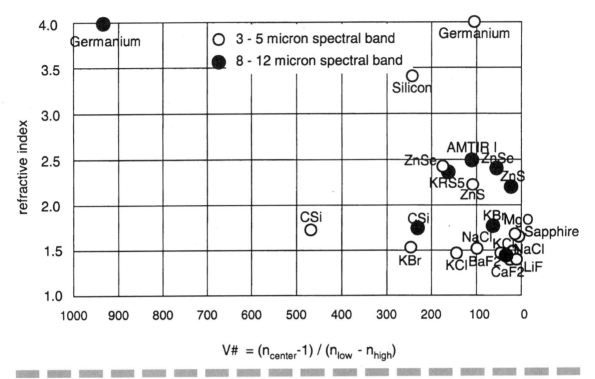

$$V\# = (n_{center} - 1) / (n_{low} - n_{high})$$

Figure 12.14
Glass Map for Common Materials in the MWIR and LWIR Spectral Bands

than polycrystalline germanium. Throughout the 1970s and 1980s there was much confusion regarding the relative need for single-crystal germanium in high-performance thermal-imaging systems. For the most part, European designers specified single-crystal material, and in the United States polycrystalline material was generally called for. Studies in the mid- to late 1980s showed that indeed polycrystalline germanium had a larger refractive index inhomogeneity, and this was due primarily to impurities at the grain boundaries. Furthermore, the presence of these impurities could be imaged onto the FPA if the material were at or near an intermediate image plane. The single-crystal germanium is preferred. Fortunately, recent advances in material manufacturing have closed the cost differential gap, and, for the most part, the optics industry uses single-crystal germanium. At high temperatures germanium becomes absorptive, with near zero transmittance at 200°C.

Single-crystalline germanium has a refractive index inhomogeneity of 0.00005 to 0.0001, whereas polycrystalline germanium is 0.0001 to 0.00015. For optical purposes germanium is generally specified as to its resistivity in ohm-centimeters, and the generally accepted value is 5 to 40 $\Omega \cdot$ cm throughout the blank. Figure 12.15 shows a typical germanium blank with an area on the right which is polycrystalline. Note that the resistivity is well behaved and slowly changing radially in the single-crystal region, whereas it changes rapidly in the polycrystalline region. If you were to look into the material with a suitable infrared camera, you would see a somewhat bizarre convoluted image resembling cobwebs, with this appearance most accentuated at the grain boundaries. This is all due to the impurities induced at these boundaries.

There is one further comment regarding germanium—its susceptibility to chipping. You must be exceptionally careful during the optical manufacturing and coating processes as well as during assembly as a nearly inconsequential tap to the edge of a germanium element could result in a chip flaking off. For this reason, germanium is often bonded into its housing using a semicompliant bonding material. Silicon and some of the other crystalline materials also have this problem, so be very careful.

Figure 12.15
Resistivity Map, in
Ohm-Centimeters,
for a Polycrystalline
Germanium Blank

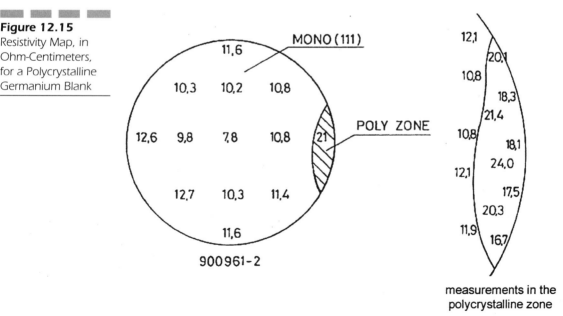

SILICON   *Silicon* is also a crystalline material much like germanium. It is used primarily in the 3 to 5 MWIR spectral band as there is absorption in the 8- to 14-µm LWIR spectral band. While the refractive index of silicon is somewhat lower than germanium (silicon is 3.4255 versus germanium which is 4.0243), it is still large enough to be advantageous with respect to aberration control. Further, the dispersion of silicon is still relatively low. Silicon can be diamond turned with great difficulty.

ZINC SULFIDE   *Zinc sulfide* is a common material used in both the MWIR and the LWIR. While its visible appearance varies greatly, it is generally rust-yellow in color and translucent in the visible. The most common process for manufacturing zinc sulfide is known as chemical vapor deposition (CVD).

If zinc sulfide is "HIP'ed" (hot isostatic pressed), it can be made to be water-clear in the visible. While available from several manufacturers, Cleartran is the most common commercially available clear zinc sulfide. Cleartran can be used for multispectral windows and lenses from the visible through the LWIR.

ZINC SELENIDE   *Zinc selenide* is in many ways similar to zinc sulfide. It has a slightly higher refractive index than zinc sulfide and is structurally weaker. Because of this, a thin layer of zinc sulfide is sometimes deposited onto a thicker zinc selenide substrate for environmental durability reasons. Perhaps the most significant advantage of zinc selenide over zinc sulfide is that it has a significantly lower absorption coefficient than zinc sulfide. For this reason zinc selenide is commonly used in high-energy CO_2 laser systems.

AMTIR I AND AMTIR III   *AMTIR I and AMTIR III* are glassy materials manufactured of germanium, arsenic, and selenium in a ratio of approximately 33:12:55. The AMTIR family of materials begins transmitting in the near IR (NIR). For this reason you can often see a very deep and faint red color transmitted through AMTIR. The *dn/dt* of AMTIR I is about 25% that of germanium, making it attractive from a thermal defocusing standpoint.

MAGNESIUM FLUORIDE   *Magnesium fluoride* is another crystalline material. In its crystalline form it transmits from the UV through the MWIR spectral bands. Magnesium fluoride is manufactured by either crystal growth or alternately by "hot pressing." In this latter process a fine powder form of the material is subjected to very high temperature and pressure in a way similar to powdered metal technology. The result is a milky looking glassy material which transmits well in the MWIR. The caution, however, is that there may be undesirable scattering which could cause contrast degradation and off-axis stray-light problems (this can be avoided using crystalline-grown material). Fortunately, small particle scattering is inversely proportional to the fourth power of the wavelength, so the milky appearance in the visible is reduced by approximately 2^4 power, or 16 times at 5 μm.

SAPPHIRE *Sapphire* is an extremely hard material (it has a 2000 Knoop hardness value as compared to 7000 for diamond). It transmits from the deep UV through the MWIR. One of the unique aspects of sapphire is that it has a very low thermal emissivity at high temperature. What this means is that the bulk material when at a high temperature will emit less thermal radiation than other materials. You might, for example, use sapphire for the window of a chamber which is subject to very high temperature, especially if you are viewing through the window in the IR. Another application for sapphire is for protective windows of super-sonic vehicles where window heating is a serious problem. The primary disadvantage of sapphire is that its hardness makes it difficult, time consuming, and expensive to optically manufacture. There is a related material whose general class is known as spinel. Spinel is, in effect, analogous to hot-pressed sapphire and can be used in place of sapphire. Spinel is also highly dispersive.

Sapphire is birefringent which means that its refractive index is a function of the plane of incident polarization.

ARSENIC TRISULFIDE *Arsenic trisulfide* is another material sometimes used in the MWIR and LWIR. It is deep red in the visible and is quite expensive.

OTHER VIABLE MATERIALS There are a number of other viable materials including *calcium fluoride, barium fluoride, sodium fluoride, lithium fluoride, potassium bromide,* and others. These materials can be used from the deep UV through the MWIR spectral bands. Their dispersion properties make them quite attractive for wide-spectral-band applications, especially from the NIR through the MWIR and on to the LWIR. Many of these materials have some undesirable properties, especially water absorption (they are hygroscopic).

Generally, optical manufacturing methods for IR materials, such as germanium, silicon, zinc sulfide, and zinc selenide, are similar to glass optics. While manufacturers clearly have their trade secrets in this area, to the designer, they can all be considered as manufacturable. Several of the crystalline materials are hygroscopic, which presents some challenges to the optical shop. Also, these materials need to be appropriately coated to prevent damage from moisture and their housings often need to be dry nitrogen purged. IR materials generally have an extremely high index of refraction, which requires antireflection coatings; otherwise the system transmission would be very low.

We have noted the hardness of sapphire previously. One further point of importance is that some IR materials can be single-point diamond turned. These include germanium, silicon, zinc sulfide, zinc selenide, AMTIR, and the fluorides. Sapphire cannot be diamond turned. Silicon can be diamond turned; however, the carbon in the silicon reacts with the carbon in the diamond and this results in a shorter tool lifetime and thus higher cost. Diamond turning can be extremely important if you need either aspheric surfaces and/or diffractive surfaces.

Reduced Aberrations with IR Materials

At the outset of this chapter, we indicated that the higher refractive indices of many of the infrared transmitting materials results in shallower and less steeply curved surfaces which, in turn, results in reduced aberrations. In order to illustrate this, consider Fig. 12.16, where we show six $f/2$ single-element lenses 25.4 mm in diameter, each bent for minimum spherical aberration. The refractive index of the lenses ranges from 1.5 to 4.0, where an index of 1.5 would be close to ordinary BK7 glass and 4.0 close to germanium. Note how the shape of the lenses is progressively changing. At index 1.5, the front is steeply convex and the rear is very slightly convex. At an index of approximately 1.62, the rear surface becomes flat. As the index keeps increasing, the lens becomes more concentric looking. While this is indeed very interesting, unfortunately it does not tell us anything about the aberrations.

Figure 12.17 shows a plot of the rms wavefront error in waves at the wavelength of 0.5 μm versus refractive index for lenses bent for

Figure 12.16

Lens Bending As a Function of Refractive Index for Minimum Spherical Aberration

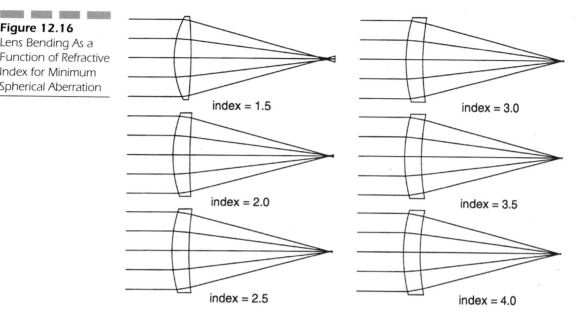

index = 1.5

index = 3.0

index = 2.0

index = 3.5

index = 2.5

index = 4.0

minimum spherical aberration. At index 1.5, we have over 10 waves rms which equates to approximately 50 waves peak to valley! This is an enormous amount of spherical aberration. Note how the aberration rapidly decreases with increasing refractive index. At an index of 2.0, which is about as high as we can find for visible glass, we have about 3 waves rms, or approximately 15 waves P-V. Note how at an index of 4.0, we have about 1 wave rms, or about 5 waves P-V. While this decrease in aberration is noteworthy, the most important point is that at an index of 4.0 we really must be thinking of the thermal infrared wavelengths, as glasses are not available with this refractive index. Let's therefore change the scale of the ordinate of the plot to indicate rms wavefront error at 10 μm. Since 10 μm is 20 times the visible wavelength of 0.5 μm, we need to reduce the values in the ordinate by 20 times. Thus, our 1-wave rms becomes 0.05 waves rms, which is approximately 0.25 wave P-V. This now meets the Rayleigh criteria and is diffraction limited! To summarize this extremely noteworthy finding: We have shown that a single $f/2$ element of glass 25.4 mm in diameter which is bent for minimum spherical aberration and has a refractive index of 1.5 similar to BK7 has approximately 50 waves P-V in the visible (200 times the diffraction limit). An equivalent single element of germanium with refractive index 4.0 when bent for minimum spherical aberration and referenced to a wavelength of 10 μm in the LWIR just meets the Rayleigh criteria with approximately 0.25 wave P-V!

It is for this reason that infrared optical designs are generally simple in form as compared with their visible counterparts. While a single germanium element may indeed suffice for an $f/2$ LWIR lens 25.4 mm in diameter, in the visible we would require three separate elements to achieve a diffraction-limited performance.

It was noted earlier that some infrared materials have very low dispersion, and for this reason color correction may not be required in some scenarios. In order to demonstrate this, we have come up with a virtual lab experiment right here in the book. Figure 12.18 shows the setup. We have a prism located 2.5 m from a vertical wall. We will now manufacture prisms similar to what is shown of various materials so that the center of the respective spectral band will be descending at a $(45-\alpha)°$ angle, where α is the prism angle. We will then measure the length of the resulting spectrum for the visible, MWIR, and LWIR spectral bands.

Figure 12.19 shows the results. Highly dispersive SF6 glass used in the visible from 0.4 to 0.7 μm has a large spectrum measuring about 120 mm. Less dispersive BK7 glass would have a spectrum measuring about 30 mm. In

Figure 12.17
Spherical Aberration As a Function of Refractive Index for $f/2$ Lens 25 mm in Diameter

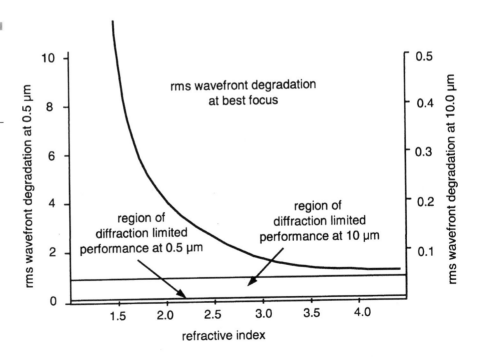

Figure 12.18
Hypothetical Experiment to Show
Length of Spectrum
for Different Materials

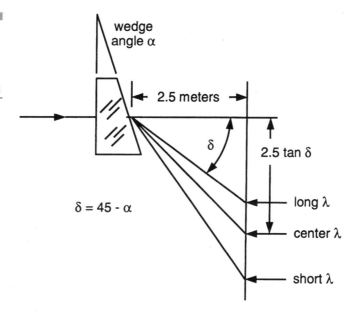

Figure 12.18
Hypothetical Experiment to Show
Length of Spectrum
for Different Materials

Figure 12.19
Length of Spectrum
on Wall 2.5 m from
Prism

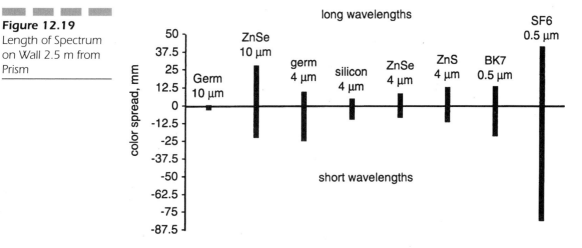

the 3- to 5-μm spectral band zinc sulfide would have a spectrum about 35 mm long, zinc selenide about 17 mm long and silicon about 12 mm. Germanium, as we know, is quite dispersive in the MWIR spectral band, and its spectrum would be about 35 mm. In the LWIR we find that the spectrum for zinc selenide is about 50 mm long and finally germanium in the LWIR is about 4 mm long. This is significantly less than any of the other materials, and this is why we can often use germanium alone in the LWIR spectral band without the need for color correction.

Image Anomalies

Thermal infrared systems often show cosmetically undesirable image anomalies which are not seen in visible optics. These effects include *narcissus, scan noise, beam wander, ghosting,* and *shading.* The effects are similar to what we generally think of as ghost images, and the resulting imagery can vary from slight brightness variations over the format to sharp bright or dark areas. While the mechanisms differ, all of these effects are due to the detector seeing more (or less) thermal energy over the field of view or through scan than dictated by the scene energy itself.

As we discussed in the section "Cold Stop Efficiency" earlier in this chapter, one of the most important methods for evaluating the properties of thermal infrared systems is to put your eye "figuratively" at the detector and look forward (into the exit pupil) and ask yourself "what do you see." This is sometimes called the *detector's eye view.* For an IR system with 100% cold stop efficiency you should see a solid angle containing only scene energy, and everything outside this solid angle should be cryogenically cold. If this is the case, then you will indeed be accurately recording or imaging the thermal radiance from the scene. However, if you can see any thermal energy outside the solid angle representing scene energy and inside the cold stop solid angle, this represents extra energy, which will behave in a similar fashion to stray light in visible systems. We will now review the primary causes of image anomalies in thermal imaging systems.

NARCISSUS *Narcissus* occurs because of a change in the magnitude of radiation reflected back into the dewar from lens surfaces within the system. Consider Fig. 12.20 where we show a scanning IR system at the center of scan (Fig. 12.20a) and at the end of scan (Fig. 12.20b). The focusing optics and detector are shown as rotated about the scan pivot point in Fig. 12.20b which is an optically valid representation of what is happening. We also show an enlarged view of the aft end of the system in Fig. 12.20c. We will be discussing only what happens from the rear surface of the last lens element. The total Narcissus effect is the radiometric summation from all lens surfaces.

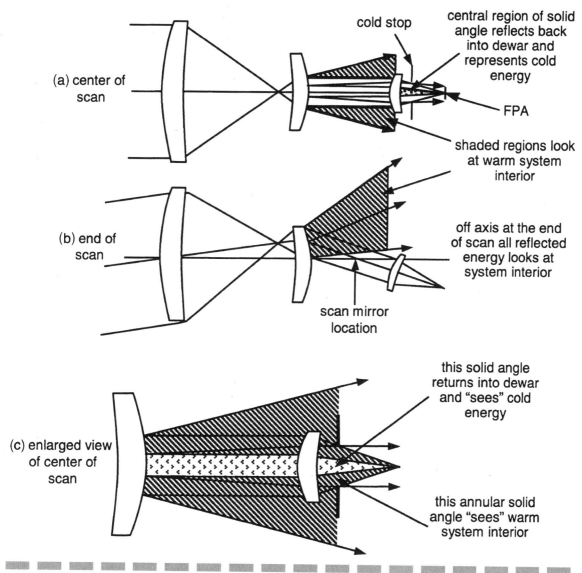

Figure 12.20
Illustration of How Narcissus Is Formed in a Scanning System

If at the center of scan (Fig. 12.20a) we look out from the center FPA pixel, this ray will travel along the optical axis and reflect right back on itself ultimately returning through the cold stop into the dewar. Thus this ray "sees" only cold radiation from inside the dewar, or perhaps better said, no radiation at all. Now let's consider the ray from the center of the FPA which just passes by the edge of the cold stop. This ray, after reflection from the lens surface, diverges on its way back and misses entirely the cold stop aperture. In fact, this ray will strike some portion of the system interior, which could be ambient temperature or it could be a hot electronics board. The total solid angle, which can return into the cold stop, is shown as the shaded solid angle in Fig. 12.20.

Now consider Fig. 12.20b, which represents the edge of scan. Here all of the energy within the entire solid angle within the cold stop reflects from the lens surface into the system interior. None of this energy will return into the cold stop and "see" cryogenically cold temperature. We thus can see how there is more "cold energy" (or lack of warm energy) at the center of scan than the edge of scan. If we now sum up the radiometric effect from each lens surface, and if we take into account the difference in solid angle returning into the dewar at the center and the edge of scan, we will find that there is more "warmth" seen at the edge of scan than at the center of scan. This is, of course, due to the summation of the dark solid angles of energy returning into the dewar at the center of scan from each lens surface. The bottom line is that the video monitor will show less thermal energy at the center of scan than the edge of scan, and this may appear as a dark porthole or disk on the monitor. The diameter of the dark disk and the gradient from the center of the monitor to the edge will depend on the specific geometry and how rapidly the cold energy solid angle reduces with scan.

Note that in nonscanning staring array systems using two-dimensional FPAs you can still have Narcissus. There is one important difference in this regard between scanning and nonscanning or staring systems—the practice in staring systems of performing a "nonuniformity correction." What this means is that the system is periodically aimed toward a uniform thermal source, and then each pixel is adjusted in its offset to yield a constant gray level over the entire format. For many thermal imaging applications where we are looking at near-ambient temperatures, simply draping a black cloth over the front of the system and then performing the nonuniformity correction will suffice. However, if the temperature of the system interior where the reflected radiation strikes at the edges of scan changes, there may be a need to perform a new nonuniformity correction.

How can we minimize the effect of Narcissus? There are two basic methods to minimize Narcissus. First we can use so-called anti-Narcissus coatings. These antireflection coatings typically have from 0.2 to 0.3% average reflectivity from 3 to 5 μm or alternatively 8 to 14 μm. The second approach is to change the relative bendings of the offending lens surfaces so as to minimize the cold solid angle of reflected radiation. This is a common technique, and to keep the desired level of optical performance, often requires the use of aspheric surfaces. Fortunately, this is not a major problem with the use of single-point diamond turning, now a mature technology. Optical systems often have a flat protective window in front of the telescope. In order to avoid the Narcissus generated by the window, the window is tilted, so that the reflected radiation falls out of the sensor's field of view.

GEOMETRICAL SCAN EFFECTS *Geometrical scan effect* (*beam wander*) is illustrated in Fig. 12.21, where we show a generic four-sided polygon scanning IR system. We will assume that the system aperture stop is on the front lens element, a likely situation. At the center of scan, the polygon will be at its nominal or neutral position as shown. As the polygon rotates to the end of scan, its geometrical position in space, where the radiation reflects from its surface, translates to the dashed position (to the left). This represents the location where the intersection of the polygon facet with the plane of the figure has translated for

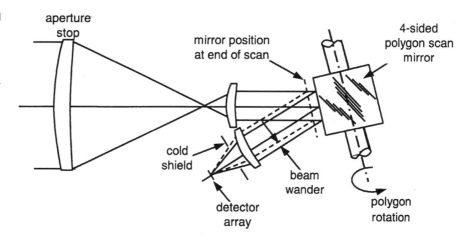

Figure 12.21
Geometrical Scan
Effects and Beam
Wander

end-of-scan imaging. Since we have stated previously that the aperture stop is on the front element, the beam of radiation reflected from the facet must translate or shift laterally by an amount called the *beam wander*.

We now have a dilemma...should the cold stop be sized for the center of scan and thus clip energy from the beam-wandered edge of scan radiation or should we increase the diameter of the cold shield (along with a lateral translation of the cold shield)? The "lesser of the two evils" is generally to increase and laterally shift the cold shield so as to eliminate clipping of the imaging radiation from any scan position.

SCAN NOISE *Scan noise* is often taken to be any undesirable change in nonscene thermal energy reaching the FPA through scan, thus causing anomalies such as bright bands and other image defects. The appearance of scan noise can often resemble flare and stray light in visible systems. One of the more common causes of scan noise is clipping or vignetting which can result if a portion of the housing infringes on the radiation bundle at the ends of scan.

Consider Fig. 12.22 where we show a representation of the solid angle of radiation reaching the detector at the center of scan (top) and at the edge of scan (bottom). At the center of scan, the solid angle is totally scene energy within the cold stop and cryogenically cold outside the cold stop. Thus, the only energy the detector will "see" or record is from the scene. At the edge of scan, the solid angle within the cold stop comes primarily from the scene; however, there is a crescent-moon–shaped shaded area which is caused by clipping or vignetting from a housing interior. The solid angle outside the cold stop is cryogenically cold as for the center of scan case. Thus the only difference from the center to the edge of scan is the clipping from the housing interior.

It can be shown that the perceived change in temperature is approximately given by

$$dt \sim \frac{\Delta A}{A}\ (temperature\ clipped\ -\ temperature\ scene)$$

The quantity, $\Delta A/A$, is the percent of area of the exit pupil within the cold stop which is clipped. Let us take some representative values for the preceding quantities. Assume that $\Delta A/A = 1\% = 0.01$, the temperature of the clipped housing is 50°, and the scene temperature is 0°C. This gives us a perceived temperature difference from the center of scan to the edge of scan of $0.01 \times 50° = 0.5°$C. If we are scanning in the azimuth direction with a polygon or alternatively with an oscillating mirror, the

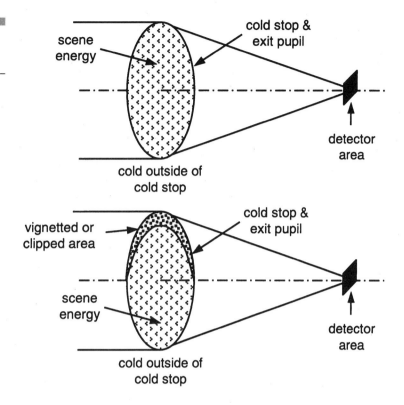

Figure 12.22
Effect of Clipping or
Vignetting

net result will likely be bright bands at the left and right of the monitor, which get brighter toward the corners of the field of view.

With today's detector technology, temperature differences far below 0.1°C can easily be seen. It is thus clear that even 0.1% clipping can often be seen and may be a problem. In fact, a piece of dirt on a lens element measuring only 0.4 mm × 0.4 mm = 0.16 mm² is equivalent to 0.1% clipping for a 25.4-mm-diameter ray bundle.

GHOSTING *Ghosting* is a term applied to an effect unique to polygon scanners. Consider Fig. 12.11 where we showed radiation incident onto a polygon scanner facet. If the polygon is in its neutral position with the facet tilted at 45° to the incoming radiation, the radiation simply reflects downward toward the FPA. Recall that to best evaluate IR systems relative to image anomalies, you should put your eye at the detector and look out and ask yourself "what do you see?" Let's do this for the facet in the central position. All we see is scene energy, as expected. However, now rotate the facet around in the counterclockwise direction until the next adjacent facet has just entered the beam of radiation. If we look out from the FPA, we will see energy reflected from the prime facet heading down to the left and a small sliver of energy from the adjacent facet heading up to the right. This sliver of radiation will be at a greater angle numerically than from the prime facet, and thus it may miss the lenses and hit an interior portion of the housing. Alternatively, this energy may make it out into object space in which case it is viewing a portion of the scene at the opposite side of the field and outside the nominal field of view. When it is hitting a housing interior, this is known as "internal ghosting" and when it views object space it is known as "external ghosting." Thus, a person holding a bright flare just outside the field of view on the right side may appear inside the field of view on the left due to external ghosting.

What is required to best understand effects such as ghosting is to perform an accurate three-dimensional modeling of your system. Every aspect of the system must be properly modeled, including the polygon axis of rotation, facet clear aperture, and the clear aperture of other lenses within the system. Only then will there be some hope of accurately predicting what is happening.

SHADING   *Shading* is very different from any of the previous phenomena, and it is more difficult to explain. Consider the exit pupil diameter at the center of scan reaching the focusing optics. If we have a 100-mm-diameter entrance pupil and a magnification of $20\times$, then the radiation bundle diameter at the exit pupil is 5 mm. Now as we proceed off axis in field of view and scan, what happens to our 5-mm beam diameter? For example, let's assume that we have positive distortion off axis, which means that the magnification increases with field of view. Thus, the 5-mm exit pupil will be smaller in scan mirror space and as it enters the focusing optics. If the aperture stop is at the front element, this will result in some area of the housing being seen at off-axis field and scan positions. The net result is a brightening of the display monitor away from the center of the field of view due to less scene energy and more housing energy.

Athermalization

As any optical system is subjected to higher or lower temperatures several things happen: the lens elements expand or contract, the housing expands or contracts, and the refractive index of the lens materials increases or decreases. While the lens and housing relative expansion can sometimes be a problem, the primary problem with infrared systems is the very large change in refractive index as a function of temperature, or dn/dt. Germanium has a dn/dt of 0.000396/°C. For comparison, BK7 glass has a dn/dt of 3.6 0E-6. It can be shown for a simple lens that

$$df = \text{change in focal length} = \frac{f}{n-1} \frac{dn}{dt} \Delta t$$

where df is the change in focal length and Δt is the change in temperature in degrees Celsius.

Consider the following example: Assume that we have a 75-mm-diameter $f/1.5$ germanium lens with a focal length of 112.5 mm. Applying the previous relationship yields a change in focal length of 0.599 mm for a 40°C Δt thermal soak. For reference, the Rayleigh criteria for one-quarter wave of defocus is ±0.046 mm, so the preceding defocus value equates to 3.3 waves of defocus, which is 13.1 Rayleigh criteria depths of focus, a huge amount! This issue can, and often is, a very serious problem in thermal infrared systems. In this example, if we were to control the temperature so as to stay within a one-quarter wave of defocus, we would need to control the temperature to within ±3°C.

It is thus apparent that unless the user can actively refocus the lens, athermalization is imperative. We could translate the FPA or the prime collecting lens; however, this is generally impractical. We could translate another lens element by a greater or lesser amount depending on the magnification, and this is often the approach. Two other approaches are also viable: negative elements with a high dn/dt can be used and use of reflective optics.

To show how effective reflective optics can be in achieving athermalization, consider Fig. 12.23. We show here two very similar systems, a fully refractive system above the optical axis and a system below the axis where a Cassegrain reflective system takes on the role of the two elements following the curved window of the refractive system. The system has a 75-mm entrance pupil diameter and is used in the LWIR at 10 μm central wavelength. Table 12.2 shows the defocus contributions for the refractive and its reflective counterpart for a 50°C thermal soak (uniform temperature change). First, look at the refractive system as in Fig. 12.23. The front curved dome has very little optical power (the power is slightly negative) and its contribution is to defocus the image outward 7.6 μm. The large first element accounts for a 1.7-mm inward image defocus, and the negative next element moves the image outward by 0.27 mm. The last two elements cause only a small defocus. The net system defocus due to the 50°C thermal soak is about the same as the large powered element at 1.71 mm inward, which is the same as the first element alone.

Looking now at the reflective Cassegrain system, we see that the two mirrors cause no thermal refocusing at all, and the total system defocus is 0.048 mm inward which just happens to be extremely close to the quarter wave Rayleigh criteria.

These data are summarized in Table 12.2.

It is important to understand just why the thermal defocus of the reflective Cassegrain is zero. In fact, there is a subtlety here—each mirror does cause a significant defocus with temperature because it changes shape with temperature. However, the total reflective optics system (the Cassegrain) has virtually zero thermal effect. Perhaps the best way to convince yourself that this is true is to consider a Cassegrain system with only a primary and a secondary mirror along with its support structure, and draw the rays coming from infinity, converging from the primary to the secondary, and finally focusing onto the image plane. Assume that the mirrors and all of the support structure supporting the mirrors and the FPA are all made of the same material such as aluminum. Under a positive thermal soak condition, the entire system will

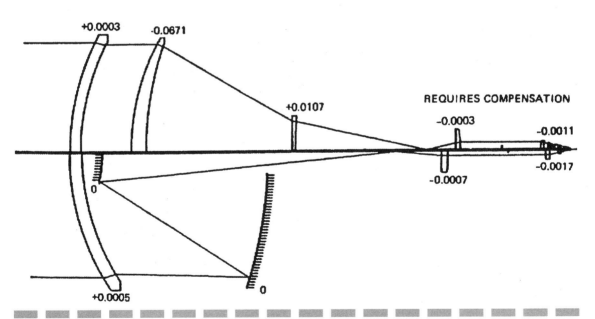

Figure 12.23 Thermal Sensitivities for Refractive System (Top) and Reflective System (Bottom)

TABLE 12.2

Thermally Induced
Focus Shifts for
Refractive and
Reflective IR Optical
Systems

Parameter	Refractive System (mm)	Reflective System (mm)
Curved dome	+0.0076	+0.0127
Large lens or mirror	−1.704	0.0
Next lens or mirror	+0.272	0.0
First small lens (collimating)	+0.0076	−0.018
Second small lens (focusing)	−0.028	−0.028
Entire system	−1.712	−0.048

expand uniformly, and the system will therefore stay in perfect focus. To be fully convinced of this conclusion, simply think about taking the drawing you have just made and enlarge it on your copy machine; then look at the enlarged drawing and notice that it is still in focus! So when we assign a zero sensitivity to the mirrors, we are assuming that this is in the context of a total reflective system of the same material.

We have shown in the preceding example how a nearly all-reflective system is athermalized to within the Rayleigh criteria. How do we restore focus in the refractive system? We could axially translate the detector or alternately translate the front element aft of the curved window. Both of these approaches are unattractive. Moving the detector/dewar assembly is just not a good idea with all of the electronic as well as cryogenic connections, and translating the massive first element is also a difficult task. The best approach is to translate another element within the system by an appropriate amount, which will be dependent on its magnification within the system. In the system shown, the best candidate is to move axially the negative element aft of the large front lens element. To first order, we will need to translate this element about the same amount as the image shifts from the large front element, and this is about ±1 mm rough order of magnitude. Thus, as the system heats up by 50°, the image from the large front element moves forward by about 1 mm, which means that the negative element must also move forward by approximately 1 mm in order to restore focus on the final image sensor. The specific magnitudes of motion need to be fully verified using your computer model. If this were the method to be employed, you could locate two to three thermisters or other temperature sensors at the outer periphery of the first element and drive the compensating element (element 2) a distance stored in a look-up table based on the temperature. The temperature sensors would be at this position in the system since the large first element is by far the most sensitive to thermal defocus, as shown earlier. Alternately, an active system approach similar to that used in 35-mm and other camera systems can be used, whereby the algorithm is based on maximizing scene contrast or some other similar criteria is used.

One of the most important points to be made here is that any element motions for athermalization reasons must be sufficiently accurate to locate the image to within approximately one Rayleigh criteria depth of focus from the nominal. In the examination of which element or elements to move, you must examine the total range of motion required. Within this range, you need to determine the finest focus that needs to be made. Make sure that this fine focus adjustment is achievable with the envisioned motion mechanism; otherwise, there will be conditions where a sharp focus cannot be achieved.

System Design Examples

To illustrate how lens designs for the thermal infrared look and work, we show in Figs. 12.24 through 12.31 designs for an $f/2$ LWIR lens 25 mm in diameter covering a field of view of $\pm 2.5°$. The materials include germanium, AMTIR 1, zinc sulfide, and zinc selenide. We show for each design a layout, the transverse ray aberration curves on axis and at 2.5° off axis. Also tabulated is the design prescription data along with the rms wavefront error, which is the average of the axis, 1.25° off axis and 2.5° off axis. With these data you can set up these designs and work with them as you wish. Do note that there is no consideration given to cold stop efficiency in these examples.

The resulting performance is shown in Table 12.3.

In Fig. 12.32 we show an $f/3$ lens with 100% cold stop efficiency along with its design data. Note that this design form is different from that shown in Fig. 12.4 where we achieved 100% cold stop efficiency by reimaging the first element, which also was the aperture stop, into the cold stop. Here we are not using any reimaging at all, and in order to make the cold stop in the dewar the aperture stop, it was simply set up that way. Notice the extreme beam motion over the front elements and the large size of these elements with respect to the entrance pupil diameter. When you can use this nonreimaging form of design, you

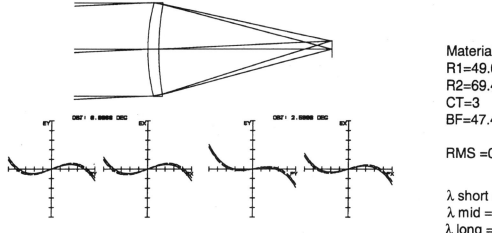

Material germanium
R1=49.030 cx
R2=69.460 cc
CT=3
BF=47.419

RMS =0.081 λ

λ short = 8 μm
λ mid = 10 μm
λ long = 12 μm

Figure 12.24
Single-Element Germanium Lens, ƒ/2 50-mm Focal Length

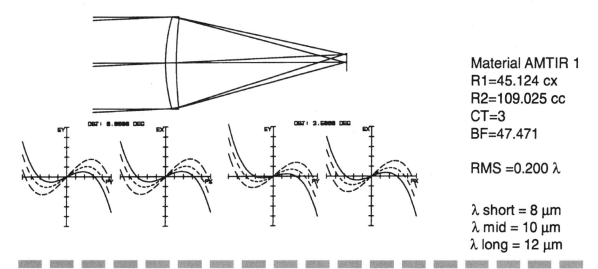

Material AMTIR 1
R1=45.124 cx
R2=109.025 cc
CT=3
BF=47.471

RMS =0.200 λ

λ short = 8 μm
λ mid = 10 μm
λ long = 12 μm

Figure 12.25
Single-Element AMTIR 1 Lens, ƒ/2 50-mm Focal Length

should do so as the design is simpler and generally performs well;
however, in some situations you must use a reimaging configuration.
This might be the case, for example, if you have a very tight packaging
requirement.

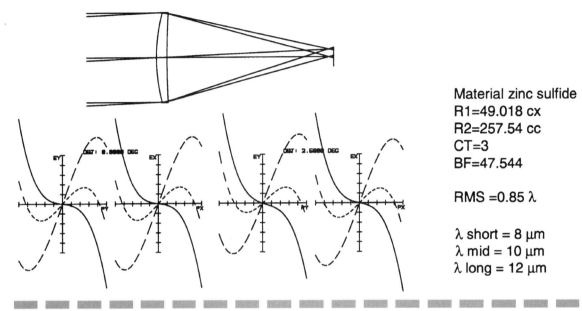

Material zinc sulfide
R1=49.018 cx
R2=257.54 cc
CT=3
BF=47.544

RMS =0.85 λ

λ short = 8 μm
λ mid = 10 μm
λ long = 12 μm

Figure 12.26 Single-Element Zinc Sulfide Lens, ƒ/2 50-mm Focal Length

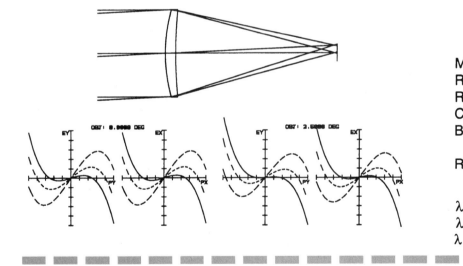

Material zinc selenide
R1=45.663 cx
R2=125.175 cc
CT=3
BF=47.483

RMS =0.35 λ

λ short = 8 μm
λ mid = 10 μm
λ long = 12 μm

Figure 12.27 Single-Element Zinc Selenide Lens, ƒ/2 50-mm Focal Length

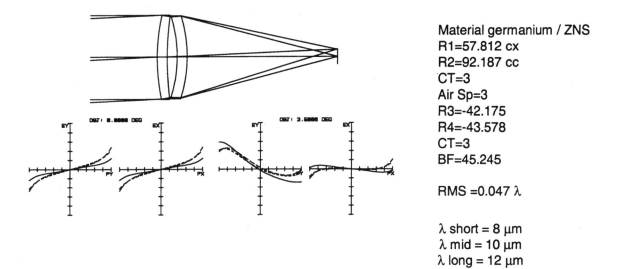

Material germanium / ZNS
R1=57.812 cx
R2=92.187 cc
CT=3
Air Sp=3
R3=-42.175
R4=-43.578
CT=3
BF=45.245

RMS =0.047 λ

λ short = 8 μm
λ mid = 10 μm
λ long = 12 μm

Figure 12.28 Germanium/Zinc Sulfide Lens, ƒ/2 50-mm Focal Length

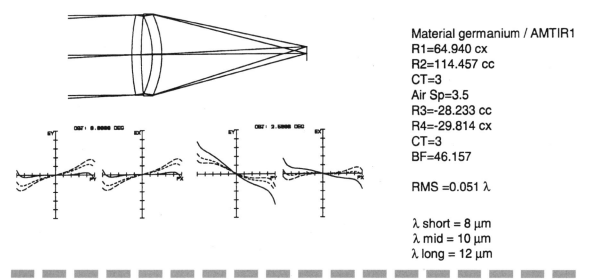

Material germanium / AMTIR1
R1=64.940 cx
R2=114.457 cc
CT=3
Air Sp=3.5
R3=-28.233 cc
R4=-29.814 cx
CT=3
BF=46.157

RMS =0.051 λ

λ short = 8 μm
λ mid = 10 μm
λ long = 12 μm

Figure 12.29 Germanium/AMTIR 1 Lens, ƒ/2 50-mm Focal Length

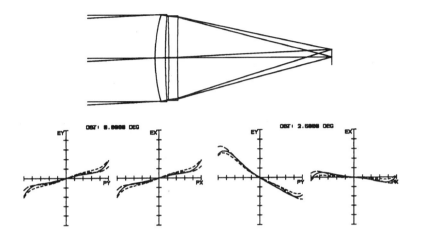

Material AMTIR 1/ZnS
R1=46.876 cx
R2=198.763 cc
CT=3
Air Sp=1.0
R3=-324.756 cc
R4=884.628 cc
CT=2.5
BF=45.566

RMS =0.053 λ

λ short = 8 μm
λ mid = 10 μm
λ long = 12 μm

Figure 12.30 AMTIR 1/Zinc Sulfide Lens, ƒ/2 50-mm Focal Length

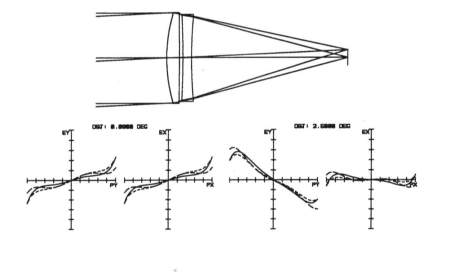

Material ZnSe/ZnS
R1=43.667 cx
R2=-1267.634 cx
CT=3.5
Air Sp=1.0
R3=-231.778 cc
R4=125.291 cc
CT=2.5
BF=44.285

RMS=0.057 λ

λ short = 8 μm
λ mid = 10 μm
λ long = 12 μm

Figure 12.31 Zinc Sulfide/Zinc Selenide Lens, ƒ/2 50-mm Focal Length

TABLE 12.3

Relative Performance of LWIR Design Examples

Lens Construction	RMS Wavefront Error	Diffraction Limited
Germanium singlet	0.08	Nearly
AMTIR 1 singlet	0.20	No
Zinc sulfide singlet	0.85	No
Zinc selenide singlet	0.35	No
Germanium/zinc sulfide doublet	0.047	Yes
Germanium/AMTIR 1 doublet	0.051	Yes
AMTIR 1/zinc sulfide doublet	0.053	Yes
Zinc selenide/zinc sulfide	0.057	Yes

Optical Systems for the UV

Working with systems in the UV is extremely challenging and demanding. In the thermal infrared, specifically the MWIR and the LWIR spectral bands, we found that the wavelength was 8 and 20 times the visible wavelength, respectively, which in some ways made IR lens systems more forgiving. In the UV we find that we now have a wavelength about one-half that of a visible system. In addition to the Airy disk expressed in micrometers being about one-half the f/number, we also find a limited number of viable optical materials are available. These include fused silica, several of the fluorides (barium fluoride, calcium fluoride, and lithium fluoride), UBK7 glass, and sapphire. The index of refraction of these materials is generally not very high. Many of these materials (especially the fluorides) are very difficult to work with and have other problems such as being hygroscopic. This leads to extreme care in manufacturing and assembly, and you may need to nitrogen purge your system to prevent moisture damage. Even sodium chloride can be used in the UV, but we recommend that you "take it with a grain of salt."

Figure 12.33 shows two deep UV lens systems, the first of which is a relatively wide-angle lens using calcium fluoride and Ultran 30 materials (the latter is no longer available). Note the relatively steep radii which is due to the inherently lower refractive indices of the materials. The second lens shown is a wafer stepper lens from a patent. This lens is similar to the Glatzel lens in Fig. 5.15, except here many more lens elements are used. Having searched the patent files extensively, this indeed is one of the more complex multielement lenses we found.

Figure 12.32
100-mm Focal
Length MWIR Lens
with 100% Cold Stop
Efficiency

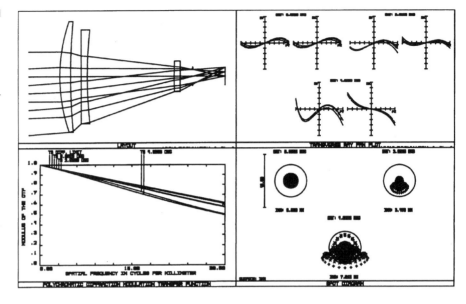

Surface		Radius	Thickness	material
1		132.661 cx	9	silicon
2		480.403 cc	8.671	
3		Infinity	5	germanium
4		340.888 cc	70.846	
5		242.243 cx	5	silicon
6		Infinity	10	
7	stop	Infinity	26.3	
8		Image, infinity		

Mirror systems offer a unique advantage in the UV. Since there are no chromatic aberrations with mirrors, these aberrations are by definition zero. And since the dispersion of refractive materials is larger at lower wavelengths, using mirrors makes good sense when you can. Two very clever reflective systems are the Schwarzschild reflective microscope objective and the Offner 1× relay for lithography, both of which are shown in Fig. 12.34. The Offner design is especially clever in that it produces a ring field of view where virtually all orders of aberrations are corrected to zero. The original patent for this system was granted in 1973 (USP 3,748,015), and the abstract reads as follows:

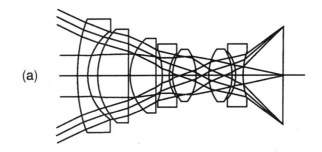

(a)

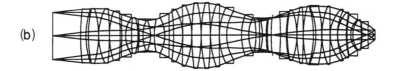

(b)

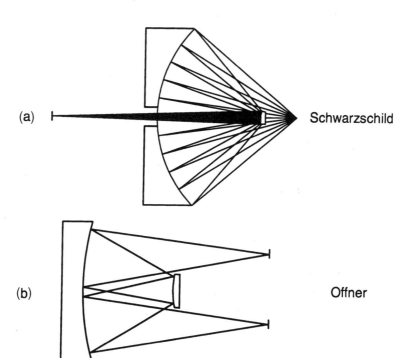

(a)   Schwarzschild

(b)   Offner

A catoptric [all reflective] system for forming in accurate micro detail an image of an object at unit magnification with high resolution is provided by convex and concave spherical mirrors arranged with their centers of curvature coinciding at a single point. The mirrors are arranged to produce at least three reflections within the system and they are used in the system with their axial conjugates at said point and to provide two off axis conjugate areas at unit magnification in a plane which contains the center of curvature, the axis of the system being an axis normal to the latter plane and through said point. This combination is free from spherical aberration, coma and distortion and, when the algebraic sum of the powers of the mirror reflecting surfaces utilized is zero, the image produced is free from third order astigmatism and field curvature.

Note that UV systems are often prone to scattering problems. The total integrated scatter (TIS) is

$$\text{TIS} \approx \left[\frac{4\pi\delta(\cos\theta)}{\lambda} \right]^2$$

where δ is the rms surface roughness, λ is the wavelength, and θ is the angle of incidence. This means that surface imperfections and finish, as well as bulk material scattering, can introduce unwanted stray light.

13

Diffractive Optics

Traditional refractive optics uses multiple glass or plastic elements consisting of spherical and/or aspheric surfaces in order to form an image. For a single material, the fact that the refractive index is higher in the blue than in the red causes the blue light to refract more severely, thereby focusing closer to the lens than the red light. This is known as *primary axial color* as we learned in Chap. 5 in the section "Axial Color." In Chap. 6 we learned how to combine a lower dispersive, positively powered element with a more highly dispersive, negatively powered element so as to bring the red and blue to a common focus position. The result of this process is the familiar *achromatic doublet,* and the further use of glasses with different dispersion characteristics is responsible for the majority of color-corrected lens systems as used in nearly all image-forming systems.

Diffractive optics allows us to take advantage of *diffraction* in addition to (or along with) *refraction* in the design of our optical systems. As you will learn, this can be an extremely powerful tool to the designer, and systems that are simpler and perform significantly better can result. However, diffractive optics is not without problems and issues, including diffraction efficiency and scattering, which can be serious in some applications.

Diffraction Gratings, Volume Holographic Elements, Kinoforms, and Binary Surfaces

Diffractive optics has as its foundation the classical *diffraction grating.* In simple terms, a transmissive diffraction grating is a series of closely ruled lines on a piece of glass or other transmissive substrate. Reflective diffraction gratings are also widely used. A collimated beam of monochromatic light incident on a grating, as shown in Fig. 13.1*a,* will result in the light scattering or diffracting from each of the lines, as shown in Fig. 13.1*b.* Emanating from each of the ruled lines will be a series of concentric wavefronts very much analogous to the water waves emanating from where the rock entered the water in our example in Chap. 3. The wavefronts from each of the ruled lines will constructively interfere when the grating equation is met:

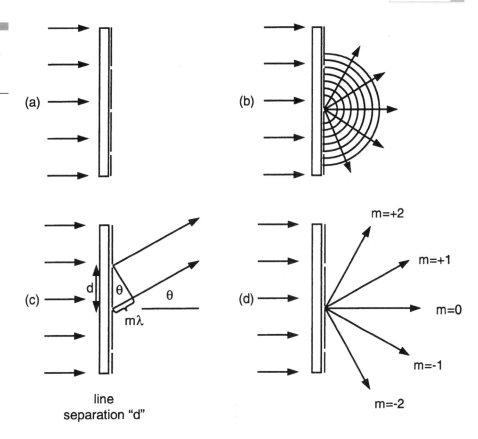

Figure 13.1
The Classical Diffraction Grating (Exaggerated)

$$\sin \theta = \frac{m\lambda}{d}$$

where m is an integer and d is the spacing of the ruled lines. This is shown in Fig. 13.1c. Thus, for example, if the line separation is $d = 20$ μm and the wavelength is $\lambda = 0.5$ μm, then for $m = 1$ we find $\theta = 1.4325°$. What this means is that if a collimated beam of light is incident onto our grating, we will see collimated light diffracted to ±1.4325° for $m = ±1$, ±2.865° for $m = ±2$, and so on as in Fig. 13.1d. In addition, there will be a so-called zero order which is straight through with $\theta = 0°$.

As discussed previously, diffractive optical elements use *diffraction* rather than *refraction* to redirect light in a predetermined manner at the interface of two different media. The first diffractive optical elements were Fresnel zone plates, which consisted of alternate transparent and nontransparent rings, shown in Fig. 13.2, built by Lord Rayleigh in the nineteenth century. Fresnel zone plates are based on constructive interference whereby the light from each successive zone constructively interferes much in the same way as described earlier for our diffraction grating. The zone spacing and the wavelength of light determine the focal length of a Fresnel zone plate.

One of the problems with gratings as described previously is that a lot of the light is not transmitted through the grating, and the light diffracts into multiple orders, resulting in significant stray-light problems if such a surface were part of an imaging optical system. In fact, it is sometimes a problem even in spectrometers where the multiple orders can overlap each other. The solution is to use a *blazed grating* or *kinoform* shape to form the grooves or lines, as shown in Fig. 13.3 where we show both a classical ruled grating (Fig. 13.3*a*) and a blazed grating (Fig. 13.3*b*). The "saw tooth" shape is commonly known as a *kinoform,* and its angle can be thought of in two ways:

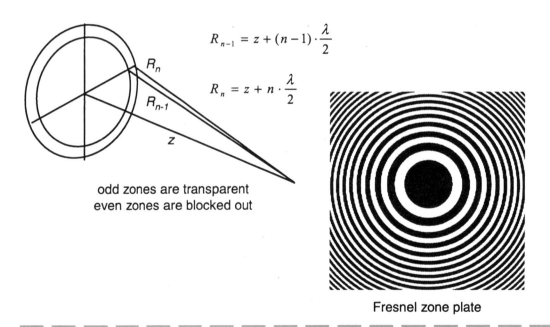

$$R_{n-1} = z + (n-1) \cdot \frac{\lambda}{2}$$

$$R_n = z + n \cdot \frac{\lambda}{2}$$

odd zones are transparent
even zones are blocked out

Fresnel zone plate

Figure 13.2
Fresnel Zone Plate

Figure 13.3
Classical and Blazed
Diffraction Grating

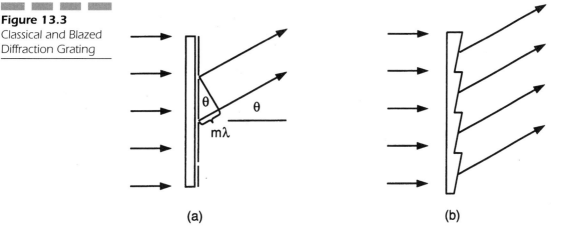

(a)   (b)

- The height of each blazed kinoform is such that it retards the wavefront by $m\lambda$, where m is an integer. For the first order of diffraction, which is normally used in diffractive optics, $m = 1$.

- Equivalent to this, the angle of each saw tooth is the angle of a prism, where the refracted light is directed into the same direction as the first order of diffraction.

Regardless of how we think of a blazed grating, its purpose is to direct the light only into the diffracted order of interest.

This process outlines the evolution of the classical diffraction grating and the blazed grating. Thus far, our discussion has, in effect, been one dimensional with straight equally spaced lines for our grating. In a lens, the bending or refraction occurs in a circularly symmetric way, and our grating analogy will be concentric rings rather than straight lines. This is analogous to the circularly symmetric Fresnel zone plate discussed earlier. Furthermore, in order to change the angle of diffraction as we proceed outward from the optical axis, the spacing between adjacent rings will change according to the grating equation and the requirements of our diffractive surface, again, much like the Fresnel zone plate.

Today many different techniques are available for manufacturing diffractive optical elements. Some of the techniques were significantly improved in recent years, making diffractive elements attractive to designers because of their potential to reduce the complexity, size, and weight of some optical systems. Diffractive optical elements (DOEs) are classified according to their method of fabrication. They can be either volume holographic optical elements (HOEs), kinoforms, or binary optical elements (BOEs).

Volume holograms are created in a photosensitive film by the exposure to two laser beams. The interference of the object beam and the reference beam creates a periodic change in the index of refraction inside the film, thus forming a phase grating, as shown in Fig. 13.4. The angles of incidence of the laser beams determine the deviation of the diffracted light incident on the hologram. The angle of diffraction can be large, for example, 45°, with high diffraction efficiency for a given wavelength.

Kinoforms, as described earlier, are a form of diffractive element whose phase modulation is introduced by a surface-relief structure. A kinoform is a surface, which consists of a set of saw tooth–shaped zones, which are concentric in most cases, as shown in Fig. 13.5. The zones are determined by the fixed optical path difference, which is a multiple of $m\lambda$:

$$OPD = n\,m\lambda \qquad n = 1,2,3,...,\text{maximum zone number}$$

where m is the diffraction order. The slope of an individual zone is optimized for the maximum diffraction efficiency for the selected order of

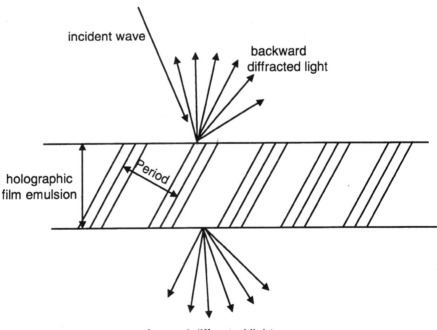

Figure 13.4
Volume Hologram

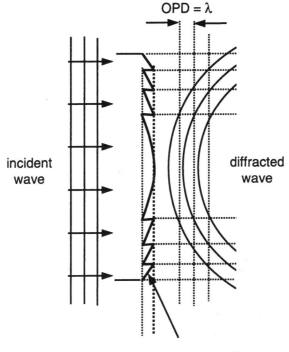

kinoform surface profile

diffraction. However, due to diffraction inefficiencies, light is distributed in more than one diffraction order, although with significantly lower intensity, forming multiple foci of the kinoform lens.

A diffractive element can be either a component as part of a simple or complex optical system or it can be a stand-alone element used for a specific application. When it is a component in a larger system, the optical designer can optimize the hybrid system comprising refractive and diffractive components to reduce the aberrations and obtain the required wavefront quality, properly determining the zone radii on the specified diffractive surface. The optical system may have more than one diffractive surface, and its optimization is done in the same way, which is to vary the coefficients in the polynomial that define zone distribution for each diffractive surface. However, it is also necessary to properly design the blaze to maximize the incident light into the chosen order. This is done separately for each kinoform surface within the system. The optical design software programs allow the designer to vary the coefficients of a polynomial describing the change or variation of kinoform spacing from the center of the element to the edge. These data are provided to the manufacturer who converts the polynomial to a surface contour to be manufactured.

There are several different methods of kinoform surface manufacturing. One method is single-point diamond turning, which can produce a linear, spherical, or aspheric blaze on a flat or curved surface. The feature size, or the smallest zone size, is limited by the tool radius. Some materials are more difficult to diamond turn than the others. Materials which can be successfully diamond turned include plastics such as acrylic and polystyrene, germanium, zinc sulfide, and zinc selenide as well as several other of the crystalline materials in the infrared. Diamond turning is often used to manufacture the inserts used to injection or compression mold plastic diffractive elements.

Another method of kinoform manufacturing is laser-writer lithography, whereby a substrate is coated with a photoresist and the photoresist is exposed to a scanning-focused laser beam. The laser beam can write continuous zone profiles by varying the exposure to shape the blaze. After etching, the surface profile created in the photoresist is transferred into the quartz substrate. Depending on the laser-scanning system, the surface profile can be either rotationally symmetric, or arbitrary. The laser-writing method requires the calibration of the photoresist, or the depth of the relief, as a function of the exposure to the laser beam.

In higher-volume production, a kinoform surface master is manufactured first, and then the kinoform components are fabricated using one of several replication techniques. Compression or injection molding is often used for plastic components. A cast and cure method is used to generate kinoforms with a very good thermal and mechanical properties.

Binary optics is diffractive elements similar to kinoforms whose phase profile is a staircase approximation to a saw tooth–shaped kinoform. These surfaces are created by photolithographic techniques using multiple masks similar to those used in the manufacture of integrated circuits. Figure 13.6 shows the manufacturing methodology. In step 1 a mask manufactured by e-beam lithography is placed on a glass or fused silica substrate and is exposed with UV light. After developing, the photoresist is removed from the exposed areas as in step 2. The part is now etched and we create a single step-height surface profile as in step 3. We then remove the photoresist and have the result in step 4. A second mask is then used for the next level as shown in steps 5 through 8. At the conclusion of step 8, we have a four-step binary optic. By using more masks, we can produce eight phase steps or more. Each sequence

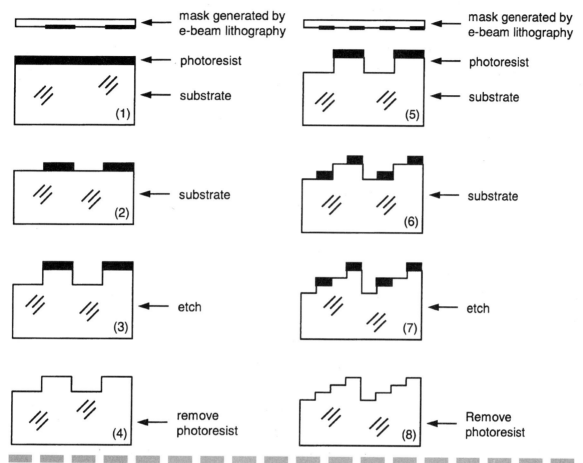

Figure 13.6
Method of Producing a Binary Optic

of the mask exposure and etching process increases the number of phase levels by a factor of 2, with the phase levels $= 2^{\text{\# of masks}}$, resulting in the name "binary" optics. The cross section of a binary optic lens is shown in Fig. 13.7. Arbitrary phase functions can also be produced with the binary technique, including nonrotationally symmetric profiles. This method is also suitable for manufacturing of lens arrays and other microoptical components. The disadvantage of this method is that it requires a flat substrate, and the diffraction efficiency is additionally limited by the staircase structure. During the manufacturing process, the masks have to be very precisely aligned with precision aligners, using fiducials on each mask.

Figure 13.7
Multilevel Approxima-
tion of a Blazed
Grating

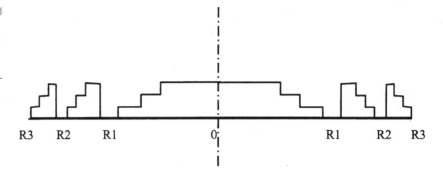

Figure 13.7
Multilevel Approxima-
tion of a Blazed
Grating

Diffraction Efficiency

A generalized diffractive surface diffracts light into many diffraction orders. The blazed surface profile is used to direct as much light into the order of preference as is possible. In the design of an optical system with one or even more diffractive surfaces, we have to consider the light distribution in other diffraction orders, not only the design diffraction order. The diffraction efficiency, or the percentage of light diffracted into the desired order, is a function of the wavelength, and it also depends on the angle of incidence, or the position of the point in the field of view. The amount of light in the design diffraction order depends on the phase introduced on each facet of the kinoform. The maximum efficiency occurs when the refracted light is in the same direction as the diffracted light. An optical system with a kinoform can be designed with an efficiency of close to 100% for a given wavelength, conjugate points, and field point. The diffraction efficiency for the mth diffraction order is

$$\text{Efficiency} = 100\% \left(\frac{\sin \{\pi[(\lambda_0/\lambda) - m]\}}{\pi[(\lambda_0/\lambda) - m]} \right)^{-2}$$

where λ_0 is the design wavelength, λ is the wavelength inside the working spectral range, and m is the diffraction order. The diffraction efficiency is equal to 100% if $\lambda_0/\lambda = m$.

For the rest of the spectral range the system is designed for, and the angular field of view over which the system must perform, the efficiency is lower than 100%. The light that does not go into the desired order of diffraction may be a source of stray light, ghost images, image color problems, and reduced contrast and resolution.

Let us look at the case of a lens with one kinoform surface, as shown in Fig. 13.8. In this example, the (+1) order is the design order. The zero-order ray direction is determined by the refraction in the basic lens surface shape. The field of view of the lens has a finite size, and we can expect to have some light from the nondesign orders to fall inside the field of view.

If the blaze, which is defined by the polynomial, is manufactured as a polynomial surface, the peak diffraction efficiency for the specific monochromatic design wavelength reaches 100%. This is shown in Fig. 13.9. When the blaze is manufactured as a linear approximation, the efficiency is 99%. However, if the diffractive surface is fabricated as a multilevel binary surface, the efficiency depends on the number of levels—the higher the number, the higher the efficiency. In the case of an eight-level approximation, the efficiency is 95%, a four-level is 81%, and a two-level is only 40%. Figure 13.10 shows the diffraction efficiency as a function of the number of binary phase steps. Note that these data are theoretical, and manufacturing errors will further degrade the diffraction efficiency.

Figure 13.8
Lens with One Kinoform Surface

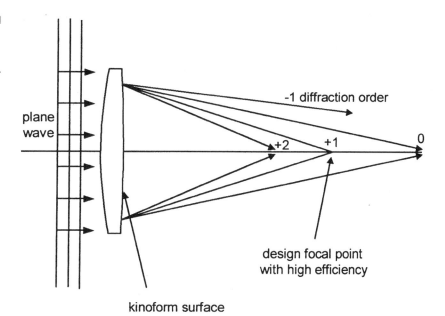

plane wave

-1 diffraction order

+2 +1 0

design focal point with high efficiency

kinoform surface

Figure 13.9
Diffractive Surface
Profiles

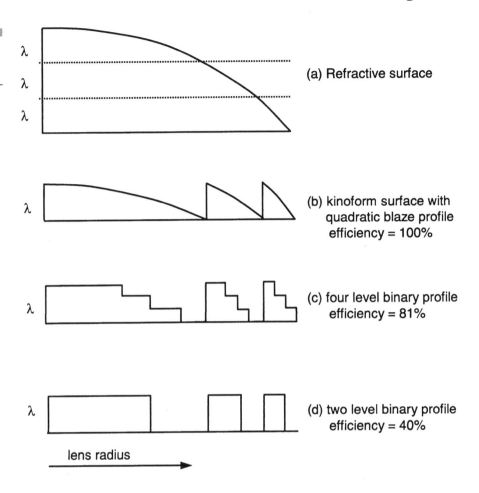

(a) Refractive surface

(b) kinoform surface with quadratic blaze profile efficiency = 100%

(c) four level binary profile efficiency = 81%

(d) two level binary profile efficiency = 40%

lens radius

Laser-based optical systems are good candidates for the use of kinoform elements. They are optimized for a specific laser wavelength and can have a high diffraction efficiency. However, systems that work inside of a specified wavelength range have to be carefully optimized for the best balance of diffraction efficiency across the spectral band. In the visible spectral range, the change in the diffraction efficiency as a function of wavelength for a few diffraction orders is shown in Fig. 13.11. A kinoform is designed with the highest efficiency for the wavelength where the human eye is the most sensitive (approximately 550 nm). The efficiency drops both in the red and in the blue region. The drop is higher in the blue region than in the red, which corresponds to the lowest sensitivity region of the human eye. As the efficiency of the design order drops, it goes up for the other orders, especially for the closest orders to the design order. If the design order is +1, we can expect the

Figure 13.10
Theoretical Efficiency
of Binary Surface As a
Function of Number
of Steps

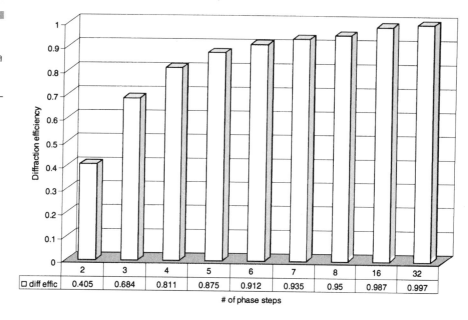

Figure 13.10
Theoretical Efficiency
of Binary Surface As a
Function of Number
of Steps

# of phase steps	2	3	4	5	6	7	8	16	32
□ diff effic	0.405	0.684	0.811	0.875	0.912	0.935	0.95	0.987	0.997

most probable "leakage" of light into the 0 and +2 orders. The diffraction efficiency of a perfect kinoform in the visible spectral band is shown in Fig. 13.12, where it can be seen that at 450 nm we have an efficiency of about 90% and at 650 nm it is about 85%. What this means is that a certain percentage of the incident light is scattered away from the preferred direction and could cause a reduction in image contrast and resolution as well as color fringing at the white-black edges in the image, similar to secondary color and/or lateral color aberrations. A very careful assessment of the diffraction efficiency and its effect on your specific system requirements is imperative.

Achromatic Doublet and the Hybrid Refractive-Diffractive Achromat

The refractive index of optical materials varies with wavelength, resulting in the presence of chromatic aberrations in the system working in a particular spectral range. Optical glasses and plastics have a higher index

■■■ ■■■ ■■■ ■■■
Figure 13.11
Diffraction Efficiency
of a Kinoform Element Designed for
the Visible Spectral
Range

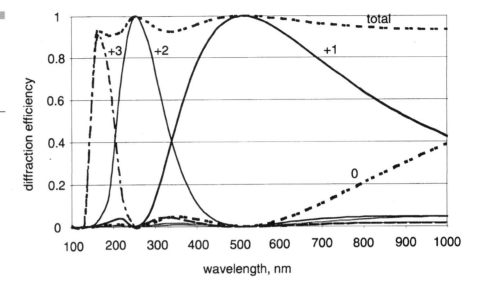

■■■ ■■■ ■■■ ■■■
Figure 13.12
Diffraction Efficiency
of Kinoform Surface
Designed for Maximum Efficiency at
Approximately
510 nm

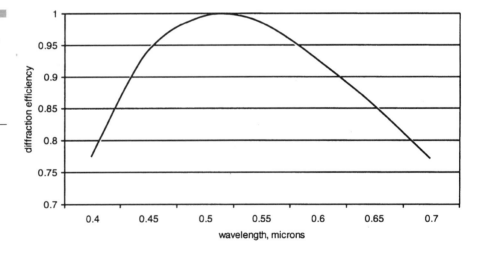

of refraction for the shorter wavelengths (blue) than for the longer wavelengths (red). This means that, in the visible spectral range, the blue light is refracted more strongly than the red light. In the case of a positive singlet lens, the blue focus is closer to the lens than the red focus. In order to correct the axial chromatism and bring two foci to the same point, two glasses with different dispersions have to be combined to design the achromatic doublet. The dispersion of optical materials in the visible spectrum is characterized by the V number or Abbe number

$$V = \frac{n_0 - 1}{n_F - n_C}$$

which is, for most glasses, between 25 and 70. The individual powers in the doublet are given by the relations

$$P_1 = \frac{V_1 P}{V_1 - V_2}$$

$$P_2 = \frac{-V_2 P}{V_1 - V_2}$$

where P is the power of the doublet. An achromatic doublet consists of one positive crown element and one negative flint element. If we combine two glasses with Abbe numbers 65 and 25, the powers of the components are

$$P_1 = 1.6 \, P$$

$$P_2 = -0.6 \, P$$

When the Abbe numbers of the components are closer to each other, the powers of both components are higher.

The other way of correcting the axial chromatism of a singlet lens is to use a diffractive surface, as shown in Fig. 13.13. In Fig. 13.13a we show a single glass element with refractive surfaces. The blue light focuses closer to the lens, as we know. In Fig. 13.13b we show a purely diffractive surface. Here the longer-wavelength red light is diffracted at a greater angle than the other wavelengths and the blue light is diffracted less. This leads us to conclude that we can use a refractive surface along with a diffractive surface to effectively bring the red and blue to a common focus position much as we can do with crown and flint glasses. This results in a single glass element which is corrected for primary axial color. This form of element is commonly known as a *refractive-diffractive hybrid*. We know that

$$f(\lambda) \sim f_0 \frac{\lambda_0}{\lambda}$$

This means that both the refracted power of the lens as well as the diffracted power can be of the same sign, with the total power being the sum of two powers. If the equivalent number to the Abbe number for glasses is calculated for the diffractive surface, it can be expressed as

Figure 13.13
Hybrid Achromatic
Singlet Lens

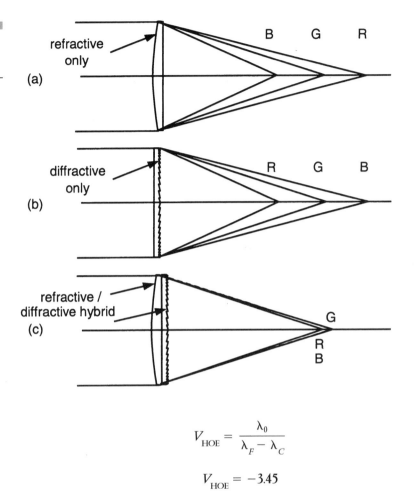

$$V_{HOE} = \frac{\lambda_0}{\lambda_F - \lambda_C}$$

$$V_{HOE} = -3.45$$

which gives the power distribution in the case of a crown glass with kinoform surface as

$$P_r = 0.95\ P$$

$$P_{HOE} = 0.05\ P$$

Note that most of the power contribution in a hybrid refractive-diffractive achromat is in the refractive part (95%) and only 5% is from the diffractive surface.

In the preceding example, we are using the quadratic polynomial term on the DOE to achromatize the hybrid singlet. It is also common to use the higher orders of the polynomial describing the kinoform period over the surface. If we use a Y^4 term in addition to the quadratic, we will also be able to minimize the third-order spherical aberration in addition to the primary color. This is well illustrated in Chap. 21 as well as later in this chapter.

Applications of Diffractive Optical Components

Diffractive optical components are used in monochromatic as well as in polychromatic systems. In monochromatic systems, such as laser systems, the maximum diffraction efficiency is obtained for the laser wavelength. With diffractive elements, the aberrations can be successfully corrected in laser beam expanders, laser collimators, null lenses or mirrors used for interferometric testing of aspheric elements, and Fourier transform lenses. Some microlens arrays are also fabricated as kinoform microlenses for laser diode optics. Diffractive elements are used in scanning systems, data storage systems, and also for athermalization in temperature-sensitive optical systems.

In high-power laser systems, there is always a danger of damaging the optical components or sensors. Measuring of the power of a laser beam may be done with the help of a kinoform surface on a fused silica substrate. A kinoform surface can diffract a very small percentage of the laser beam, which is then brought to the detector, without the danger of damaging it, and measure the energy of the beam. Diffractive elements can be used to split the laser beam into a few diffraction orders to form beams with specified energy distribution.

It is very important to know how much optical power is put into the diffractive surface, or what are the angles of diffraction resulting from the optical design. The small diffractive angles, which require the smallest feature size of a few micrometers or larger, are suitable for production. If the zone width is, for example, 70 μm or larger, such an element is a good candidate for molding in plastic. Zones in the order of 1 μm can be done with a direct laser writer in quartz or some other materials. However, if larger diffractive angles are required, volume holograms are a convenient and a more efficient technique to use. These volume holograms are, however, produced in the laboratory by the appropriate interference of laser beams and this can become costly in production.

Diffractive surfaces are often used in the MWIR and LWIR thermal infrared spectral regions. The kinoform surface is easily diamond turned in germanium. A very good state of color correction can be achieved for fast lenses with only a few kinoform zones. This is because of the extremely high Abbe number of almost 1000 for germanium in the 8- to 12-μm spectral region. This way the number of components, the size, and the weight of the system as well as its cost is generally reduced.

Holographic diffusers find application in projection systems, where the light has to be redirected into a specified volume in space, called the *eye-box*. They are not cheap components, but by using a holographic diffuser, the brightness of the image in the visible spectral range can often be increased by an order of magnitude.

Diffractive components also find applications in polychromatic imaging and illumination systems. In wide field-of-view systems, like projection TV optics, where the lateral color correction can be obtained only with the use of expensive optical materials, one kinoform surface, whose location in the system is carefully chosen, may correct the lateral color more successfully than the combination of several exotic glass elements. The use of a kinoform surface in an eyepiece reduces significantly the size and the weight of the eyepiece.

There are other application areas which use the same photolithographic techniques as used in manufacturing diffractive optics. These surface structures, with periods smaller than the wavelength of light, can be used to make antireflection structured surfaces as well as polarization components. Rigorous electromagnetic theory describes the phenomena of light transmission through a surface with a microstructure smaller than the wavelength of light, dividing two media with index of refraction n_1 and n_2. The surface structure synthesizes an effective index of refraction. Antireflection surfaces of this type are successfully used in the systems working in the longer-wavelength regions. The structure is a two-dimensional array of either pyramids or some other binary or multilevel structure. This is shown in Fig. 13.14.

Note that with the extremely small feature sizes used in this case, the surfaces need to be kept extremely clean to avoid damage and unwanted scattering.

When a subwavelength structure is made as a one-dimensional array or grating, the light averages the properties of the grating region and the surface acts as a polarizer, as shown in Fig. 13.15. The surface, in effect, acts as a birefringent film. Wire-grid polarizers work on the same

Figure 13.14
Two-Dimensional
Antireflection Grating
with Subwavelength
Structure

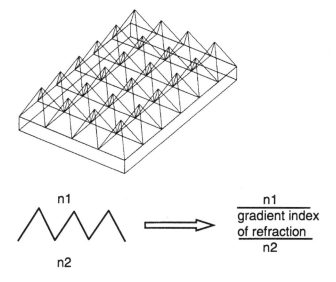

principle, and a good contrast ratio can be obtained with the wire-grid polarizers in the visible spectral range.

Parametric Examples of Diffractive Optics Designs

In order to illustrate how we design systems with diffractive optics, as well as to demonstrate the relative merits of these designs, we will show several representative examples. The specifications for the design example is as follows:

- *Entrance pupil diameter*: 25 mm
- *Field of view*: on axis only
- *Wavelengths*: C (656.3 nm), d (587.6 nm), F (486.1 nm)
- f/numbers: f/10, f/5, f/2.5, f/1.25

Figure 13.16*a* shows the transverse ray aberration curves for an f/10 hybrid singlet using the preceding specifications. The material used was BK7 glass, and very similar results would result for acrylic, which would be a fine material choice if the element were to be molded. The diffractive surface is located on the second surface of the lens. The spacing, or

Figure 13.15
Polarization Beam-
splitting with Sub-
wavelength Structure

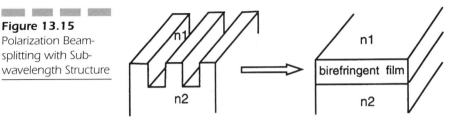

separation, between adjacent kinoforms, or rings, was allowed to vary with respect to the square as well as the fourth power of the aperture radius, or y^2 and y^4, where y is the vertical distance from the vertex of the surface perpendicular to the optical axis. The quadratic term allows for the correction of the primary axial color whereby the F and C light (blue and red) are brought to a common focus. The fourth-order term allows correction of the third-order spherical aberration as well. The resulting ray trace curves show the classical performance typical of an achromatic doublet. Since the lens is of a relatively high f/number, the spherical aberration at the center wavelength is fully corrected. There is a residual of spherochromatism which is the variation of spherical aberration with wavelength. This residual aberration is due to the fact that the dispersion of the kinoform is linear with wavelength, whereas the dispersion of the BK7 glass used is nonlinear. The surface has 28 rings with a minimum period of 229.7 μm. The insert for injection molding this surface could be easily diamond turned. Figures 13.16*b* to *d* show similar results for hybrid singlets that are $f/5$, $f/2.5$, and $f/1.25$, respectively. Note that as the f/number gets lower and lower, the higher-order spherical aberration increases and the spherochromatism increases as well, to a point where the spherochromatism is the predominant aberration. The number of rings and the minimum kinoform period are listed.

Note that these data do not scale directly as we can do with conventional optical designs. As a DOE is scaled down in focal length while maintaining its f/number, a true linear scaling of all parameters (except, of course, the refractive index which is unitless and thus does not scale) is not correct. This is because we need to maintain the kinoform height or depth to create the one wave of OPD between adjacent kinoforms, and a linear scaling would result in one-half the kinoform height and thus one-half wave of OPD between adjacent kinoforms. Thus, we find that for a 0.5 scaling, we end up with one-half of the number of rings with approximately the same minimum kinoform period. For example, if we were interested in a 12.5-mm-diameter $f/2.5$ hybrid, we would scale

Figure 13.16
Hybrid Refractive-
Diffractive Achromatic
Singlet As a Function
of f/Number

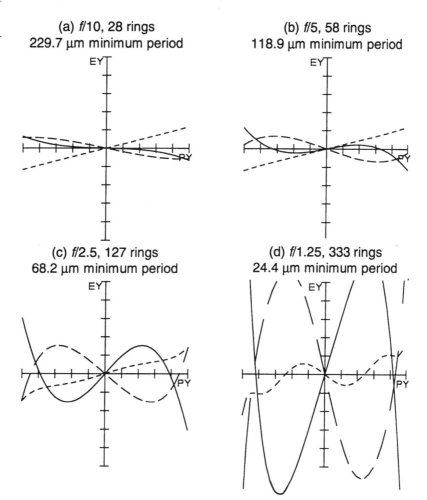

MAXIMUM SCALE: +/- 50.000 MICRONS
0.486 0.588 0.656

(a) f/10, 28 rings
229.7 µm minimum period

(b) f/5, 58 rings
118.9 µm minimum period

(c) f/2.5, 127 rings
68.2 µm minimum period

(d) f/1.25, 333 rings
24.4 µm minimum period

the radii, thickness, and diameter by 0.5× from the 25-mm starting design. However, the number of kinoform rings will decrease by a factor of 2, with essentially the same minimum kinoform period. It is highly recommended that you reoptimize any lens or lens system containing one or more DOEs after scaling in order to assure that your surface prescription is correct.

It is feasible to manufacture kinoform as well as binary surfaces with minimum periods of several micrometers or less; however, it is best to discuss your specific requirements with your manufacturer prior to finalizing your design. And do remember the issues with respect to diffraction efficiency presented earlier in this chapter. Figure 13.17 shows, for comparison, the performance of classical $f/10$ and $f/2.5$ achromatic doublets using BK7 glass. The results are somewhat improved over the hybrid solution. Note that, at the lower f/number of $f/2.5$ (Fig. 13.17b), the spherical aberration of the achromatic doublet is becoming a problem, and we show an aspherized design in Fig. 13.17c.

Figure 13.18 shows graphically the relationship between the lens f/number, the number of rings, and the minimum kinoform period for

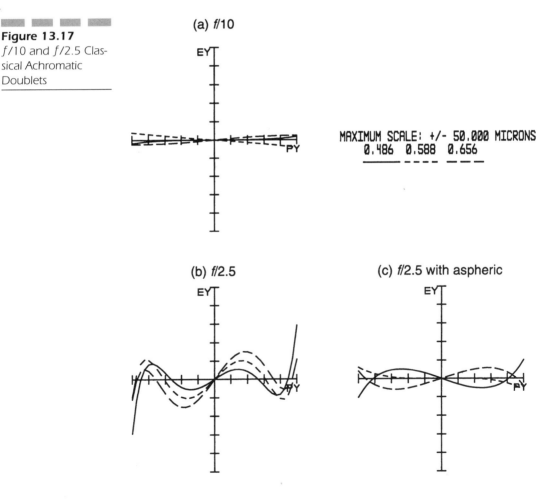

Figure 13.17
$f/10$ and $f/2.5$ Classical Achromatic Doublets

(a) $f/10$

the hybrid designs in Fig. 13.16. As we would expect, the lower the f/number, the larger the number of rings and the smaller the minimum kinoform period.

Figure 13.19 shows several design scenarios, all for a constant f/5 single-element lens. For reference, Figs. 13.19a and b both have no diffractive surface; however Fig. 13.19b does have an aspheric surface for correction of spherical aberration. The primary axial color is the same in both lenses, and quite large as expected. Figure 13.19c has an aspheric surface for spherical aberration correction and a quadratic diffractive surface for correction of the primary axial color. Figure 13.19d is all spherical with a quadratic and a fourth-order kinoform variation. It is interesting that this solution is very similar to that in Fig. 13.19c, except that this solution has more spherochromatism than the solution with the aspheric surface. Finally, Fig. 13.19e allows both an asphere as well as a quadratic and fourth-order kinoform variation. The aspheric, along with the quadratic kinoform variation, of Fig. 13.19c was so well corrected that no further improvement is possible here.

One of the more interesting observations is that the aspheric surface, along with the diffractive surface, allows for the correction of both the spherical aberration as well as the spherochromatism. The all-diffractive surface with the quadratic and the fourth-order kinoform variation has

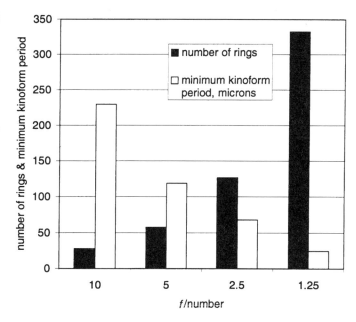

Figure 13.18
Number of Rings and Minimum Kinoform Period As a Function of f/Number for Hybrid Singlet Example

Figure 13.19
Performance of ƒ/5
Hybrid Singlets with
Different Surface
Descriptions

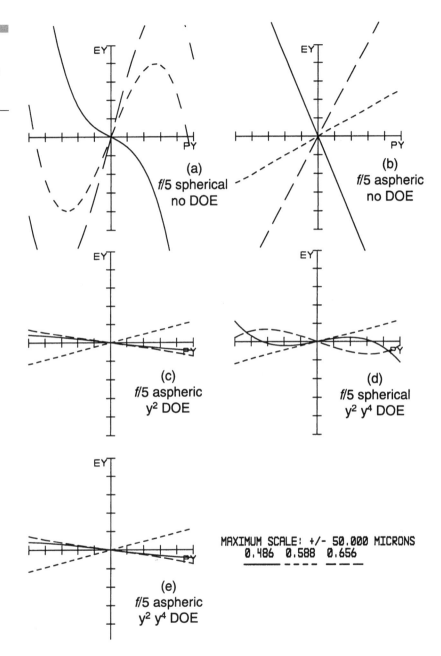

(a)
f/5 spherical
no DOE

(b)
f/5 aspheric
no DOE

(c)
f/5 aspheric
y² DOE

(d)
f/5 spherical
y² y⁴ DOE

(e)
f/5 aspheric
y² y⁴ DOE

MAXIMUM SCALE: +/- 50.000 MICRONS
0.486 0.588 0.656

a residual of spherochromatism. The reason for this subtle difference is that in the aspheric case the spherical aberration correction and the chromatic aberration correction are totally separate from one another, thereby allowing better performance. In the all-diffractive design we are more constrained and do not have sufficient variables to eliminate the spherochromatism as well as the spherical aberration and the primary axial color.

Summary of Diffractive Optics

There are a number of important conclusions from our discussion of diffractive optics, and they are presented here in outline form:

- If your diffractive surface is immersed within a more complex lens, it may need to do more "work" than if it is simply used as a hybrid refractive-diffractive singlet, as shown in the previous section. This may, and probably will, result in many more rings and significantly smaller kinoform periods. These parameters, especially the minimum kinoform period, should be closely monitored during the design process to assure cost-effective manufacturability.

- The theoretical change in diffraction efficiency with wavelength must be dealt with—it cannot be ignored. The scattered light can be a serious problem in certain applications. A good example is a camera lens using a DOE on one of its outer lens elements. If the sun is outside of the field of view, and if sunlight is permitted to directly illuminate the diffractive surface, then you may find a highly objectionable spike of stray light as well as an overall loss in contrast.

- Binary optical surfaces have a further diffraction efficiency problem due to the staircase approximation to a kinoform.

- Manufacturing errors, as well as ray obliquity on a diffractive surface, can further degrade performance with unwanted scattered light.

- Remember that one of the fundamental relationships in optics is that scattering is inversely proportional to the fourth power of the wavelength. This is one of the reasons we often have problems with diffractive optics at short wavelengths. In the thermal infrared we generally find very few rings with huge kinoform periods are often the case. Furthermore, the scattering and diffraction efficiency is rarely a problem due to the long wavelength.

■ The different lens design software programs do not all use the same mathematical representation for diffractive surfaces. Make sure that you are fluent with what your program's convention is and that your manufacturer is also fluent with it.

Diffractive optical elements can offer very significant design benefits. However, "user beware." The issues of diffraction efficiency, manufacturability, and the other issues discussed in this chapter may be of major significance in your application. Finally, we once again urge you to work very closely with your manufacturer to assure that your paper design is producible and will yield the results you predict.

14

Design of Illumination Systems

Introduction

Illumination optics is required in many varied system applications, including, for example, microscopes, projection systems, machine vision systems, industrial lighting. In optical systems where a light source is illuminating an object that has to be projected onto a screen as in a desktop projector, the design often requires a high brightness and uniformity across the image. High brightness implies a high collection efficiency of the light emitted from the source. Furthermore, these systems often require small packaging of the optical system.

Light sources have a wide range of types, shapes, and sizes, and the choice of the design of the illumination optics is very dependent on the source. Sources can be tungsten halogen lamps of different shapes, metal halide lamps, light-emitting diodes (LEDs), xenon lamps, frosted bulbs, different forms of arc lamps, or fusion (sulfur) lamps. Some sources have sufficient brightness and uniformity across their emitting area, and these can be imaged directly onto the object that has to be illuminated, but in most cases the sources need some kind of homogenization in the illumination optics to achieve the required brightness uniformity, while simultaneously minimizing throughput losses. The most common physical parameters that are used in photometry to characterize the source and the illumination system are *flux, intensity, illuminance,* and *brightness. Photometry* deals with visual systems, and it is implicit that all relationships are weighted by the spectral sensitivity of the human eye. Clearly, if an object being viewed visually appears to be of a certain brightness, this must take into account the eye's spectral sensitivity. *Radiometry* is a closely related subject, which deals directly with power as it is emitted by a source, and ultimately irradiates a surface. The concepts and the basic theory are identical to those used in photometry; however, the units are totally different. We will define the basic photometric parameters and their corresponding units:

- *Flux* corresponds to power in radiometry, and it is the total power emitted by the source. The unit for flux is the lumen.

- *Intensity* is the flux per unit solid angle, or a solid angle flux density or the flux angular distribution, which assumes that the flux comes from a point source. The unit for intensity is the candela = lumen/steradian.

■ *Illuminance* is the flux per unit area incident onto the surface being illuminated, or the area flux density. It does not relate to the angular flux distribution incident on the surface. The unit for illuminance is the foot-candela = lumen/ft^2, or lux = lumen/m^2.

■ *Brightness* (*luminance*) is the flux per solid angle per unit area, or the area and solid angle flux density. The unit for brightness is the nit = candela/m^2.

Köhler and Abbe Illumination

There are two classic approaches to the design of illumination systems. Most modern illumination system designs are modifications of one of these basic concepts.

Abbe illumination, which is sometimes called *critical illumination,* images the source directly onto the object to be illuminated, as shown in the paraxial model in Fig. 14.1. It is used in the cases when the source is sufficiently uniform for the system requirements. An illumination relay lens images the source onto the object. The object being illuminated can be film as in a movie projector, a 35-mm slide, or other similar transparency, or a transmissive or reflective LCD panel as in a desktop projector. Either the source must be inherently uniform as noted previously, or the condenser must have sufficient aberrations that blur the image of the source enough to eliminate the structure of the source at the transparency plane. The projection lens then images the object (transparency) onto the screen. At the same time, it images the image of the source onto the screen. This type of illumination works well with frosted bulbs or large sources, but it does not work with high brightness sources where high throughput is required and where there is significant structure in the source as with filament lamps and arc lamps. Generally, the resulting brightness and uniformity across the screen depends on the brightness and uniformity of the source, which means both spatial and the angular source uniformity.

In the case of a small bright, highly nonuniform source, such as an arc or a filament lamp, uniform brightness at the image is achieved with another form of illumination known as *Köhler illumination,* as shown in Fig. 14.2. Here, an illumination relay lens or condenser lens images the source into the *pupil* of the projection lens rather than onto the object being projected. This illumination relay system has its aperture stop at

Figure 14.1
Abbe Illumination

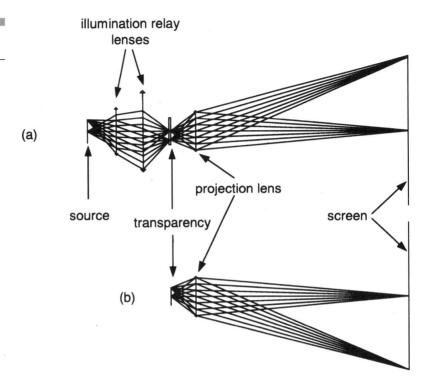

illumination relay
lenses

(a)

projection lens

source   transparency

screen

(b)

the location of the film, or the object, that is to be projected onto the screen, or into the eye. Since each point of the source illuminates the entire surface of the film, the film is by definition uniformly illuminated. Brightness nonuniformity with Köhler illumination can be caused by significant intense nonuniformity of the source.

Optical Invariant and Etendue

When a given light source emits a certain flux, not all of the emitted light reaches the screen or the detector. Some of that light is lost immediately after leaving the source. Since most sources emit into a large solid angle, often into a full sphere, we find that the first optical element, which may be a condenser system, has a difficult job of collecting and orienting or directing that light toward the aperture of the projection optical system. Some of the light is lost in the optical system as absorbed, scattered, diffracted, or vignetted light. There may also be some compo-

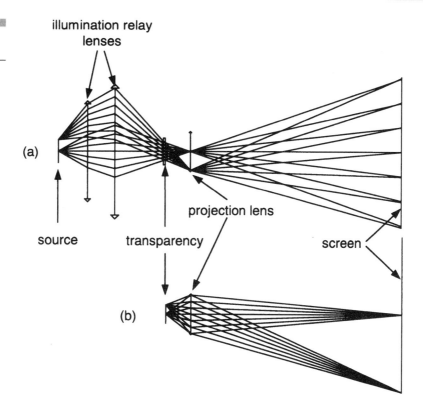

Figure 14.2
Köhler Illumination

illumination relay
lenses

(a)

source   transparency   screen

projection lens

(b)

nents, such as the beamsplitters, filters, or polarizers, whose purpose in the system is to transmit only a certain type of light and block the rest.

There are many different terms that are related to the coupling of the light from the source to the screen. These terms are the *optical invariant, etendue, light-gathering power, throughput, angle to area,* and *area-solid angle product.*

Let us first take an example of an imaging system such as a telescope. The flux that is transmitted through a telescope goes through the aperture stop of the telescope shown in Fig. 14.3. If our telescope has an entrance pupil diameter D_{in} and the field of view is $2\omega_{in}$, the exit pupil diameter is D_{out} and the exiting field of view is $2\omega_{out}$, then

$$D_{in} \sin(\omega_{in}) = D_{out} \sin(\omega_{out})$$

In other words, optical invariant in the entrance pupil has the same value as in the exit pupil. This product maintains its value through the entire optical system. If we square the previous equation, we get

Figure 14.3
Flux at the Stop of a
Telescope and Tele-
scope Throughput

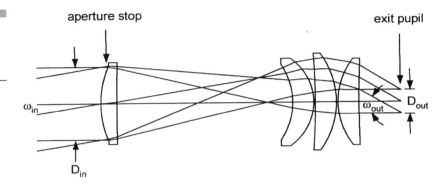

$$D_{in}^2 \sin^2(\omega_{in}) = D_{out}^2 \sin^2(\omega)_{out}$$

The total flux that passes through the entrance pupil of the telescope will pass through the exit pupil if there are no losses within the system from absorption, vignetting, beamsplitting, filters, or scattering. This product $D_{in}^2 \sin^2(\omega_{in})$ is the measure of the throughput of the optical system and it is proportional to the area − solid angle product, or etendue. Etendue is, in effect, the conservation of flux within an optical system. In a system with a light source, the goal is to use as much of the emitted light from the source as possible, and couple it into the optical system such as a projection lens for example. The etendue of the source (area of the source times the solid angle into which the source emits light) should be the same or slightly larger than the etendue of the projection lens. If the etendue of the source is significantly larger than the etendue of the projection lens, there are a lot of rays that are stopped by the apertures in the optical train, and these rays are lost and never reach the screen. On the other hand, if the projection lens is designed with the etendue larger than the etendue of the source, then the lens might be unnecessarily complex, with an underfilled aperture stop or underfilled field of view.

There are cases when the etendue is increased along the optical train. For example, if there is a need to polarize the light, then the unwanted polarization, instead of throwing it away, can be rotated at the expense of an increase in etendue. Another example is the presence of diffractive components in the system. If more than one diffractive order is used, the etendue in the system after the diffractive element is larger than before the diffractive element. Unfortunately, it is not possible to decrease the etendue without the light loss in an optical system.

The best correction of aberrations in most imaging optical systems is generally done at the center of the field of view. In illumination systems, on the contrary, the best aberration correction must be done at the edge of the field to get the sharp edge of the illuminated patch of light on the film, LCD, or transparency.

Light pipes, as shown in Fig. 14.4, are commonly used as light homogenization components in illumination systems. Light pipes can be either hollow structures with reflective inside surfaces or they can be solid structures inside of which the light is totally internally reflected. Light pipes have two important roles in illumination systems. The first is that they are used as light homogenizers to change a spatial nonuniform distribution of the light at the input of the pipe into a uniform output. By definition, the etendue at the input surface of the light pipe is equal to the etendue at the output of the pipe.

The second very important function of light pipes is that tapered light pipes can change the angle of the input cone of light into a cone that can be accepted by the system into which the light from the source is injected. The optical invariant at the input of the pipe is equal to the invariant at the output

$$\frac{D_{out}}{D_{in}} = \frac{\sin(\omega)_{in}}{\sin(\omega)_{out}}$$

This formula gives the angle conversion in the case of a tapered light pipe and this is shown in Fig. 14.5. A tapered light pipe with multiple inside reflections is exactly like a kaleidoscope! Changing the angle between the rays and the optical axis after each internal reflection, it creates an array of virtual sources. This helps to spatially homogenize

Figure 14.4
Light Pipes

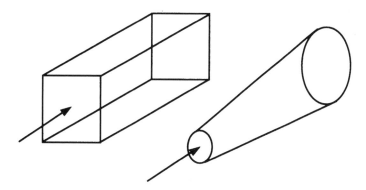

the brightness at the output of the pipe. Some shapes of the input (output) pipe surfaces spatially homogenize the light better than the others. Shapes such as a rectangle, a triangle, or a hexagon are the tileable shapes after unfolding due to internal reflections, and they spatially homogenize the light very well. Different shapes are shown in Fig. 14.6. Straight walls of these pipes are good enough to give a uniform output. Other types that are not tileable, such as a circular input surface, do not homogenize the light as well, even when the pipe is very long. In the case of a circular input, improvement in uniformity is achieved with a pipe wall that is parabolic rather than a cone. This type of pipe is called a *compound parabolic concentrator* (CPC), as shown in Fig. 14.7.

Other Types of Illumination Systems

The most efficient way to collect the light from a small source is to place the source at the focus of a parabolic or elliptical reflector. These reflectors collect the light emitted by the source into a large solid angle, and either collimate the light in the case of a parabolic reflector, or focus it to the second focal point of the ellipse shown in Fig. 14.8. There are different ways to reduce the nonuniformity of the image of the source and get a smooth illumination. One way of smoothing in

Figure 14.5
Angle Conversion in a Tapered Light Pipe

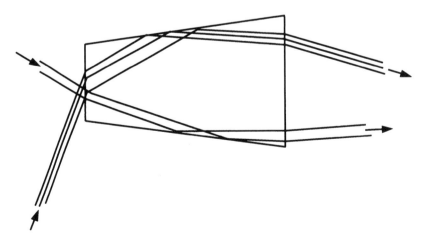

Figure 14.6
Different Shapes of
Light Pipes

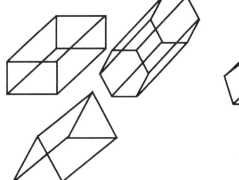
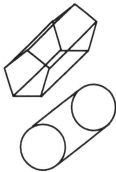

shapes that are good
spatial homogenizers of
light at the output

shapes that are bad
spatial homogenizers of
light at the output

Figure 14.7
Compound Parabolic
Concentrator (CPC)

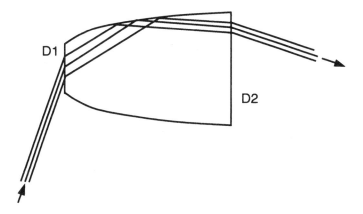

collimating reflectors, such as flash lamps or street or automobile lighting, is the wedged reflector, where the reflector has a basic parabolic form, but it is divided into a lot of wedge-shaped segments. The other way is to combine the collimated light from the reflector with a smoothing plastic element located in front of the reflector, consisting of the extruded array of prisms or structures with combined prismatic and sinusoidal profiles.

In the case of a film projector, the goal is to send the light from the source into the rectangular area of the film, with the numerical aperture

Figure 14.8
Parabolic and
Elliptical Reflectors

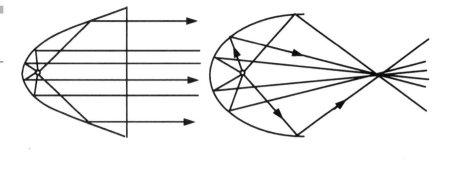

parabolic reflector　　　　　elliptical reflector

of the illumination matched to the numerical aperture of the projection lens. There are two common ways of getting a rectangular uniformly illuminated area in the film plane from a highly nonuniform, nonrectangular source.

The first method is to initially collect the light from the source with a reflector, then focus it down onto the input surface of the rectangular light pipe, as in Fig. 14.9. The magnification of the tapered light pipe is chosen such that the etendue at the output from the pipe is equal to the etendue of the optical system that relays the rectangular pipe output surface onto the plane of the film. The pipe has to be sufficiently long to get the spatially uniform output. A good rule of thumb is that the output will be uniform if there are at least three ray reflections along the pipe. The larger the magnification of the pipe, the smaller the minimum length of the pipe needed to achieve the uniform output.

The second method is often used in the illumination systems of front desktop projectors with transmissive image panels and with projection lenses of $f/3$ and above. It uses lenslet arrays along with a polarization recapture plate. This is shown in Fig. 14.10. The light from the source is collimated with the parabolic reflector. The expanded and collimated beam then goes through the lenslet array (lenslet array is shown in Fig. 14.11). Each lenslet focuses the beam inside the lenslet of the second lenslet array, producing an image of the source. The light that illuminates the image panels has to be linearly polarized. After the second array, there is an array of polarization beamsplitters, which splits the light into two orthogonal linearly polarized beams. One of the beams then undergoes the rotation of its plane of polarization, and two beams emerge in the same polarization state out of the polarization

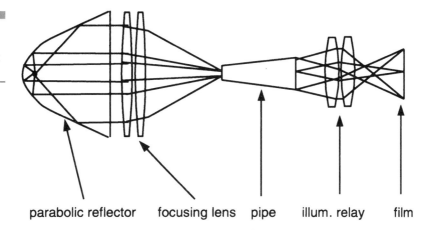

Figure 14.9
Illumination System
with a Tapered Light
Pipe

parabolic reflector   focusing lens   pipe   illum. relay   film

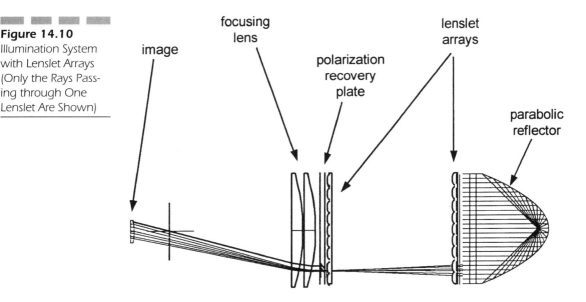

Figure 14.10
Illumination System
with Lenslet Arrays
(Only the Rays Pass-
ing through One
Lenslet Are Shown)

image

focusing
lens

polarization
recovery
plate

lenslet
arrays

parabolic
reflector

recovery array. The second lenslet array, together with the focusing lens, images the lenslets (which are rectangular) of the first array onto the image panel. The focusing lens superimposes images of the lenslets of the first array in the plane of the image panel. In summary, a circular distribution of flux from the parabola is sampled by rectangular elements of the lens arrays and is then superimposed at the image panel, thus homogenizing the nonuniform parabolic output and transforming

Figure 14.11
Lenslet Arrays

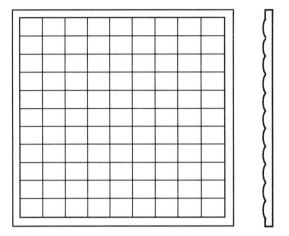

the geometry to match that of the image panel. An illumination system with a light pipe has a high throughput and small packaging. However, mounting of the light pipe is a problem, since all pipe surfaces are the optical surfaces, and any contact with the sides of the pipe frustrates the total internal reflection, resulting in a loss of light. Although the system with lenslet arrays requires more space, it accommodates the polarization recovery, and increases the throughput 30 to 40%.

15

Performance
Evaluation
and
Optical Testing

The quantitative characterization of optical performance, or image quality, is extremely important. Generally, the optical design engineer plays a key role in system testing, and for this reason we feel it is important to include the basics of optical testing in this book. Testing can range from the somewhat simplistic bar target to the more sophisticated means for characterizing the modulation transfer function (MTF).

Testing with the Standard 1951 U.S. Air Force Target

The simplest form of resolution target is perhaps the white picket fence shown in Fig. 15.1. The image of the fence consists of alternating bright and dark bars as formed by the white pickets and the dark background between the pickets. If we image this fence with a camera lens, the image will be demagnified by approximately the ratio of the camera focal length to the distance to the fence. Let's assume the fence pickets are 75-mm wide and that we are imaging the fence with a 50-mm focal length lens from a distance of 20 m. The magnification is therefore $50/20,000 = 0.0025\times$. The fence is 150 mm/picket pair, or equivalently 150 mm/line pair. This equates to 0.006667 line pairs/mm. At the image formed by our lens, this becomes 0.375 mm/line pair, or 2.667 line pair/mm. Most camera lenses will resolve this spatial frequency just fine, as it is a rather low spatial frequency. Let's now move our lens to a distance of 200 m. Here the magnification is $0.00025\times$ and the spatial frequency becomes 26.67 line pair/mm. A reasonably good camera lens will have a contrast of approximately 50% or higher at this spatial frequency. Needless to say, as the lens moves further from the fence the spatial frequency, in line pairs per millimeter increases and the contrast decreases as a result of aberrations, diffraction, assembly and alignment errors, and other factors.

If we had no other metric, a white picket fence would be a reasonable target to use for lens performance testing and characterization. In the laboratory, the most basic means for measuring image quality is through the use of the so-called 1951 U.S. Air Force Target. This form of target is readily available, low in cost, and easy to use. A typical Air Force target is shown in Fig. 15.2. The legend for the target is shown in Table 15.1 where it is evident that the target is divided into groups and

Figure 15.1
The White Picket
Fence Analogy to a
Bar Target for Optical
Testing

Figure 15.1
The White Picket
Fence Analogy to a
Bar Target for Optical
Testing

elements so, for example, group 2, element 4 is a bar pattern of 5.66 line pair/mm.

How do we use an Air Force type target? Let us consider the example of measuring the performance of a 100-mm focal length 35-mm camera lens. Let us further assume that we must "resolve" 50 line pair/mm. As will be discussed in Chap. 21, this is a reasonable value for such a lens. We now construct a test setup shown in Fig. 15.3. We locate an Air Force target at the focus of a collimator lens. The collimator is used to simulate an infinite object distance. It is *critical* that the quality of the collimator *must* be independently validated and demonstrated to be better than the level of performance we are looking for. Generally, the focal length of the collimator lens should be at least a few times longer than the focal length of the lens under test (a factor of 3 is the minimum). The collimated light now enters the lens under test and the image of the target comes to focus in the image plane as shown.

In order to compute which pattern corresponds to 50 line pair/mm, we can simply multiply 50 by the magnification from the target to the final image. This magnification is the focal length of the camera lens divided by the focal length of the collimator. Assume we have a collimator with a 1-m focal length. This gives us a magnification of 0.1×, which

Figure 15.2
The Standard 1951
U.S. Air Force Target

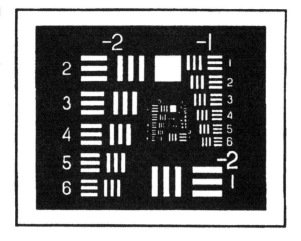

means that 5 line pair/mm is the spatial frequency on the target corresponding to 50 line pair/mm at the focus of our camera lens. Group 2, element 3 is 5.04 line pair/mm and is a sufficiently close match.

We now magnify this image with a microscope of suitable magnification so that the 50 line pair/mm is sufficiently well magnified so as not to be limited by the resolution of the eye. In order to compute the required magnification, first we need to find what angular subtense our 50 line pair/mm equates to: 50 line pair/mm = 0.02 mm/line pair, which when viewed at a nominal distance of 254 mm corresponds to 0.02/254 = 0.00008 rad. The eye resolves approximately 1 min of arc, which is 0.0003 rad. Our target is thus about 4 times smaller than what the eye can resolve, and this is not acceptable. If we magnify the target by 40× with the microscope, it will be seen at 10 times the resolution limit of the eye, which should be just fine. Ten arc minutes equates to 1 cm at a distance of 3 m. Thus, a 40× microscope should be used. To determine the magnification of the microscope, simply multiply the magnification of the objective lens by the magnification of the eyepiece. Thus, a 4× objective and a 10× eyepiece will do the job for us. We may want to use a higher magnification to be sure that the eye is in no way limiting and for more comfortable viewing of the image. The microscope objective has to have a numerical aperture that is larger than the numerical aperture of the lens under test.

The person performing the test now visually views the bar pattern through the microscope and judges whether he or she can distinguish three separate bars in the pattern in group 2, element 3 (for both

TABLE 15.1

Legend for 1951
Standard U.S. Air
Force Resolution
Target, lp/mm

Element Number	Group Number									
	-2	-1	0	1	2	3	4	5	6	7
1	0.25	0.5	1.0	2.0	4.0	8.0	16.0	32.0	64.0	128.0
2	0.28	0.561	1.12	2.24	4.49	8.98	17.95	36.0	71.8	144.0
3	0.315	0.63	1.26	2.52	5.04	10.1	20.16	40.3	80.6	161.0
4	0.353	0.707	1.41	2.83	5.66	11.3	22.62	45.3	90.5	181.0
5	0.397	0.793	1.59	3.17	6.35	12.7	25.39	50.8	102.0	203.0
6	0.445	0.891	1.78	3.56	7.13	14.3	28.51	57.0	114.0	228.0

Figure 15.3
Using the Air Force
Target to Test a
Camera Lens

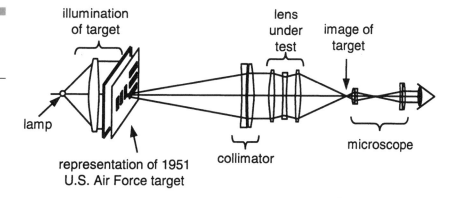

horizontal and vertical orientations). If the answer is "yes," the lens passes, if the answer is "no," the lens fails. This test is indeed quite simple and easy to set up. The disadvantage of this form of test is that a real person can only judge whether the bar pattern can or cannot be resolved, not the level of contrast or sharpness of the image. Also, the test is somewhat subjective and sometimes different people will get different answers. A common rule of thumb is that the eye can resolve a modulation of approximately 5%, or 0.05. This is a useful metric, but what about the lens' ability to produce a good image with a higher contrast, say at 20 line pair/mm? The lens will likely resolve this spatial frequency fine; however, the user cannot judge the level of modulation or contrast. The only thing the user can judge is whether a specific spatial frequency is or is not resolvable or distinguishable into three bars.

The Modulation Transfer Function

The modulation transfer function provides a more quantitative measure of the image quality of an imaging lens system than a bar target. We discussed the basics of MTF in Chap. 10. From a laboratory standpoint, there are two principal ways of measuring MTF. First, we can use sinusoidal patterns of different spatial frequencies and image them with the lens under test onto a CCD sensor to characterize the resulting modulation as a function of spatial frequency. If you know the modulation of your target pattern, you can compute directly the modulation transfer function.

A more robust and easier way to measure MTF is by computing the Fourier transform of the line-spread function, which gives the MTF directly. The Fourier transform is a widely used mathematical transformation whereby we can compute the transformation from the spatial domain to the frequency domain, thereby giving the MTF directly. A lab setup, such as in Fig. 15.4 is used, and is explained here. The test setup shown is for a finite conjugate lens, which is $f/20$ on its object side and $f/2.8$ on its image side:

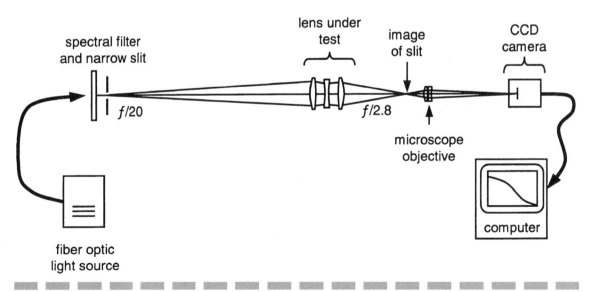

Figure 15.4
Example of Lab Setup to Measure MTF

▪ A light source with an appropriate spectral filter illuminates a narrow slit. The slit width should be narrow enough so that its finite width does not affect the results. As we know, the diameter of the Airy disk diffraction pattern is approximately equal to the f/number, in micrometers. A good rule of thumb is that if our slit width is 25% or less of this value, then its effect on the resulting Fourier transform will be negligible. For our f/20 object cone, the slit width required is approximately 5 μm or less since the Airy disk diameter would be about 20 μm. If we cannot find a sufficiently narrow slit, the software can divide the resulting MTF by the Fourier transform of the rectangular slit, which is the MTF of the slit itself. This, in effect, divides out the effect of the finite slit width.

▪ The lens under test is now positioned on the optical bench.

▪ The image of the slit is now magnified with a high-quality microscope. The profile of the intensity distribution in the image of the slit, in the direction normal to the slit, is called the *line-spread function* (LSF). The magnified image of the slit is imaged onto a CCD array.

▪ The analog or digital output of the camera is input into a computer with a frame grabber, where the real-time video image of the slit is displayed in one of the windows, as shown in Fig. 15.5. A scan through the gray levels (line-spread function) is displayed in another window.

▪ The Fourier transform of the LSF is computed and is displayed in the remaining window. The perfect system MTF or diffraction-limited MTF is shown for reference and is computed based on user input of the central wavelength and the f/number.

The MTF is one of the most useful means for characterizing the optical performance of an imaging system. Most scenes consist of objects of many spatial frequencies, and the MTF data tell us how the modulation of the object is transferred from the object to the image as a function of the many varied spatial frequencies in the scene.

All lens design computer programs allow for the computation of the MTF for a given design. It is important to remember that these results are for the design prescription, or paper design, and do not include the effects of fabrication, assembly, and alignment errors. This nominal design thus needs to have sufficiently higher MTF to account for the MTF degradations due to these manufacturing errors.

Figure 15.5
Example of MTF
Measurement
Software

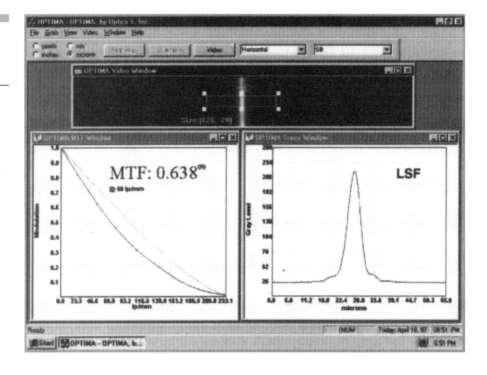

Interferometry

Interferometry allows us to measure quantitatively the optical path difference of a lens or, alternatively, the departure of a curved or flat surface from its nominal shape. While the geometry of many interferometers is different, their basic functionality is similar in that the wavefront from a perfect reference surface, such as an optical flat, is compared and interferes with the wavefront from the lens or surface under test. To illustrate just how an interferometer works, consider Fig. 15.6, where a Twyman-Green interferometer is shown:

■ Light from a monochromatic and coherent light source, such as a HeNe laser, is input into a beam expander. The beam expander expands the approximately 0.8-mm-diameter beam to a diameter in the order of 25 to 50 mm. While the beam expander should be of high quality, it is not necessary that it be *perfect* as any small errors will cancel since the beam goes to both the reference arm as well as the test arm of the interferometer, and small wavefront errors will cancel each other.

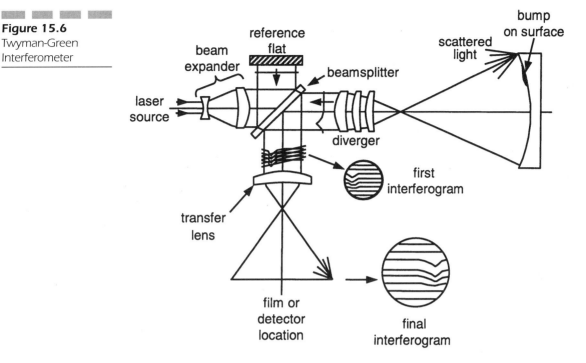

Figure 15.6
Twyman-Green
Interferometer

▣ The expanded laser beam is incident onto a nominally 50–50 beamsplitter plate. Let us assume that the first (left-hand) surface has the beamsplitter coating and the opposite side is antireflection coated.

▣ Fifty percent of the light now travels upward to the reference flat. This mirror must be near perfect and is often in the order of $\lambda/40$ wave P-V. The light reflects downward from the reference flat and 50% of it passes through the beamsplitter, with the rest going back toward the laser. This is the so-called reference arm of the interferometer.

▣ The remaining 50% of the light, which passes through the beamsplitter, enters what is known as a *diverger* or *transmission sphere*. This is the test arm of the interferometer. We will assume that we are testing here a spherical mirror. The diverger is, in effect, a perfect lens that creates a perfect Airy disk at its focus and creates perfectly diverging wavefronts following the intermediate image. The purpose of the diverger is to create a wavefront that perfectly matches and nests into the nominal shape of the mirror under test, which means it appears that the wavefront emerges from the point coincident with the center of curvature of the mirror under test. If the mirror were a paraboloid or some other nonspherical shape, then a *null lens* is used instead of a diverger, with the same goal of creating a wavefront which nests into the nominal surface under test. We will show the use of a null lens in Chap. 22.

- We will assume that the mirror under test has a small bump on it, as shown, but is otherwise perfect. A portion of the wavefront will first hit the top of this bump and reverse its direction. The rest of the wavefront will now travel to the mirror surface and then turn around. Meanwhile the part of the wavefront that hits the top of the bump is already heading back toward the diverger. The bump in the wavefront will thus be a factor of 2 times as large as the physical height of the bump itself.

- The return wavefront passes through the diverger and is once again a series of plane parallel wavefronts, with the exception of the bump.

- Fifty percent of the light incident onto the beamsplitter will now reflect downward, with the rest going toward the laser. At this point, we have two wavefronts that are located between the beamsplitter and the transfer lens, one from the reference flat and the other from the mirror under test. Since the light is monochromatic and coherent, the light will interfere. What this means is that for regions where the two wavefronts are in phase we see a bright fringe and for regions where they are 180° out of phase we see a dark fringe. Since the wavefront from the reference arm of the interferometer is essentially perfect, any deviations are from the test arm, specifically from the mirror under test. If we were to place a white card in the beam here, we would see interference fringes indicative of the departure of the mirror under test from perfect sphericity. There is only one problem—due to scattering from the edge of the mirror under test we may see artificially curved fringes at the outer periphery of the interferogram. This effect could be interpreted as a turned up or down edge on the mirror, which is false.

- In order to eliminate the false turned up or down edge phenomena described, a *transfer lens* is used. This lens, or lens group, serves a dual role. It images the surface under test onto the sensor (CCD, film, etc.), thus bringing the scattered light back to the edge of the interferogram where it belongs to eliminate false fringes. It also matches the size of the image to the sensor. If this is done properly, the final interferogram will show crisp edges with no artificial edge effects. Figure 15.6 shows how the scattered light at the edge of the surface under test is reimaged onto the interferogram plane by the transfer lens.

Another common form of interferometer is the Fizeau interferometer, as shown in Fig. 15.7. This system is similar to the Twyman-Green interferometer; only the reference beam is created from the partial reflection from a reference flat, which is also used in transmission.

It is important in any of the interferometer tests that the two beams (the reference and the test beams) are of approximately the same intensity in order to maximize the fringe contrast.

The interpretation of an interferogram is very much like interpreting a topographical map with contours of equal altitude or elevation. In an interferogram, the fringes are lines or contours of equal height just as with a topographical map. Consider the topographical map shown in Fig. 15.8. The lake at the lower left is flat or level; hence there are no

Figure 15.7
Fizeau Interferometer

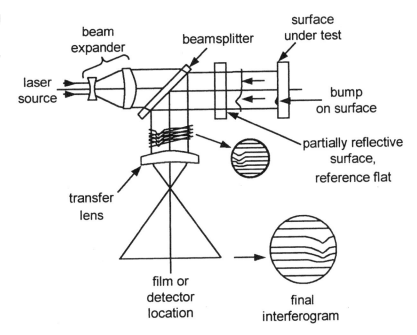

Figure 15.8
Topographical Map
Analogy to Interferogram

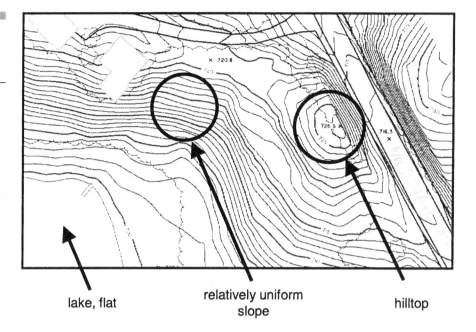

lake, flat   relatively uniform
                                  slope   hilltop

contour lines. The top center circle contains nearly straight contours, which indicates a flat region which is also tilted or sloped, and the circle at the right of the hilltop shows a somewhat dome-shaped hilltop which has a rapid falloff to the right. The interpretation of an interferogram is exactly the same as that of a topographical map, with the only difference being the scale of the contours. In a topographical map the contour levels may be in units of meters or feet, and in an interferogram the units are in wavelengths of light.

Now consider Fig. 15.9 where we show four interferograms of a nominally flat mirror. Each fringe is due to one-half wave of surface departure from flatness, which results in one wave to the reflected wavefront, hence one fringe. The upper left interferogram shows a surface, which for the most part is tilted left to right. The upper right interferogram is quite flat, with a residual tilt from an 11 o'clock to a 5 o'clock direction. And the lower left interferogram shows a saddle-shaped residual. It is important to note that interferograms contain no sign information whatsoever with respect to what is high and what is low on the surface. This information *must* be obtained when the part is being tested, as it is impossible to derive the sign information at a later time. The saddle could be down in the 10 o'clock and 4 o'clock directions and up in the

1 o'clock and 7 o'clock directions. Alternatively, the up and down directions could be reversed. In fact, it is physically possible (although unlikely) that the surface only departs up (or down) in the four regions.

Other Tests

There are many other optical system testing methods, and the more important of these will be briefly outlined here:

■ The *star test* is where you view the image of a point object (a pinhole) similar to a star through a microscope visually (or via a video camera). If a reasonably narrow-band filter is used, the image should be that

Figure 15.9
Typical Interferograms of Nominally Flat Mirrors

interferograms of thin aluminum mirror
(both interferorgrams are of same mirror)

interferogram of thin
aluminum mirror with
astigmatism or saddle

interferogram of
thin beryllium
mirror

of an Airy disk if the system is diffraction limited. This is an extremely sensitive test since very small asymmetries in the Airy disk are quite evident. The human eye is sensitive to many orders of magnitude in dynamic range, which makes the test quite robust. If the microscope is now focused inside and then outside of best focus, you will see a disk with concentric rings which becomes larger as you move further from best focus. This ringed disk should appear similar both inside and outside of best focus. If it is not, then you have a residual of spherical aberration. While not quantitative, the star test is an extremely sensitive test.

- The *Hartman test* is often used at observatories to test telescopes. A mask with a grid of small apertures is placed in front of the primary mirror, and the resulting imagery is recorded at several through-focus locations. This is the analogy of creating a series of through-focus spot diagrams, and the residual aberrations can be derived using appropriate software.

- Other forms of optical system testing relate more to alignment of the system as a whole as well as its subcomponents or groups. Measurement of focal length, distortion, and other lens metrics is also important. We often use alignment telescopes, laser beams, and other methods for assuring that our optics is sufficiently well aligned.

16

Tolerancing and Producibility

Introduction

Most of the material presented in the earlier chapters of this book is associated with achieving the optimum lens design. There has been a lot said over the years about lens design optimization theory and algorithms, global search algorithms, aberration theory, and other related topics. These are all directed toward achieving the optimum lens performance for what we sometimes call the "paper design." The performance of the paper design is that of the design prescription, or the theoretical design, with the effects of absolutely no manufacturing errors of any kind included. While the performance of the paper design is indeed important, the effects of real-world hardware-related manufacturing errors and tolerances can and will, by definition, alter and degrade the level of performance from the theoretical paper design performance. We often find that the degradations due to manufacturing errors can many times even surpass the image degradations of the nominal design itself.

Tolerancing is the science (and art, to some extent) of distributing and error budgeting the manufacturing tolerances of all optical and optomechanical components and dimensions throughout the system to assure that your system will meet its required level of optical performance at a reasonable cost.

Unfortunately, there is little, if any, direct correlation between the performance of the paper design and the robustness or insensitivity of the design with respect to the level of manufacturing errors or tolerances. For example, we may have a design where steep bendings and high angles of incidence has allowed for the effective balancing of higher orders of aberration, thus yielding a high level of performance for the paper design. Unfortunately, this design may be extremely sensitive to tolerances due to the higher angles of incidence and the presence of higher-order aberrations. On the other hand, a different design configuration for the same lens requirements may have significantly reduced angles of incidence, reduced higher-order aberrations, and may result in a somewhat lower level of performance for the paper design. However, due to the reduced angles of incidence, this design may be less sensitive to manufacturing errors and tolerances. An example of this was shown in Chap. 9 where we discussed several designs submitted to the 1980 International Lens Design Conference. One of these designs was a very compact lens with small angles of incidence at the surfaces, while one of the other designs had extremely large angles of incidence on several surfaces. This latter design will, in all likelihood, have tighter manufacturing tolerances.

The key point is that manufacturing errors in the form of fabrication, assembly, and alignment errors can be extremely important and are often a major contributor to the overall level of performance of an optical system, even if the paper design is excellent.

In tolerancing an optical system, we need to assign tolerances to all optical and mechanical components within the system. This includes all lenses and/or mirrors as well as the mechanical components, which directly or indirectly support the optics. The overall goal for the system is that the optical performance is met (MTF and/or other image quality criteria), optical component costs are minimized, assembly and alignment costs are minimized, and yields are maximized. Tolerancing is necessary whether you are producing a lens in a high-production environment or a one-of-a-kind lens.

Ultimately, the goal of the tolerancing effort is to aid in establishing a performance error budget whereby you can, with confidence, predict the expected level of optical performance.

What Are Testplates and Why Are They Important?

Prior to embarking on an extensive tolerance analysis, your design needs to be completed and finalized. One of the very last steps in this process is that of matching radii to existing testplates or tooling. Virtually all optical shops have in their inventory hundreds of so-called testplates. These testplates (sometimes called *test glasses*) are a convex and concave mating pair of tooling with radii ranging from very short to very long. They are often made of low-expansion Pyrex glass, they have very low surface irregularity, and their radii have been measured to a high level of precision.

Once a design is finalized and the shop to manufacture the optics has been selected, the designer should proceed to match as many radii as possible to existing testplates. Let's take an example: Assume that we have a six-element double Gauss lens with 10 different radii. Further, assume that after we compared the radii in the final design to the testplate list of the selected lens vendor, we found that the closest radius was 25.21 mm, and our vendor has a testplate of radius 25.235 mm, a difference of only 2.5 μm. Five sections of the testplate list from OPTIMAX is shown here to give you an indication of the number of plates available and the density, or closeness, of radii in the different radii regions.

Plate ID	Radius	Diameter	CC	CX
1	1.5000	2.5000	X	X
2	2.0000	3.2000	X	X
3	2.0000	3.9000	X	X
4	2.0470	3.5000	X	X
5	2.0470	4.0000	X	X
6	2.5150	3.2000	X	X
7	2.5150	5.0000	X	X
168	10.0150	14.2000	X	X
169	10.0150	14.9000	X	X
170	10.0600	13.2000	X	X
171	10.0600	14.9000	X	X
172	10.1800	13.9000	X	X
173	10.1800	14.7000	X	X
174	10.2240	13.5000	X	X
175	10.2240	16.6000	X	X
176	10.3100	17.0000	X	X
177	10.3100	15.0000	X	X
178	10.3900	13.6000	X	X
179	10.3900	14.9000	X	X
180	10.5000	13.4000	X	X
181	10.5000	13.7000	X	X
483	25.0550	40.0000	X	X

Plate ID	Radius	Diameter	CC	CX
484	25.0550	41.1000	X	X
485	25.1000	17.8000	X	X
486	25.1000	16.8000	X	X
487	25.2350	38.3000	X	X
488	**25.2350**	**44.3000**	**X**	**X**
489	25.2600	33.5000	X	X
490	25.2600	44.8000	X	X
491	25.3800	38.7000	X	X
492	25.3800	41.7000	X	X
493	25.4050	37.7000	X	X
494	25.4050	41.8000	X	X
495	25.4460	33.1000	X	X
496	25.4460	35.0000	X	X
948	75.3200	57.6000	X	X
949	75.3200	61.2000	X	X
950	75.5000	56.2000	X	X
951	75.5000	44.7000	X	X
952	75.8260	70.0000	X	X
953	75.8260	71.5000	X	X
954	76.2950	63.0000	X	X
955	76.2950	63.0000	X	X
956	76.4350	62.0000	X	X
957	76.4350	63.7000	X	X
958	76.6150	49.9000	X	X
959	76.6150	52.0000	X	X
960	77.0930	69.0000	X	X
1491	500.2100	85.0000	X	X
1492	500.2100	87.0000	X	X

Plate ID	Radius	Diameter	CC	CX
1493	501.5200	78.3000	X	X
1494	501.5200	56.6000	X	X
1495	508.1400	87.3000	X	X
1496	511.8840	88.0000	X	X
1497	511.8840	88.0000	X	X
1498	514.5900	69.3000	X	X
1499	514.5900	69.8000	X	X
1500	520.7050	68.3000	X	X
1501	520.7050	68.9000	X	X
1502	528.5250	102.4000	X	X
1503	528.5250	103.5000	X	X
1504	537.4110	73.6000	X	X
1505	537.4110	75.4000	X	X

We then proceed to change the radius in the design to the testplate value of 25.235 mm and then freeze it from any further changes. In effect, we will reoptimize the lens while constraining this radius to exactly match the testplate. All other variables and constraints of the design are the same as for our final optimization cycles, and our error function remains the same. After the first radius has been matched, we once again search for the closest radius of those remaining to any on the testplate list and match it to this testplate radius. Note that during the reoptimization process following the first testplate insertion, all of the remaining radii will change or "shuffle" a small amount, giving a whole new scenario with respect to which surface is now closest to an existing testplate. This process continues until we have matched all of the radii to existing testplate radii.

It is our experience that in most cases 100% of all radii should be able to be matched to existing testplates. If for some reason you have problems with the last one, two, or three radii, you can release several radii that you have already matched and then match the problem radius or radii, with the intent of matching 100% of all radii to existing testplates. A very important point to keep in mind is that when we are matching our last one or two testplates, there are barely enough variables to correct the aberrations, much less any constraints such as focal length. Thus, for the last few testplates we highly recommend either to allow some or all of the element thicknesses to vary (in addition to the already varying air spaces), or alternatively we may need to release or relax one or more of the system constraints such as focal length. Any further changes in the design will likely cause a negligible change in focal length. We had a system some years ago where we inadvertently left the constraint on focal length during the match of the final testplate. In order to meet this constraint, the design took on a highly degraded level of optical performance, which would have caused it to fail its performance specification. During the testplate fit, you may elect to match the closest or the farthest radius first; this is your choice. Our experience shows the closest algorithm to work just fine.

Most software programs have an automated testplate fitting routine, and these are quite robust and work well. But use them with care, as they are quite automated and you could run into one of the problems cited previously if you are not careful (such as the match of the last one or two testplates, producing poor performance due to overconstraining a first-order parameter such as focal length). If you do have a particular problem with matching a specific radius, you might consider running your optimization and varying this surface radius along with a constraint to match it to the testplate. By working the match into the optimization, your chances are best.

Why do we emphasize so strongly matching all radii to existing vendors, testplates, and how does this relate to tolerancing? The reasons are several:

■ Testplates cost several hundred dollars or more each, and they take time to manufacture, perhaps up to several weeks.

■ Testplate radii can be measured more accurately than they can be manufactured. For example, a radius of 25 mm can be *manufactured* to within about ±0.025 mm (precision level for manufacturing); however, it can be *measured* to within about ±0.00625 mm or less. We all realize that, given enough time and money, we could of course manufacture the radius to within the ±0.00625-mm tolerance or even better; however, this is generally not economically feasible. Thus, matching testplates is like an insurance policy whereby your level of confidence in your lens working as predicted is enhanced after matching the radii to existing testplates.

When you perform your final tolerance analysis, the way you can best model the real-world situation with respect to the surface radii is to first assign a power fit to the testplate in units of fringes. You will then also assign a radius tolerance, which in this case means the *accuracy* to which the testplate radius is known to have been measured. This is a measurement of accuracy or capability tolerance, and it should be discussed with your lens manufacturer. And if you really want to assure that your system works well, have the shop remeasure the testplates you have matched and incorporate these newly measured results into your design. After all, equipment and techniques constantly improve, so why depend on a testplate radius measurement from 20 years ago?

On a different, yet related matter, for extremely high-precision systems the designer often incorporates the measured refractive indices of the glasses into the design. This is called a *melt design,* and is yet a further assurance that the design will perform as predicted. As with testplates, glass refractive indices can be measured more accurately than they can be manufactured.

How to Tolerance an Optical System

The basic procedure for tolerancing an optical system is shown in outline form in Fig. 16.1 and is described as follows:

1. We first assign viable tolerances to all toleranceable parameters within our system. This includes all optical as well as mechanical components. This first candidate set of tolerances should be reasonably achievable at a rational cost. If we know we have a very sensitive system and/or we require a very high level of performance, a somewhat tighter set of tolerances should be used; conversely if we have an insensitive system or a poorer level of performance, then looser tolerances can be used.

For an optical system with a reasonable level of performance, a candidate tolerance set may be derived from Table 16.1.

TABLE 16.1

Candidate Tolerances for a Reasonable Performance Lens

Parameter	Tolerance	Parameter	Tolerance
Radius	Testplate measurement accuracy	Tilt	0.05-mm TIR
Power fit to testplate	Three fringes	Decenter	±0.05 mm
Surface irregularity	One fringe	Refractive index	±0.001
Thickness	±0.05 mm	Abbe number	±0.8%
Air space	±0.05 mm	Glass inhomogeneity	±0.0001
Wedge/centration	0.025-mm TIR		

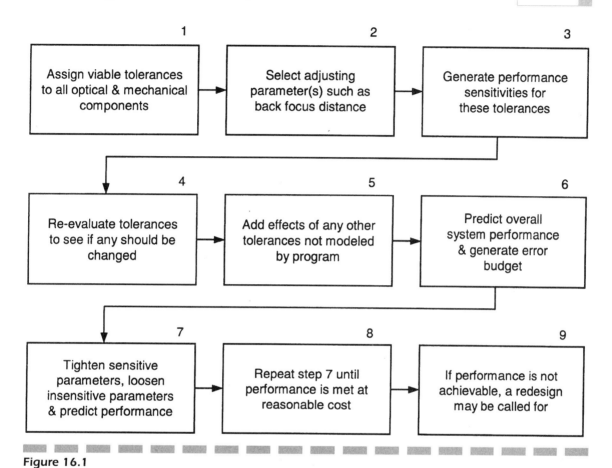

Figure 16.1
Tolerancing Procedure

Note in Table 16.1 that the radius tolerance is listed as testplate measurement accuracy. This is the accuracy to which the testplates have been measured at the shop. Using a value of ±0.01 mm is probably a reasonable assumption, but check with your shop to be sure. For long radii, a better assumption might be the radius change corresponding to ±0.25 wave of sag at the outer periphery of the element clear aperture, or the testplate, whichever is smaller. This is because for long radii testplates the effective f/number of the light cone from the center of curvature of the surface is high, thus yielding a sizable depth of focus, which ultimately impacts the ability of determining the radius.

2. We then generate performance degradation sensitivities for all toleranceable parameters within the system using the tolerance routine in our lens design computer program. In other words, each and every fabrication, assembly, and/or alignment-related tolerance on all components is evaluated and sensitivities are determined. The tolerance forms include radius, power fit to testplate, surface irregularity, element thickness, airspace, element wedge, element tilt, element decentration, refractive index, Abbe number, and glass inhomogeneity. Other specific factors, such as the effects to performance due to the thermal environment, may also need to be included here.

3. As part of the preceding tolerance sensitivity analysis, we need to select the appropriate *adjusting parameter* or *parameters*. An adjusting parameter is an adjustment which you plan to allow for during the final lens assembly and testing. Back focus is the most common adjusting parameter and will almost always be used as a final adjustment during final system assembly. There can, in principle, be other adjusting parameters ranging from airspaces to tilts and decentrations. More on adjusting parameters later.

4. We now generate the performance sensitivities for this initial set of candidate tolerances. Thus, if our performance criteria were the MTF at 30 line pair/mm, we would determine the drop in MTF at this spatial frequency for each toleranceable parameter within the system.

5. We now go back and look carefully at each tolerance to determine if it should be changed from a manufacturing or assembly point of view. For example, larger airspaces may require somewhat looser tolerances. This is the all-important time to talk with your optical shop and your mechanical designer as well as your machine shop to reach a mutual understanding of the optical components and the mechanical design with respect to the anticipated levels of tolerances, which will affect directly the lens elements.

6. We now add the performance degradations from any other effects, which may not have been covered by our computer model. These may include atmospheric turbulence, surface irregularities that may not have been modeled in the computer analysis, and/or other effects. The net result is to predict the expected level of optical performance based on the assumed tolerance set.

7. We then predict the overall system performance and generate a performance error budget.

8. We now proceed to tighten any sensitive parameters and loosen insensitive ones and, again, predict performance. Here you need to look carefully at the effect on your performance metric such as MTF for each toleranced parameter to determine its effect. You will sometimes find that only a small subset of the overall tolerance set is sensitive, and the majority of the tolerances are insensitive. In these situations, tightening only a small number of tolerances can have a big payoff in improved performance. It is important to review the tolerances you intend to tighten with your optical shop and/or machine shop.

9. Step 8 is repeated as necessary until we meet the performance goal at a reasonable level of cost. The results of step 8 comprise the final set of manufacturing and assembly tolerances which, if applied and adhered to, will result in a system which meets your performance goals and objectives at a reasonable cost.

10. If the required performance is not achievable at a reasonable level of cost, you may need to return to your initial specifications to see which ones you may be able to relax. Also, a redesign may be called for with the specific goal of loosening the manufacturing tolerances.

How Image Degradations from Different Tolerances Are Summed

When you compute tolerance sensitivities as outlined earlier, you will ultimately need to predict the net system performance. Ideally, this should represent the predicted level of performance for some reasonably high percentage of manufactured systems such as a 95% cumulative probability or confidence level. But how do we add performance degradations in order to accurately predict performance? In other words, if we have, for example, 10 tolerances and each one degrades the performance by a different amount, how do we use these data to predict the net result of assembling a large number of systems?

One common method of adding degradations is known as *RSS addition*, which stands for root sum square. In this method, we take the *square root of the sum of the squares* of each of the individual degradations, and if the degradations are of the same form, this should lead to a 95% confidence level. The degradations could be a peak-to-valley optical path difference (OPD), RMS OPD, drop in MTF, or some other criteria. While this can all be done, the criteria are not all mathematically correct, as will be discussed later.

To illustrate RSS addition and how powerful the method is, let us assume that we have a stack of pennies, and each penny has thickness T ±0.1 mm. If we have 100 pennies, the predicted thickness of the stack of 100 pennies is $100 \cdot T$ ±10 mm worst case or $100\ T$ ±1 mm for a 95% confidence based on an RSS addition, since $\sqrt{0.1^2 \times 100} = 1$. The tolerance on the thickness of a penny is of a simple form in that the penny is either thicker or thinner than nominal; in other words, the degradations are all of the same type—they are all thickness errors.

In lenses, the degradations to performance are of different forms and are caused by different types of parameter errors. For example, performance degradation introduced by center thickness and airspace errors will likely introduce defocus plus spherical aberration on axis. On the other hand, element wedges, tilts, and decentrations will likely introduce coma and/or astigmatism on axis and across the field of view. The question now becomes: can we use an RSS addition to predict the net expected performance due to axial spacing errors as well as asymmetrical errors such as decentrations? The answer is a qualified "yes," we can indeed RSS performance degradations caused by unlike errors (such as thickness errors and decentrations), but the more important question is whether the results are accurate and meaningful. The bottom line is that RSS addition simply does not handle properly the mix of aberrations that we often encounter in a lens system. For certain cases, RSS addition is quite appropriate and valid, and one such case is in computing the expected refocusing required following assembly. If we take the RSS of the refocusing required for each and every individual tolerance, the result should be valid since all of the refocusing is of the same form.

Fortunately, there is another approach known as *Monte Carlo tolerancing*, which is a method of simulating the performance statistics of a lens system in a high level of production. In a Monte Carlo simulation, tolerances are assigned to all toleranceable parameters along with a likely probability function such as normal, uniform, spiked, or a skewed gaussian, as shown in Fig. 16.2. While a normal distribution seems to be the best model, we often find that tolerances sometimes tend to end up toward one end of the allowable range. Element center thickness is a good example, as the optician often leaves extra thickness just in case a scratch occurs and the element needs to be reground. Thus, the probability distribution for CTs is a skewed gaussian distribution with the highest probability on the thick side of nominal as shown in Fig. 16.2.

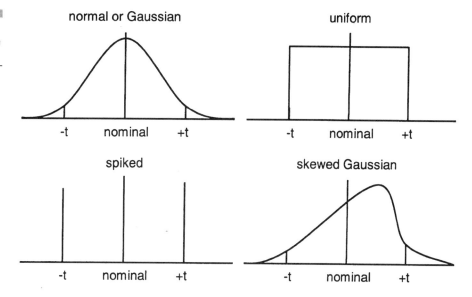

Figure 16.2
Tolerance Distribution
Models

The following is the procedure used in computing a Monte Carlo tolerance analysis:

- *Every* parameter is independently and randomly perturbed according to its assigned tolerance, based on a likely probability distribution. At this point, we have created a random lens assembly on the computer, based on our current set of tolerances.

- The performance is now computed after applying any compensators such as refocusing. The system model at this point is a simulation of a single manufactured system with its tolerances randomly distributed as discussed previously.

- The preceding process is repeated 25 or more times, and for each Monte Carlo sample, we have, in effect, a simulated system that has been manufactured. We can use the resulting output to compute the level of performance versus the cumulative probability. With these results, we can easily determine the level of performance for, say, 90 or 95% cumulative probability of occurrence.

The beauty of the Monte Carlo approach is that it is fully valid regardless of the relative nature, mix, or form of the aberrations because we are simulating a manufacturing environment. Each and every Monte Carlo sample is, in effect, a new and different manufactured system.

Forms of Tolerances

We have discussed the various tolerances, along with ways and means for adding the performance degradations. Next we will show the various forms of the tolerance parameters:

■ *Symmetrical errors* relating to fabrication and assembly include radius, power fit to testplate, thickness of elements, airspaces, refractive index, and Abbe number. Figure 16.3 shows how a testplate is used to characterize a lens element in manufacturing. The testplate is used as a standard of a known radius to high precision. The surface under test is placed, or "nested," into the testplate as shown, and every time the airspace or gap between it and the surface under test changes by λ/2 we see one full fringe. In the example shown, we see approximately two rings, or fringes, which means that approximately 1.0 wave of optical power or mismatch exists between the testplate and the surface under test. If the gap between the testplate and the surface under test is rotationally symmetrical, the interference pattern is indeed round, with circularly symmetric rings, as shown in the upper part of Fig. 16.3. If, on the other hand, we tilt the surface under test (this is the same as moving the ring center way off the part as shown), we see curved fringes as in the lower part of Fig. 16.3. The dashed vertical line is a reference to tell us how many fringes of power we have, and, as you can see, we have about two fringes of power.

■ *Asymmetrical errors* in assembly and alignment include element wedge, element decentration, element tilt, surface irregularity, and inhomogeneity of refractive index. In effect, an element with a wedge is really the same as an element with its optical centerline tilted with respect to its mechanical centerline. Further, the element has an *edge thickness difference* as we rotate the element around. Consider Fig. 16.4 where on the left we show a nominal perfect lens element. The two centers of curvature, when connected, represent a line straight through the mechanical centerline of the element. This line is, by definition, the optical axis of the element. On the right is an element with a severe tilt on its top surface. Note that this surface tilt or element wedge results in an edge thickness difference from left to right (the edge is thinner on the left, thicker on the right). The wedge, in radians, is the edge thickness difference, 2δ, divided by the element diameter, d. Thus, the wedge, in radians, equals $2\delta/d$.

Figure 16.3
Power Fit to Testplate

appearance
of fringe pattern

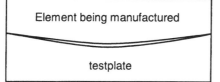

Element being manufactured

testplate

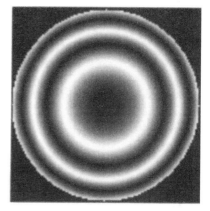

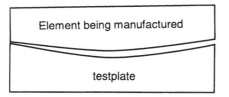

Element being manufactured

testplate

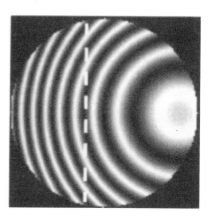

As we have shown, a lens element with spherical surfaces will, by definition, have a single optical axis, and it is the process of lens centering which will bring the mechanical edge of the element to be concentric with the optical axis of the lens. Figure 16.5 shows two mechanical methods. In Fig. 16.5a the lens is located on a precision spindle so that its lower surface runs true. Spinning the element will result in a large total indicator runout (TIR) reading on the dial indicator. Moving the element to the left in its shown rotational position will enable the top surface to run true, and the element is then edged using a diamond wheel. Figure 16.5b shows a method of centering called "cupping," in which the upper and lower rings (similar to the edge on a cup) will only fully contact the lens

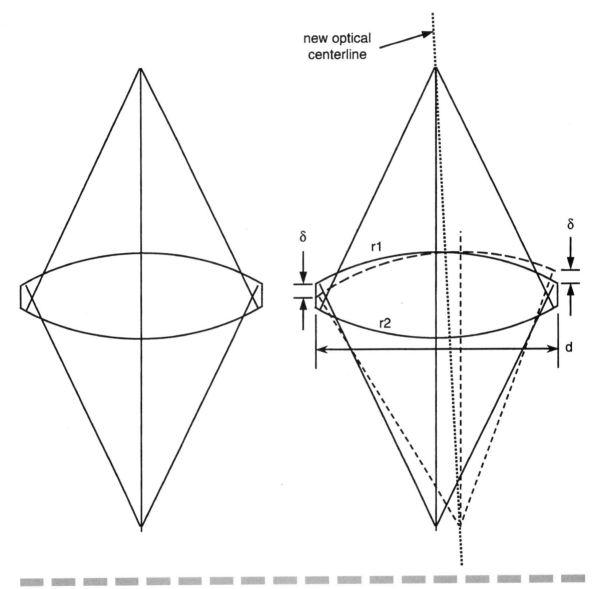

Figure 16.4
Nominal Element (Left) and Wedged Element (Right)

surfaces when the two surfaces are on a common and vertical axis. When this alignment condition is achieved, the element is edged to the proper diameter.

In many of today's optical shops techniques using HeNe lasers are often used to aid in centering lens elements. With one surface running true, the laser beam is either reflected off of the other surface or transmitted through the element, while the element is being rotated. The laser beam will nutate, or "wobbulate," until the element is properly centered. By using sensors such as CCDs or quad cells, extremely accurate indications of the element centration can be made.

Element decentration can be either a simple lateral decenter (up and down) or it can be a "roll" whereby the element maintains contact with a housing seat, as shown in Fig. 16.6. Note that while the net effect is quite similar, the two decentration models are actually quite different. In the roll situation the left-hand radius, which is in perfect contact with the housing seat, ends up perfectly aligned, with the surface tilt

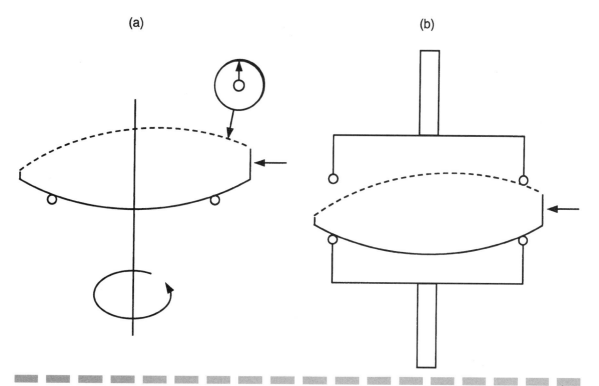

Figure 16.5
Centering on a Spindle (Left) and by Cupping (Right)

occurring only on the right-hand surface. Many of the computer software packages can model both pure decentrations as well as roll.

Finally, element tilt is self-explanatory, and it is expressed either as a total indicator runout (TIR) or sometimes as the tilt, in minutes of arc. Since 1 min of arc is 0.0003 rad, we simply multiply the tilt, in radians, by the diameter to derive the TIR.

Adjusting Parameters

Adjusting parameters, sometimes called *compensators,* are those parameters used to optimize the optical performance in the laboratory or on the assembly line at some time during the final lens assembly and testing.

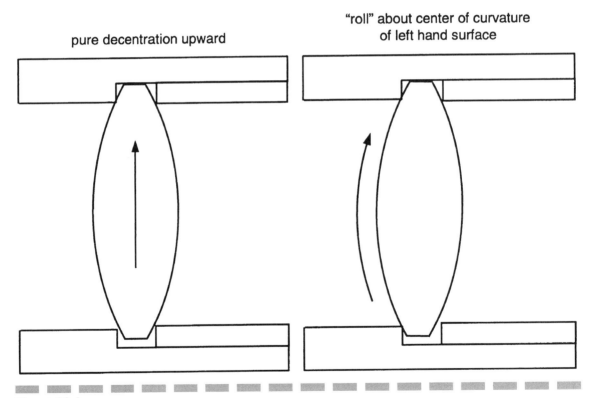

pure decentration upward

"roll" about center of curvature
of left hand surface

Figure 16.6
Element Decentration Models

The most common adjusting parameter is refocusing during the final assembly and testing procedure. Let's consider the manufacturing of a 35-mm camera lens or a similar high-quality lens for a CCD camera. The hypothetical example shown in Fig. 16.7 illustrates the situation. The lens will be mounted to the camera using a bayonet or similar accurate mounting methodology, and the distance from the rear flange to the image plane (the film or the CCD chip) is a tightly held mechanical dimension. Needless to say, it is imperative that when the completed lens is fastened to the flange on the camera body, the image of an infinite object is in perfect focus on the film or CCD when the lens is focused at infinity.

Referring to Fig. 16.7, we show a nominal lens imaging onto a sensor at the nominal flange back-focal distance. We also show in dashed lines the last radius of the lens with a steeper or more powerful radius than nominal by two fringes of power. The effect of this is to move the image inward, and the image will be out of focus on the sensor.

The optical path difference (OPD) = $(n − 1)$ t, where n = the refractive index and t = the separation between the nominal surface and the manufactured surface at the edge of the aperture as shown. If we have two fringes of power, this is due to one wave of surface error, or sag, at the edge of the element. And since OPD = $(n − 1)$ t, and t = 1.0λ, the OPD = 0.5λ peak to valley. If the flange back-focus distance is absolutely correct for the nominal camera and for the nominal lens assembly, the image will be out of focus by 0.5λ peak to valley, which is a factor of 2 from the Rayleigh criteria, due *only* to the last radius being two fringes of power from nominal!

The entire lens in this simplified example of course has six radii as well as the two airspaces, element thicknesses, and other tolerances, all of which will further contribute to defocus errors. If we RSS the effect of

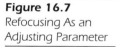

Figure 16.7
Refocusing As an
Adjusting Parameter

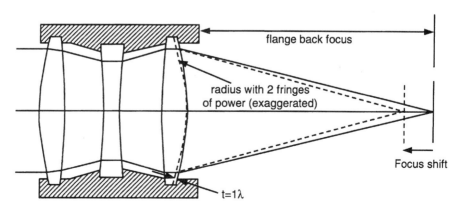

the six radii being two fringes of power each, we find that the final image could be out of focus by 1.22 waves of defocus, which is of major significance. What this means is that if we consider only the power fit to testplate, then with a tolerance of two fringes per surface, we predict a focus error of 1.22 waves of defocus with a 95% confidence level. This is about 5 times the Rayleigh criteria, and the imagery will be poor relative to the diffraction limit. We *must* refocus the lens, and this is accomplished by using an adjusting parameter, in this case the back-focus distance, during final assembly.

It is not uncommon to sometimes use an element decentration as an adjusting parameter. This is done when the system is quite sensitive to what is sometimes called *axial coma*, or coma which occurs on axis (and carries somewhat uniformly across the field of view) due to asymmetrical tolerances such as element wedge, tilt, and decentration. By allowing a strategically located element within the system to be adjusted in *X* and *Y* during the final assembly and testing, you can often cancel the coma introduced by all of the other tolerances within the system. Not only will this permit you to produce a system which otherwise may not work properly, but you may be able to relax some of your other tolerances from their otherwise tight levels, thus lowering the cost and enhancing the producibility. The trade-off often results in producing a system with extremely tight tolerances versus looser tolerances and one adjustment made during final assembly and testing.

In large field-of-view projection optics, it is sometimes imperative to have a lens or a group of lenses that are adjusted in centration during final assembly of the system in order to relax the tolerances on the other components, and have a reasonable cost to the system.

In microscope objective manufacturing a common test is the "star test" discussed in Chap. 15. In some high-precision shops, if the objective shows axial coma, it is sometimes tapped or "banged" lightly on a table and retested one or more times until the performance is met. While somewhat of a "brute force" method of compensator usage, it does work.

Typical Tolerances for Various Cost Models

There have been several papers presented over the years showing the effect on cost associated with various levels of tolerances. One of these, presented by John Plummer, is shown in Fig. 16.8. Note that there is a similar updated table in Chap. 17. You may find it interesting to compare the two sets of data. Plummer's company, at the time, was primarily involved in high production of reasonable quality optics which we might call "riflescope quality" in other words not ultratight tolerances and not "loosey goosey" either. For 10 manufacturing parameters, the level of tolerance is shown on the first line and the relative cost on the second line for various levels of tolerances. Let's look at several examples:

diameter tolerance	± 0.1	± 0.05	± 0.025	± 0.0125	± 0.0075		
in millimeters	100	100	103	115	150		
center thickness	± 0.2	± 0.1	± 0.05	± 0.025	± 0.0125		
tolerance in mm	100	105	115	150	300		
stain charac.	0	1	2	3	4	5	5+
of the glass	100	100	103	110	150	250	500
# of lenses	25	18	11	6	3	1	
per block 100	105	115	130	175	300		
eccen. Toler.	6 min.	3 min.	2 min.	1 min.	30 sec.	15 sec.	
In light dev.	100	103	108	115	140	200	
Figure tolerance	10-5	5-2	3-1	2 1/2	2 1/4	1 1/8	
in λ (pow/irreg.)	100	105	120	140	175	300	
Dia. to thick.	9-1	15-1	20-1	30-1	40-1	50-1	
ratio (fig. 3-1)	100	120	150	200	300	500	
Beauty defects	80-50	60-4	40-30	20-10	10-5		
(MIL-C-13830A)	100	110	125	175	350		
Raw glass cost	$3.00	$5.00	$8.00	$15.00	$25.00	$50.00	$100.00
in 1000 lb lots	100	108	115	125	135	200	350
Coating speci-	unctd.	Mg. Fl.	3-4 layer	>4 layers			
fications	100	115	150	200-500			

Figure 16.8
Relative Cost of Manufacturing As a Function of the Level of Tolerance

■ The standard diameter tolerance is ±0.05 mm. If we require ±0.025, the cost increase is only about 3%; however, if we need ±0.01, then the cost increase is about 25 to 30%.

■ Meeting tight tolerances for element thicknesses is difficult and costly. Achieving a tolerance of ±0.05 costs 15% above the standard level of ±0.2, and achieving ±0.025 will cost an additional 50%.

■ Even stain characteristics on glass will affect cost. The reason for this is that a glass type with a 5 stain will acquire a stain within minutes, and for this reason the elements must be sent almost immediately into the vacuum chamber for coating after they have been polished. This, of course, affects the workflow in the shop, and for this reason it becomes costly.

You should take the time to read through the table in Fig. 16.8 and become familiar with the various tolerances and their relative cost versus quality trade-offs. More in-depth material relating to optical manufacturing and tolerances is presented in Chap. 17.

Example of Tolerance Analysis

In order to best show how to tolerance a lens system, we will go through the tolerances for a 10× reduction lens Cooke triplet for a machine vision application. Our basic specifications are as follows:

Parameter	Specification
Object distance	200 mm
Object full diagonal	60 mm
Magnification	0.1×
Image full diagonal	6 mm
f/number	f/3.5 at used conjugate
Focal length	≈20 mm
Full field of view	≈17.06°
Spectral band	550 to 650 nm, uniform weights
Sensor	$1/_3$-in CCD

Parameter	Specification
Number of pixels	640 × 480
Pixel pitch at image	7.5 × 7.5 μm
Pixel pitch at object	75 μm
Nyquist frequency at image	66.67 line pair/mm
Nyquist frequency at object	6.667 line pair/mm
MTF spec at image	>0.3 at Nyquist

Figure 16.9 shows the layout and performance of the final design for this lens. Note that the design is quite close to diffraction limited as evidenced from the MTF. The geometrically based spot diagrams show square boxes measuring 7.5 × 7.5 μm, which is one pixel at the CCD sensor. Figure 16.10 is a listing of the specifications and prescription data for the lens in case you want to set it up and work with it.

We will now apply a standard set of manufacturing tolerances to the lens. We show in the following the resulting tolerance sensitivities. In

Figure 16.9
Cooke Triplet for Machine Vision Application

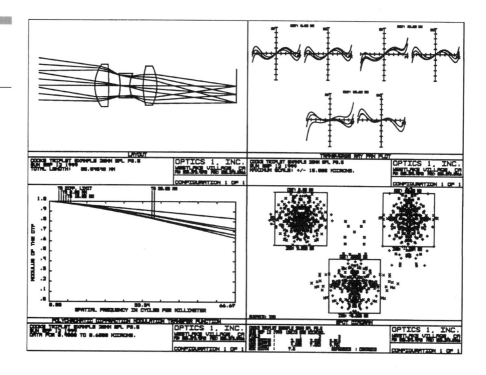

Figure 16.10
Prescription of Cooke
Triplet for Machine
Vision Application

```
System/Prescription Data

GENERAL LENS DATA:

Surfaces          :              8
Stop              :              4
System Aperture : Image Space F/# = 3.5
Eff. Focal Len.  :      19.00051 (in image space)
Back Focal Len.  :      12.4271
Total Track       :      35.48153
Image Space F/#  :       3.5
Stop Radius       :       1.777314
Parax. Ima. Hgt.:       3.018941
Parax. Mag.       :      -0.1006314
Entr. Pup. Dia.  :       5.428717
Entr. Pup. Pos.  :      15.4601
Exit Pupil Dia.  :       6.196183
Exit Pupil Pos.  :     -23.54149
Field Type        : Object height in Millimeters
Maximum Field    :             30
Primary Wave      :       0.55
Lens Units        :    Millimeters
```

SURFACE DATA SUMMARY:

Surf	Type	Radius	Thickness	Glass	Diameter
OBJ	STANDARD	Infinity	190		60
1	STANDARD	Infinity	10		0
2	STANDARD	6.572511	2.5	SK5	7.303856
3	STANDARD	-51.05087	2.041158		6.332172
STO	STANDARD	-8.89512	1.75	SF2	3.696617
5	STANDARD	5.048865	2.408424		4.172604
6	STANDARD	12.99962	2.5	SK5	6.31239
7	STANDARD	-8.044117	14.28195		6.824242
IMA	STANDARD	Infinity			6.020948

order to conserve space, data will be shown for the central element only (this is for most of the tolerances, the most sensitive). In the data, we see in the "Field" column "All," which is the average of all of the fields of view. Below that are each of the individual fields (field 1 is on axis, 2 and 3 are ±70% of the full diagonal, and 4 and 5 are ±full field). The "MTF" columns are the MTF after applying the tolerance and refocusing for compensation (by the "Change in Focus" value), and the "Change" columns are the change in MTF.

In the following data, the tolerance designations are:

- TFRN is the number of fringes of power fit to the testplate.
- TTHI is the airspace or element thickness, in millimeters.
- TEDY is the decentration of an element, in millimeters.

▪ TETY is the tolerance of element tilt, in degrees.

▪ TIRY is the tolerance on the total indicator runout or surface tilt for wedge or element centration.

▪ TIRR is the tolerance on surface irregularity in fringes.

Type	Sf1	Sf2	Field	Minimum Value	MTF	Change	Maximum Value	MTF	Change	
TFRN		3	All	-4.00	0.713	-0.027	4.00	0.710	-0.031	4 fringes power
			1		0.739	-0.023		0.724	-0.038	to testplate
			2		0.659	-0.060		0.726	0.006	
			3		0.659	-0.060		0.726	0.006	
			4		0.751	0.008		0.681	-0.061	
			5		0.751	0.008		0.681	-0.061	
Change in Focus				:		0.126			-0.124	
TFRN		4	All	-4.00	0.732	-0.008	4.00	0.727	-0.013	4 fringes power
			1		0.754	-0.008		0.759	-0.003	to testplate
			2		0.731	0.011		0.688	-0.031	
			3		0.731	0.011		0.688	-0.031	
			4		0.714	-0.028		0.740	-0.002	
			5		0.714	-0.028		0.740	-0.002	
Change in Focus				:		-0.096			0.097	
TTHI	3	4	All	-0.050	0.720	-0.020	0.05	0.720	-0.021	thickness
			1		0.764	0.001		0.761	-0.001	
			2		0.746	0.026		0.679	-0.040	
			3		0.746	0.026		0.679	-0.040	
			4		0.662	-0.080		0.726	-0.016	
			5		0.662	-0.080		0.726	-0.016	
Change in Focus				:		0.045			-0.045	
TTHI	4	6	All	-0.050	0.729	-0.011	0.050	0.732	-0.008	airspace
			1		0.763	0.000		0.762	-0.000	
			2		0.694	-0.025		0.736	0.015	
			3		0.694	-0.025		0.736	0.015	
			4		0.736	-0.006		0.702	-0.040	
			5		0.736	-0.006		0.702	-0.040	
Change in Focus				:		-0.042			0.042	
TEDX	3	4	All	-0.025	0.607	-0.134	0.025	0.607	-0.134	element
			1		0.633	-0.129		0.633	-0.129	decenter x
			2		0.587	-0.132		0.587	-0.132	
			3		0.587	-0.132		0.587	-0.132	
			4		0.602	-0.140		0.602	-0.140	
			5		0.602	-0.140		0.602	-0.140	
Change in Focus				:		0.000			0.000	
TEDY	3	4	All	-0.025	0.589	-0.151	0.025	0.589	-0.151	element
			1		0.633	-0.129		0.633	-0.129	decenter y
			2		0.550	-0.169		0.607	-0.112	
			3		0.607	-0.112		0.550	-0.169	
			4		0.563	-0.179		0.559	-0.183	
			5		0.559	-0.183		0.563	-0.179	
Change in Focus				:		0.000			0.000	
TETX	3	4	All	-0.150	0.584	-0.157	0.150	0.584	-0.157	element tilt x
			1		0.757	-0.005		0.757	-0.005	
			2		0.703	-0.016		0.477	-0.242	
			3		0.477	-0.242		0.703	-0.016	
			4		0.417	-0.325		0.532	-0.210	
			5		0.532	-0.210		0.417	-0.325	
Change in Focus				:		0.000			0.000	

```
TETY      3   4   All   -0.15    0.661   -0.080    0.150    0.661   -0.080   element tilt y
                  1               0.757   -0.005             0.757   -0.005
                  2               0.648   -0.071             0.648   -0.071
                  3               0.648   -0.071             0.648   -0.071
                  4               0.598   -0.144             0.598   -0.144
                  5               0.598   -0.144             0.598   -0.144
Change in Focus                :  0.000                              0.000
TIRX      3       All   -0.015   0.566   -0.174    0.015    0.566   -0.174   wedge TIR x
                  1               0.638   -0.124             0.638   -0.124
                  2               0.552   -0.167             0.552   -0.167
                  3               0.552   -0.167             0.552   -0.167
                  4               0.518   -0.224             0.518   -0.224
                  5               0.518   -0.224             0.518   -0.224
Change in Focus                :  0.000                              0.000
TIRY      3       All   -0.015   0.570   -0.170    0.015    0.570   -0.170   wedge TIR y
                  1               0.638   -0.124             0.638   -0.124
                  2               0.629   -0.090             0.508   -0.211
                  3               0.508   -0.211             0.629   -0.090
                  4               0.482   -0.260             0.555   -0.186
                  5               0.555   -0.186             0.482   -0.260
Change in Focus                :  0.000                              0.000
TIRX      4       All   -0.015   0.683   -0.058    0.015    0.683   -0.058   wedge TIR x
                  1               0.736   -0.026             0.736   -0.026
                  2               0.664   -0.055             0.664   -0.055
                  3               0.664   -0.055             0.664   -0.055
                  4               0.654   -0.088             0.654   -0.088
                  5               0.654   -0.088             0.654   -0.088
Change in Focus                :  0.000                              0.000
TIRY      4       All   -0.015   0.627   -0.114    0.015    0.627   -0.114   wedge TIR y
                  1               0.736   -0.02              0.736   -0.026
                  2               0.708   -0.011             0.539   -0.180
                  3               0.539   -0.180             0.708   -0.011
                  4               0.507   -0.235             0.604   -0.138
                  5               0.604   -0.138             0.507   -0.235
Change in Focus                :  0.000                              0.000
TIRR      3       All   -1.000   0.636   -0.104    1.000    0.652   -0.089   surface
                  1               0.669   -0.093             0.651   -0.111   irregularity
                  2               0.554   -0.165             0.721    0.001
                  3               0.554   -0.165             0.721    0.001
                  4               0.704   -0.038             0.595   -0.147
                  5               0.704   -0.038             0.595   -0.147
Change in Focus                :  0.000                              0.000
TIRR      4       All   -1.00    0.709   -0.032    1.000    0.691   -0.050   surface
                  1               0.719   -0.043             0.730   -0.032   irregularity
                  2               0.748    0.028             0.624   -0.095
                  3               0.748    0.028             0.624   -0.095
                  4               0.665   -0.077             0.730   -0.012
                  5               0.665   -0.077             0.730   -0.012
Change in Focus                :  0.000                              0.000
```

The preceding are the tolerance sensitivities for the central element. There are many numbers...what do they all mean?

At this point, you should look over all of the sensitivities to see if any are especially sensitive. For example, decentration of the central element by 0.025 mm drops the average MTF about 0.134 for an x decentration (in and out of the figure) and 0.151 for a y decentration. The effect over the field is reasonably uniform (this is not always the case). A less sensitive tolerance is the four fringes of power on surface 4 (the rear of the second element), where a maximum MTF drop of only 0.013 average (0.031 maximum drop) has resulted. We should be able to loosen this tolerance if our shop feels it is worthwhile.

In Fig. 16.11 we show in a graphical form those parameters which drop the average MTF over the field of view by 0.02 or more for our lens. There are many tolerances not even represented in the data. Note that there are only about five to six tolerances that are most sensitive, and these are primarily tilts, decentrations, and wedges. The most sensitive tolerances are for element 2, as discussed earlier.

The best measure of the overall lens performance and the most reliable means for predicting performance is via the Monte Carlo analysis. Here we computed 20 Monte Carlo samples and the resulting statistics are shown in Table 16.2. Recall that each Monte Carlo sample is, in effect, a simulated fabricated system. Each individual parameter is changed   according   to a normal probability distribution between its minimum and maximum values.

TABLE 16.2

Monte Carlo
Results

		Field				
	Average	**1**	**2**	**3**	**4**	**5**
Nominal	0.741	0.762	0.720	0.720	0.742	0.742
Best	0.602	0.760	0.664	0.649	0.616	0.641
Worst	0.326	0.360	0.297	0.327	0.196	0.165
Mean	0.491	0.585	0.484	0.498	0.444	0.410
Standard deviation	0.068	0.108	0.101	0.086	0.105	0.107

Compensator Statistics	
Change in back focus:	
Minimum	−0.232282
Maximum	0.222737
Mean	0.031034
Standard deviation	0.138210

Ninety percent of Monte Carlo lenses have an MTF above 0.306.
Fifty percent of Monte Carlo lenses have an MTF above 0.393.

Figure 16.11
All Cooke Triplet Tolerances which Drop the MTF by >0.02

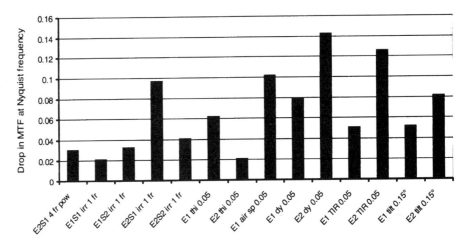

Ten percent of Monte Carlo lenses have an MTF above 0.556.

The Monte Carlo results show that 90% of our lenses should have an MTF at 66.7 line pairs/mm of 0.306 or better. This just meets our MTF goal of 0.3 or better! It still may be beneficial to review each tolerance to see if it can be loosened, and the most sensitive tolerances should be tightened if possible. Once these changes have been made, another Monte Carlo analysis is in order as this is our best way of modeling the predicted lens performance.

It appears that 20 Monte Carlo samples is a small number to represent accurately the statistics of assembling lenses in production. So, we ran 500 Monte Carlo samples with the same tolerances. Here 90% of the lenses are predicted to have an MTF at Nyquist greater than or equal to 0.243. While this result is, indeed, slightly less than 0.306 for 20 samples, it does tell us that 20 samples give a reasonably accurate answer. Indeed, we will need to tighten some of the sensitive tolerances after all.

The compensator is the back focus, and a total range of ±0.23 mm was encountered with a standard deviation of 0.14 mm.

Surface Irregularities

In Chap. 4 we discussed the concept of optical path difference and its influence on image quality. One of the most important and influential rules of thumb is the Rayleigh criteria which tells us that if the peak-to-valley optical path difference is less than or equal to one-quarter wave, then the image quality will be nearly indistinguishable from perfect diffraction-limited performance.

Optical path difference (OPD) can be introduced by the following factors:

- The fundamental *aberrations present in the basic lens design* such as spherical aberration, coma, astigmatism, defocus, and other image-degrading aberrations.

- *Assembly and alignment errors.* Included in this category are the various tolerance forms such as element thicknesses, airspaces, wedges, tilts, decentrations, and other tolerance types.

- *Environmental effects* such as thermal soaks and/or gradients.

- *Effects external to the system* such as atmospheric turbulence.

- *Surface irregularities and other wavefront errors not included in the previous errors.* Included here are the residual manufacturing errors, which cause a surface to deviate or depart from its nominal shape. Most often, this is a deviation from sphericity or flatness, but it can, of course, also be a deviation from a prescribed aspheric profile if the nominal surface is aspheric.

The basic lens design performance residual, which is due to the various forms of aberrations, will consist of the orders of aberration present in the design such as defocus, spherical aberration, coma, astigmatism, field curvature, and the chromatic aberrations of axial and lateral color which are changes in the basic aberrations with wavelength. Thus, the design of a double Gauss camera lens may have a mix of third, fifth, and higher orders of aberration, both on axis with spherical aberration, as well as off axis with other aberrations.

Recall that the OPD polynomial is one exponent higher than the transverse ray aberrations. In other words, third-order spherical aberration affects the wavefront proportional to the fourth power of the aperture, third-order astigmatism is linear with aperture and the effect to the wavefront is quadratic with aperture, and so on. The net result is that the basic lens design will have a mix of all of the residual aberrations present in the prescription. Fortunately, lens design software programs are quite robust and can model these aberrations and accurately predict the MTF and other measures of performance. The optical path difference introduced by our design prescription is accurate and quantifiable.

The optical path difference introduced by assembly and alignment errors include symmetrical errors such as power fit to testplates, element thicknesses, and airspaces, as well as asymmetrical errors such as element wedge, tilt, and decentration. In most cases, the effects of these tolerances will be the introduction of primarily low-order aberration and, to a reasonable extent, this will be true throughout the system. For example, the resulting effect of the designated number of fringes of the power fit to testplate will, to a very large degree, contribute to the on-axis image defocus and, to a lesser extent, third-order spherical aberration. The axial image will also show the effects of the asymmetrical tolerances, including wedge, tilt, and decentration. These perturbations will contribute primarily low-order coma and astigmatism to the axial image. Similarly, off axis we will see the introduction of mostly lower-order aberrations due to the manufacturing and alignment errors.

The results of environmental effects, such as thermal soaks and gradients, will also be primarily low order.

This brings us to *surface irregularities*, which we will divide into several categories:

- *Conventional optical manufacturing.* Here we are talking about spherically surfaced lenses in approximately the 10- to 30-mm-diameter range manufactured in a production environment using conventional machinery. For longer radii, elements are blocked for cost economies. The residual surface irregularities for these elements are primarily cylindrical in form. Thus, the surface is literally a toroid which is best thought of as a cylindrical departure from sphericity. Clearly, to the axial imagery, this will introduce astigmatism, and the aberration is low order.

- *Larger elements.* As lens elements become larger, the surface irregularities tend to depart some from the classical cylindrical shape. We can have asymmetries, which can in many cases become more highly asymmetric and nondescript and less well correlated.

- *Larger surfaces such as telescope mirrors.* Whenever larger surfaces are involved, especially if there are aspheric surfaces, we often encounter much higher-order surface irregularities. These effects can, for example, be due to the manufacturing process where subdiameter polishing tools are often used. In the process of reaching the desired surface profile (often a paraboloid or similar conic section, sometimes with an intentional higher-order residual), high-frequency irregularities are often left in the surface.

■ *Thin lenses, windows, mirrors, and plastic optics.* Very often, we require very thin lenses, windows, and mirrors, and also in this category we have injection-molded plastic lenses (as well as compression-molded glass lenses). We may, for example, use flat glass manufactured by a process called *float glass.* In many of these cases the residual surface irregularity is less straightforward to predict and can often have nonrotational symmetrical residuals from the nominal surface shape.

The primary difference between small, high-production lenses, where a cylindrical departure from sphericity results, and some of the latter examples, such as injection-molded plastic lenses and thin-float glass mirrors or beamsplitters, is that in these latter scenarios we may have both larger departures from the ideal surface profile as well as a lower "correlation" surface due to more "bumpiness" on the surfaces and therefore to the wavefront.

How Does Correlation Relate to Performance?

Consider Fig. 16.12 where we show three sinusoidal wavefront profile models: Fig. 16.12a is plane or flat, Fig. 16.12b has a given peak-to-valley surface irregularity with 1.5 bumps across the surface, and Fig. 16.12c has the same P-V irregularity with five bumps across the surface. The term "correlation," in very simple terms, is the inverse number of bumps across the surface. Thus, the surface in Fig. 16.12b has a correlation of 0.666 and the surface in Fig. 16.12c is less correlated with a correlation of 0.2.

If the preceding represents the deviation of the wavefront at the exit pupil from its ideal spherical shape, then multiplying the total angular ray deviation by the focal length of our optical system will give the maximum image blur diameter or extent at the image. The larger the slope errors, the larger the image blur. Unfortunately, these effects are often random and change from part to part or system to system, so some form of modeling and approximation is in order.

Figure 16.12
Sinusoidal Wavefront
Shapes

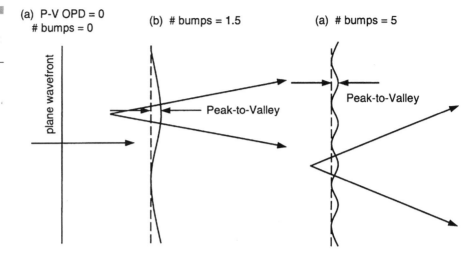

Effect to Spot Diameter

Assume that we have a wavefront that departs from perfect flatness or sphericity sinusoidally, as shown in Fig. 16.12. Assume the following:

A = the peak-to-valley variation of the wavefront
n = number of bumps over the wavefront extent
D = the total extent of the wavefront (exit pupil diameter)

The maximum slope of the wavefront can be shown to be

$$\text{Maximum slope} = \frac{\pi A \lambda n}{2D}$$

Thus, the total angular spread of the rays proceeding to the image is twice the preceding result. If we assume that the rms wavefront error is one-fifth of the peak to valley (a reasonable assumption), the angular spot diameter containing 100% of the light is

$$100\% \text{ spot diameter (rad)} = \frac{5\pi\sigma\lambda n}{D}$$

where σ is the rms wavefront error. If we now assume that the energy is uniformly distributed in the image, we can rewrite the previous relationship in its final and most useful form:

$$\text{Spot diameter} = \sqrt{p} \left(\frac{5\pi\sigma\lambda n}{D} \right)$$

where p is the fraction of the total energy. This gives us the diameter of the spot inside of which there is a given percent energy.

Let's do a quick sanity check. Assume we have a wavefront with three bumps, which just meets the Rayleigh criteria exiting an $f/10$ lens with a 100-mm exit pupil diameter. The rms wavefront error is thus 0.05 wave, and the wavelength is 0.5 μm. Our focal length is 1000 mm. Thus, the diameter of the spot containing 80% of the energy is

$$\text{80\% spot diameter} = \sqrt{0.8} \left\{ \frac{5 \times 3.141 \times 0.05 \times 0.5 \times 3}{100{,}000} \right\} \times 1000$$

$$= 0.0105 \text{ mm}$$

The diffraction-limited Airy disk diameter for the previous $f/10$ lens is approximately 0.012 mm, which is very close to the derived value of 0.0105 for a situation which should be essentially diffraction limited. As the number of bumps increases or the other parameters change, we can compute the approximate predicted blur diameter. While this derivation is not rigorous, and it is based solely on geometry, it is extremely useful when you really do not know the exact form of the wavefront error but you do have an idea of the correlation and the wavefront error.

Figure 16.13 shows graphically the 80% energy blur diameter, in radians, as a function of rms wavefront error and the number of bumps on the wavefront. These data are based on the derivation shown earlier. Results smaller than the Airy disk diffraction diameter are fictitious and in these situations the prediction should revert back to the Airy disk diameter.

Effect to MTF: The Optical Quality Factor

A number of years ago, Hufnagle of Perkin-Elmer developed an empirical relationship whereby the MTF degradation can be derived as a function of the rms wavefront error and the correlation of the wavefront.

Figure 16.13
Predicted Blur
Diameter As a
Function of rms
Wavefront Error and
Number of Bumps

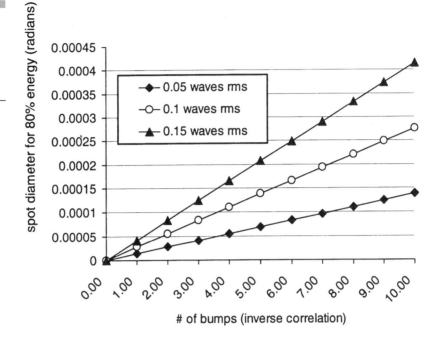

This relationship yields what has become known as the optical quality factor (OQF), and is given by

$$\mathrm{OQF} = e^{-(2\pi\mathrm{rms})^2}\left(1 - e^{-2n^2 s^2}\right)$$

where n is the number of bumps over the exit pupil (the inverse correlation) and s is the normalized spatial frequency relative to cutoff. Figure 16.14 shows the OQF as a function of normalized spatial frequency and rms wavefront error for a correlation of 0.333, or three bumps across the pupil. If we multiply these data by the perfect system MTF, we get the results in Fig. 16.15.

There is an empirically derived approximation by Shannon which can also be used, and this is shown in Fig. 16.16. The equation for this data is

$$\mathrm{MTF}\,(v) = \frac{2}{\pi}\left[arc\,\cos(v) - v\,\sqrt{1 - v^2}\,\right]$$

$$\mathrm{ATF}\,(v) = \left\{ 1 - \left[\left(\frac{W_{\mathrm{rms}}}{0.18}\right)^2\right] [1 - 4(v - 0.5)^2] \right\}$$

Tolerancing and Producibility

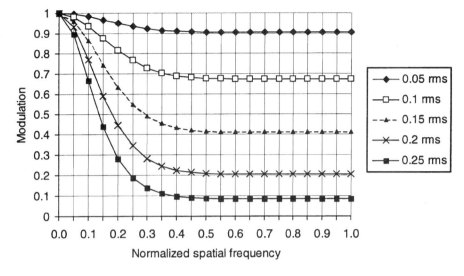

Figure 16.14
OQF As a Function of rms Wavefront Error and Number of Bumps

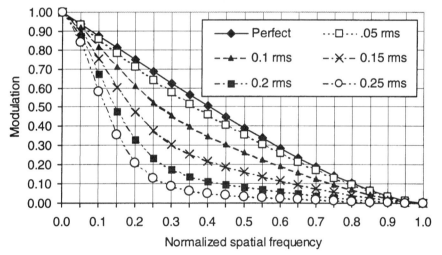

Figure 16.15
Predicted MTF As Function of rms Wavefront Error and Number of Bumps

$$v = \frac{\text{spatial frequency}}{\text{cutoff frequency}} = \frac{N}{[1/(\lambda)\, f/\#]}$$

In this relationship, the diffraction-limited MTF is given by the first equation as a function of the normalized spatial frequency, v. The equivalent to the OQF described earlier is given by the second equation as a function of the normalized spatial frequency v, and the rms wavefront error.

Figure 16.16
MTF Versus rms
Wavefront Error

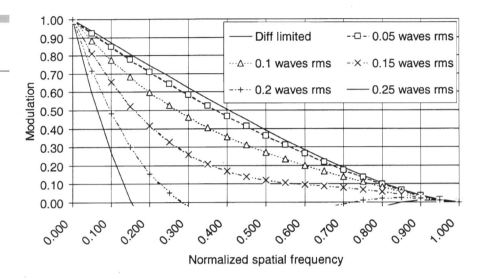

Another form of MTF degradation, or OQF, is shown in Fig. 16.17*a*, where we show the effect of a mix of third-, fifth-, and seventh-order aberrations. The wavefront degradations are in the form of rms wavefront error. In Fig. 16.17*b* and *c* we show the OQF for 0.1 wave rms of nonrotationally symmetric aberrations and rotationally symmetric aberrations, respectively. It is interesting to note that for 0.1 wave rms the MTF drops to approximately 0.6 of its nominal value, regardless of the form of aberration.

The data in this section can be very useful for predicting the MTF drop due to wavefront errors which you may have derived from a performance error budget, but you may not know the specific form of the error so you cannot model it directly on your computer program. For example, if you were to have approximately one-half wave P-V of random irregularity due to a mirror in your system, this equates to approximately 0.1 wave rms, and the OQF at midfrequency would be in the order of 0.6. These data allow us to quickly and easily assess and predict the effect of wavefront errors on the optical system.

Beam Diameter and Surface Irregularity

There is one additional important point, which needs to be discussed—how the beam diameter relates to the surface and wavefront irregularity.

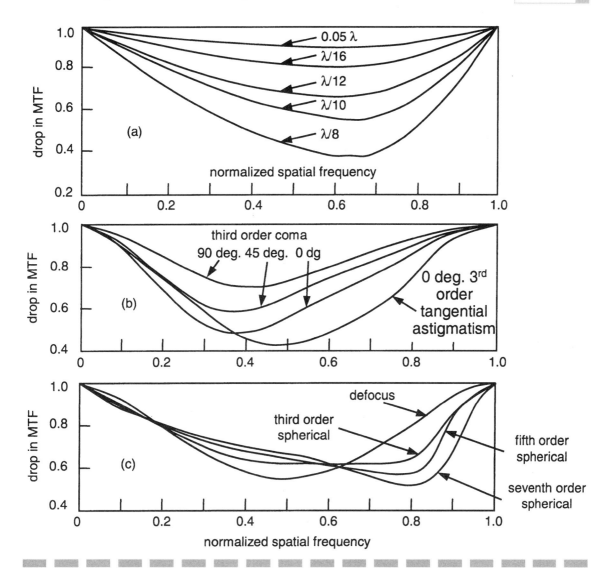

Figure 16.17
MTF Drop As a Function of Wavefront Error

The critical factor is how much wavefront error is introduced to the wavefront as it proceeds through our system, surface by surface. A parameter that becomes extremely important is the beam diameter or the "footprint" of light at each surface. Consider the lens in Fig. 16.18. Note how almost the entire aperture of the cemented doublet is used regardless of the field

Figure 16.18
Beam Diameter

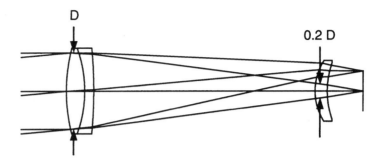

position. The element closest to the image is nearly the same diameter as the doublet; however, the beam diameter going to any given image point uses only about 20% of its diameter.

The impact of this is clear from Fig. 16.19a, where we show a simulated interferogram with five waves of astigmatism. Let us assume that an equivalent wavefront error is introduced by the final lens element over its full diameter. Since the beam diameter reaching any point in the field of view is only 20% of the lens diameter, what we have is the effect of the same interferogram but only within the white circle. If we have astigmatism, as shown, the net wavefront error to the system will be reduced from 5 waves to 0.2 wave since the wavefront polynomial for astigmatism is quadratic with aperture, and $0.2^2 = 0.04$, which means that only one-twenty-fifth of the astigmatism is introduced to the wavefront.

Figure 16.19b shows a random wavefront where we can also see that, over a reduced beam footprint, the net wavefront error can be significantly reduced. In this case, the full diameter has a residual of approximately 0.8λ peak to valley, and over 20% of its aperture (the white circle) we find approximately 0.02 wave, a reduction of 40 times.

The Final Results

Once we predict the net system degradation, the performance can be shown in various formats. In order to show how these results all compare, consider Figs. 16.20 and 16.21, where we show the performance for a perfect $f/5$ system with no diffraction or aberrations (a), as weel as a diffraction-limited system (b), 0.25λ $P - V$ (c), 0.5λ $P - V$ (d), and 0.75λ (e). The aberrations are all third-order shperical elements. The following performance metrics are shown in Figs. 16.20 abd 16.21:

Figure 16.19
Reduction of OPD
for Smaller Beam
Diameters

(a) Astigmatism

(b) Random wavefront

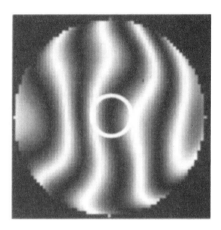

White circle is 20% of full diameter

■ *A graphical representation of a three-bar Air Force–type target whose bar pattern is selected to be at 50% of the cutoff spatial frequency at 250 line pairs/mm.* Note that for these data, as well as the other data presented here, the 0.25-wave situation is only slightly degraded from the diffraction-limited metric, as predicted by the Rayleigh criteria. It is interesting to note that if there were no diffraction effects, and we had an aberration-free system, then the three-bar pattern would image as shown in Fig. 16.20a. This corresponds to a modulation of unity and, for all practical purposes, represents the target for the data in Figs. 16.20 b through d.

■ *A plot of the MTF.* The reduction in contrast is quite evident, reaching zero for the 0.75-wave case. Note how the bar pattern imagery correlates with the MTF data.

■ *Geometrical spot diagrams, with the Airy disk size shown for reference.* Note in the case of the spherical aberration that we have here somewhat of an intense center to the pattern rather than a uniform intensity pattern which would be evident from pure defocus. As the pattern grows from the Airy disk diffraction pattern diameter, then the MTF drops and the contrast degrades, as already shown.

■ *A plot of encircled energy.* Here, too, we can see the increasing blur diameters as the spherical aberration increases. As before, the 0.25-wave case is close to diffraction limited, as predicted by Rayleigh.

Figure 16.20
Image Quality for
Different Amounts of
Peak-to-Valley Third-
Order Spherical
Aberration

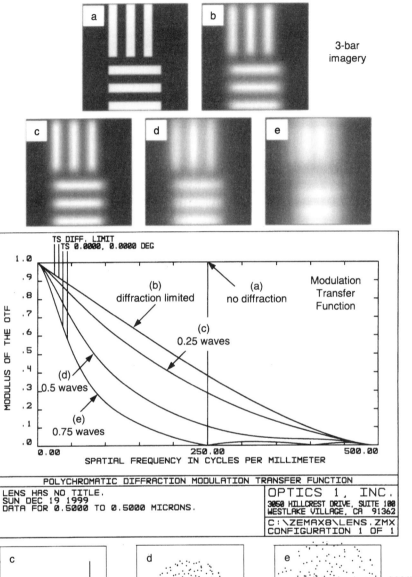

3-bar
imagery

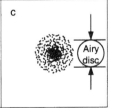

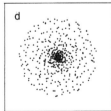

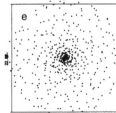

geometrical
spot
diagrams

Figure 16.21
Image Quality for
Different Amounts of
Peak-to-Valley Third-
Order Spherical
Aberration

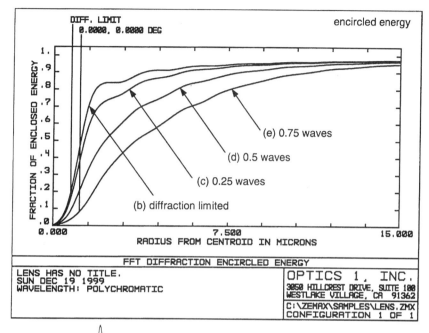

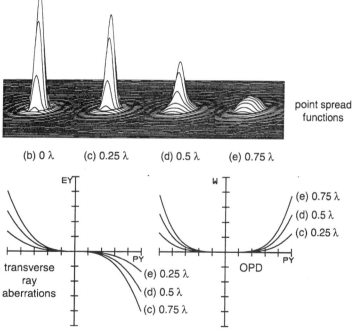

▪ *Plots of the point spread function (PSF).* This is one of the more instructive examples of the Rayleigh criteria. The PSF for the perfect system and the 0.25-wave system are very nearly identical in overall appearance. While we do see a drop of intensity at the center of the pattern, the overall appearance of the PSF is nearly the same as for the perfect system. The first ring is still quite evident. However, as soon as we reach 0.5 wave and above, the whole character and appearance of the pattern degrades, and more energy is thrown out from the central maximum of the pattern.

▪ *Plots of the transverse ray aberrations.* Here we show the transverse ray aberrations for the different amounts of wavefront error. These data are cubic with aperture.

▪ *Plots of the optical path difference.* Finally, we show the optical path difference for the different amounts of wavefront error. These data are proportional to the fourth power of the aperture.

All of the data here are for *precisely* the same system with the same third-order spherical aberration. These data, along with the OQF and other relationships discussed earlier in this chapter, can be used to help predict the performance of imaging optical systems for various amounts of wavefront error.

Optical
Manufacturing
Considerations

From the point of view of a lens manufacturer, what design attributes have the most influence on manufacturing efficiency? The primary design considerations are optical material, component size, shape, and manufacturing tolerances. All of these attributes are variable at the design phase and can have significant impact on lens manufacturing costs.

In order to narrow the scope of this chapter, the text assumes the manufacture of a precision glass lens of approximately 50-mm diameter using grinding and polishing techniques. The information is presented in the following order:

1. *Material.* A summary of manufacturing considerations for optical glasses

2. *Manufacturing.* An overview of conventional and advanced process technologies

3. *Special fabrication considerations.* A review of tolerancing trade-offs and finishing options

4. *Relative manufacturing cost.* An analysis of manufacturing variables

5. *Sourcing considerations.* Suggestions for achieving project goals

6. *Conclusion.* A summary table for quick reference

While this analysis is based on a 50-mm-diameter glass lens, it can also be adapted to include specific market niches such as microoptics (diameters smaller than 5 mm), macrooptics (diameters larger than 300 mm), prisms and flats, molded glass and plastic optics, diamond-turned crystal and metal optics, and diffractive optical elements. These niches are addressed in additional chapters of this book.

Material

There are more than 100 different optical glasses available worldwide, and each has a unique set of optical, chemical, and thermal characteristics. Only a few glass manufacturers in the world produce these optical glasses, and each manufacturer has a company-specific glass-naming convention. Cross-referencing the glasses is possible via a six-digit glass code (ABCXYZ) that is derived from the index of refraction ($n_d = 1.ABC$) and the Abbe value ($v_d = XY.Z$). For the vast majority of optical applications, glasses from differing manufacturers can be direct substitutes. Lens designers should be aware, however, that equivalent glasses having the same six-digit glass code might not have exactly the same optical, chemical, and mechanical properties. For example, Schott's SK-16 (620603) has slightly different characteristics than Ohara's S-BSM-16 (620603). Be aware that optical design software will define glasses that can achieve a desired optical performance, but it cannot determine the glasses' current availability in the market. Nor will the software give consideration for the glasses' chemical and thermal properties. For example, it may be important to consider that the index of refraction of a glass changes with temperature at a known rate. Other parameters that are important to consider are spectral transmission, dispersion, material quality, and mechanical, chemical, and thermal properties.

Design Considerations

Material quality is defined by tolerances of optical properties, striae grades, homogeneity, and birefringence. Optical properties include spectral transmission, index of refraction, and dispersion. Data for each glass type is available from its manufacturer. If tighter than standard optical properties are required, then additional cost and time are usually associated with obtaining the material. Specification of glass based on material quality is provided in the International Standard ISO 10110 and the U.S. military specification MIL-G-174B. A brief summary of glass material specifications using nomenclature from Schott optical glass is shown in Fig. 17.1.

Before finalizing an optical design, some consideration should be given to glass cost and availability. Glass prices vary from a few dollars per pound to several hundred dollars per pound. In some cases, it may be more economical to add a lens to the design in order to avoid expensive glasses. In addition, many glasses are not regularly stocked; instead they are melted to order, which can take several months. Pricing and melt frequencies are available from glass manufacturers. Each manufacturer has a list of "preferred" glasses that are most frequently melted and usually available from stock. It's important to note that preferred does not imply "best glass type available." From a manufacturing perspective, preferred refers only to the availability of the glass in stock. For example, BK-7 is readily available from stock and is among the most economical of glass types. On the other hand, a glass like SF-59 is not made as frequently and may not be as readily available. If delivery is a concern, the

Figure 17.1
Glass Material
Specifications

Striae Grade AA (P) is classified as "precision striae" and has no visible striae. Grade A only has striae that are light and scattered when viewed in the direction of maximum visibility. Grade B has only striae that are light when viewed in direction of maximum visibility and parallel to the face of the plate.

Birefringence is the amount of residual stress in the glass and depends on annealing conditions, type of glass, and dimensions. The birefringence is stated as nanometers per centimeter difference in optical path measured at a distance from the edge equaling 5% of the diameter or width of the blank. Normal quality is defined as (except for diameters larger than 600 mm and thicker than 100 mm):

1. Standard is less than or equal to 10 nm/cm.
2. Special annealing (NSK) or precision annealing is less than or equal to 6 nm/cm.
3. Special annealing (NSSK) or precision quality after special annealing (PSSK) is less than or equal to 4 nm/cm.

Homogeneity is the degree to which refractive index varies within a piece of glass. The smaller the variation, the better the homogeneity. Each block of glass is tested for homogeneity grade.

Normal grade	$\pm 1 \times 10^{-4}$
H1 grade	$\pm 2 \times 10^{-5}$
H2 grade	$\pm 5 \times 10^{-6}$
H3 grade	$\pm 2 \times 10^{-6}$
H4 grade	$\pm 1 \times 10^{-6}$

Tolerances of optical properties consist of deviations of refractive index for a melt from values stated in the catalog. Normal tolerance is ±0.001 for most glass types. Glasses with n_d greater than 1.83 may vary by as much as ±0.002 from catalog values. Tolerances for n_d are ±0.0002 for grade 1, ±0.0003 for grade 2, and ±0.0005 for grade 3.

The dispersion of a melt may vary from catalog values by ±0.8%. Tolerances for v_d are ±0.2% for grade 1, ±0.3% for grade 2, and ±0.5% for grade 3.

designer may want to use only glasses from the frequently melted glass list.

Fabrication Considerations

Since the mechanical, chemical, and thermal properties of glass are what determine the ease or difficulty of making optics from the material, these properties are of particular interest to the optical fabricator.

MECHANICAL PROPERTIES *Mechanical properties* include hardness and abrasion resistance. These properties determine the rate at which material is removed, and should be among the first to consider.

Hardness is measured in accordance with International Standards Organization (ISO) 9385. It is measured with a microhardness tester that utilizes a precision diamond point applied with a specific amount of force. This probe contacts and penetrates the polished glass sample at room temperature. Carefully measuring the resultant indentation yields a calculation known as the "Knoop hardness" of the material. Knoop hardness ranges from 300 to 700 for most optical glasses, where 300 represents a soft glass and 700 harder glasses. In general, the harder the glass, the longer the time required to grind and polish the lens.

Abrasion resistance describes how fast the glass will process. Abrasion resistance is the ratio of material removed on a test piece of glass to the material removed from a BK-7 sample. The abrasion resistance of BK-7 is set to equal 100. The higher the number, the faster the material will be removed. The values range from about 60 to 400. Compared to BK-7, a glass with a value of 60 will take almost twice as long to process. Conversely, glass with a value of 400 will take only one-quarter of the time. The process time seems to imply that softer glasses are cheaper to fabricate. One must remember, however, that other factors, such as cosmetic finish, may offset potential savings. Soft glasses are more difficult to polish to achieve very good cosmetics and low rms surface roughness. As a general rule of thumb, for lenses with identical specifications, except for material, a BK-7 lens will be cheaper to produce. The cost of a lens increases as the abrasion resistance value moves away from that of BK-7. For example, glasses that have high abrasion resistance can require significantly longer grinding and polishing times. On the other hand, glasses with a low abrasion resistance are more difficult to achieve tight thickness tolerance, especially when good cosmetics are required.

RELATIVE COST AND DENSITY *Relative cost and density* are also important factors to consider. The density of glass is described in grams per cubic centimeter. Multiplying this number by the blank volume (including cutting allowances) and cost yields the approximate cost of a blank. It is important to remember dollars per pound of glass is not the only factor that determines the cost of the optic. For example, SF6 and SFL6 are virtually identical optically. SFL6 costs 63% more per pound, but its density is only 65% of the density of SF6, offsetting the higher per pound cost. In addition, SFL6 is much easier to process, which ultimately results in lower manufacturing costs.

CHEMICAL PROPERTIES *Chemical properties* are also of interest to the optician. There are several tests that characterize the chemical behavior of glass with regard to humidity, acid, alkali, and phosphate stainability. The values reported from these tests reflect the degree of processing difficulty and special handling a glass will require. Designers should, therefore, refer to chemical property test values when making lens design decisions. The chemical properties tests for glass are explained in more detail in Fig. 17.2.

To summarize the chemical properties listed in Fig. 17.2, if a glass is low in all categories, then it is stable and unlikely to stain during standard manufacturing processes and storage. If a glass is high in one or more categories, it is very likely to cause problems *if* special care is not taken. As a general rule, any glass with a stain coefficient of 3 or more must be handled with special care. Glasses with stain designations in the 50s (for example, SK-55 or S-FPL53) tend to be very troublesome. The poor chemical properties of these glasses can lead to residual stain from deblocking, cleaning, and/or handling of the lens. If stained, the lens may require repolishing to remove the stain. This causes more risk to the part, either from handling or missing the mechanical tolerances. For example, if tight thickness control is required and the glass is prone to

Figure 17.2
Chemical Property Tests for Glass

Climate resistance (CR) is a test that evaluates the material's resistance to water vapor. Glasses are rated and segregated into classes, CR 1 to CR 4. The higher the class, the more likely the material will be affected by high relative humidity. In general, all optically polished surfaces should be properly protected before storing. Class 4 glasses should be processed and handled with extra care.

Resistance to acid (SR) is a test that measures the time taken to dissolve a 0.1-μm layer in an aggressive acidic solution. Classes range from SR 1 to SR 53. Glasses of classes SR 51 to SR 53 are especially susceptible to staining during processing and require special consideration.

Resistance to alkali (AR) is similar to resistance to acid because it also measures the time taken to dissolve a 0.1-μm layer, in this case, in an aggressive alkaline solution. Classes range from SR 1 to SR 4, with SR 4 being most susceptible to stain from exposure to alkalis. This is of particular interest to the optician because most grinding and polishing solutions become increasingly alkaline due to the chemical reaction between the water and the abraded glass particle. For this reason most optical shops monitor the pH of their slurries and adjust them to neutral as needed.

Resistance to staining (FR) is a test that measures the stain resistance to slightly acidic water. The classes range from FR 0 to FR 5, with the higher classes being less resistant. The resultant stain from this type of exposure is a bluish-brown discoloration of the polished surface. FR 5 class lenses need to be processed with particular care since the stain will form in less than 12 min of exposure. Hence, any perspiration or acid condensation must be removed from the polished surface immediately to avoid staining. The surface should be protected from the environment during processing and storage.

staining, it is more difficult to achieve a stain-free surface within the desired thickness tolerance.

THERMAL PROPERTIES *Thermal properties* of glass may also affect optimal process methods. Thermal expansion coefficients range from 4 to 16×10^{-6}/K. Glasses with a coefficient over 10 must be handled very carefully during any operation involving rapid thermal change. In fact, even body heat that is transferred by touching the glass may cause subsurface microfractures. Glasses with high thermal coefficients of expansion are more susceptible to surface distortion and catastrophic fractures during blocking and handling. If the coefficient is over 10, then the process should not include any rapid thermal processes. Due to the difficulty in handling these glasses, they should be avoided whenever possible.

Optical glasses can be segregated into groups by their material properties. It may be helpful to contact the preferred glass manufacturer for a particular material to get summary data. As an example, Table 17.1 is a quick reference chart for selecting the more favorable glasses.

Glass material is available in various forms of supply. It can be in block, rod, or slab form, requiring sawing or core drilling operations to make it into disks or it can be purchased as a disk. In any case, the desired form is defined by a diameter and a thickness, or, in other words, a cylinder that totally contains the final lens geometry with some oversize allowance for processing. This approach is the quickest but not the most cost effective. Buying the glass as a molded blank results in the lowest cost. Material efficiency is achieved by taking a piece of glass of the appropriate weight, heating it, and pressing it into a metal mold to make the shape (slightly larger) of the final lens. This approach requires several weeks for the glass to be delivered; however, it minimizes glass cost for higher-volume projects.

TABLE 17.1

Optical Glasses Categorized by Material Properties

Stable Glasses	Climate Stainable	Alkaline Stainable	Acid Stainable	Heat Sensitive	Soft Material
BK7	PSK50	LaK11	FK3	FK52	FK3
BaK2	SK16	LaK21	PSK52	PSK53A	PSK54
SFL6	LaK21	All KzFS	SK16	SF59	SF6
SF11	KzFS1	SK16	SSKN5	TiF6	TiF6

Manufacturing

For more than 100 years, the manufacture of lenses has remained essentially unchanged. While these conventional methods utilize relatively low-cost machinery, they are also very labor intensive and require highly skilled craftsmen. With recent innovations in computer numerically controlled (CNC) machines, faster and less labor-intensive manufacturing methods are now viable options over conventional methods. From prototyping to high-volume production, automated grinding and polishing technologies are now available for lens fabrication.

Although these new technologies are more efficient and provide more reliable production, they require a significant initial capital investment. In addition, there are some situations where conventional methods are simpler to use and more cost effective. To understand the practical applications and benefits of each type of lens fabrication, brief descriptions of the manufacturing methods follow.

Conventional Lens Fabrication

Conventional lens fabrication (Fig. 17.3) begins with a plano-plano disk of glass or a near form–molded lens blank. The blank is placed into a chuck that rotates around the mechanical center of the glass disk. A ring tool with embedded diamonds removes bulk material and grinds down the top surface of the blank. This process gives the lens blank a spherical shape and a coarse surface finish. This surface has significant subsurface microfractures, which must be removed by loose abrasive lapping at a later stage in the manufacturing process. The lens blank can then be flipped and its second side ground to near net shape using the same process. This overall process is called *generating* because the end result is the generation of a blank in the shape of the final lens.

Fine Grinding and Polishing

To prepare for the fine-grinding and polishing process, the perimeter of the lens blank is wrapped with tape to create a reservoir. Molten pitch is poured onto the surface of the lens, filling the reservoir. The pitch is allowed to cool at room temperature until a solidified pitch layer, called a *pitch button,* is developed.

Figure 17.3
Conventional
Generation

The next step is to arrange the lenses in a circular pattern to be processed as a group called a *multiple block*. Multiple blocks are assembled by laying the buttoned lenses into a tool, which has a radius approximately equal to the design radius. Now, with the generated spherical surface down and the pitch button side facing up, the array of lenses is ready to receive the blocking tool. This heated metal tool is placed in contact with the pitch buttons, allowed to melt into the pitch, and then quickly cooled to room temperature. The resultant "block of lenses" is then ready for loose abrasive lapping.

The purpose of loose abrasive lapping, often referred to as grinding, is to remove the residual subsurface damage that was incurred during the generating process. The block of lenses (Fig. 17.4) is fine ground, with loose abrasive grains mixed with water. Grinding is a step-down process that begins with large grains and continues with sequentially smaller and smaller grains. Grain sizes typically range from 30 to 5 μm. At this point in the manufacturing process, the operator is trying to achieve two goals: (1) a spherical surface very close to the design radius and (2) no subsurface damage. It is important for the optician to be aware of the abrasion resistance of the glass in order to control center thickness while minimizing subsurface damage. To achieve a thickness within the center

Figure 17.4
A Block of Lenses

thickness tolerance, a certain amount of material (on the order of tens of micrometers) is left on the lens for removal during polishing.

The lens is polished to the specified radius of curvature, spherical irregularity, and cosmetic finish by using a soft-pitch lap pressed to the desired radius and rotated about the spherical lens surface while a cerium oxide–polishing slurry is applied. The radius of the lens is controlled with a test glass or test plate of known radius. The lens is compared to the test plate by direct contact and/or evaluating the fringes of the Fizeau interferometric test. This test also gives the optician the ability to measure spherical irregularity, which is the maximum allowable perturbation of the spherical wavefront. The cosmetic requirements for the lens dictate the maximum allowable surface imperfections such as scratches, digs, and chips.

Conventional Centering

Once both sides of the lens are polished, the lens is centered by precision grinding the edge of the lens on a special lathe (Fig. 17.5). This process accomplishes two tasks. First, the lens is ground to its final diameter. Second, the optical and mechanical axes of the lens are made coaxial with one another. This is also the point at which any flats or special mounting bevels are ground onto the lens.

Once the lens is centered, manufacturing is complete. The lens is cleaned and inspected for quality. If it is satisfactory, the lens will be delivered either uncoated or with an antireflection coating. If the lens is not satisfactory, it is returned to one or more of the steps in the process to be corrected. If the lens cannot be reworked to meet the required specifications, it is scrapped.

CNC Lens Fabrication

Recent advancements spearheaded by the Center for Optics Manufacturing (COM) at the University of Rochester in Rochester, New York, have led to the development of equipment and processes that enable the optician to perform a variety of operations on computer-controlled machines—processes called *CNC lens fabrication.*

Figure 17.5
Conventional
Centering

The equipment combines the accuracy of multiaxis CNC motion control with robust machine designs that are faster, more versatile, and more precise than conventional machines. This automated process minimizes part-handling and transfer errors, which are prone to happen with the more manual conventional process. It also enables the optician to generate precision surfaces that are precentered to final diameter and ready to polish. The precision spindles yield little subsurface damage, which reduces polishing time and shortens overall production time. An added benefit of using this equipment is that the lens can be shaped to precise complex dimensions during the generation sequence.

Once the lens has been precision generated, it is polished to meet all the requirements for surface accuracy and cosmetics. Polishing may be done utilizing the conventional process described earlier in the "Conventional Manufacturing" section of this chapter or with a new CNC polishing machine (Fig. 17.6). Determining which polishing method is most appropriate depends on geometry of the lens as well as the quantity being produced. For example, if the lens has a relatively long radius of curvature, then conventional polishing of a multiple block may be most cost effective.

It is important to note that this is a two-machine process with very good process control. For most applications the lens fabrication is

Figure 17.6
Deterministic Grinding with CNC Machine

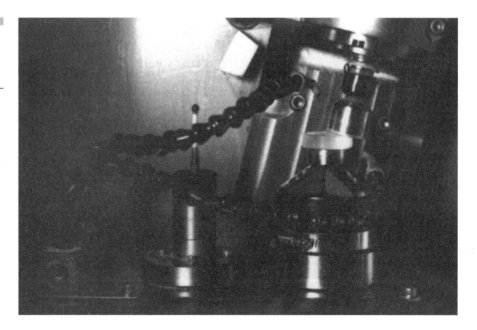

complete. However, for high-precision applications COM has developed a complementary machining technology that can significantly improve the surface figure of the lens.

This new technology uses a unique fluid that is magnetically manipulated to deterministically remove material from the lens. The process is called magnetorheological finishing (MRF) (Fig. 17.7). The magnetorheological (MR) fluid stiffens as it passes through a magnetic field, thus forming a temporary finishing surface or polishing pad. The MR fluid carries polishing slurry that is presented to the lens surface in a precisely controlled pattern by varying the magnetic field's strength and direction. Since fresh abrasive is continuously delivered to the polishing zone, heat and debris are constantly removed. This process reduces cycle times and is capable of producing fractional wave-surface irregularity.

The development of CNC machine technologies led directly to the capability to fabricate precision aspheric lenses in brittle materials, for example, optical glass. Robust CNC machines are able to profile grind complex rotationally symmetric shapes defined by polynomial equations (Fig. 17.8). This development effort continues today. Commercially viable processing methods are being developed for conformal optics. Conformal optics is loosely defined as nonrotationally symmetric, such as a saddle or a toroid. In fact, processing methods for conformal topics

Figure 17.7
Magnetorheological
Finishing (MRF)

have progressed so far that testing the finished optic is often more challenging than making it.

For more information regarding these new technologies please visit one or more of the following web sites:

www.Optipro.com

www.QED.com

www.opticam.Rochester.edu

www.photonicsonline.com

Special Fabrication Considerations

Centering Tolerance

Centering tolerance is a complex optomechanical parameter that is frequently misinterpreted. For example, 1-arc min edge thickness difference (ETD) may be reasonable for a 50-mm-diameter lens, but a 6-mm-diameter lens with this tolerance requires centering to 0.003 mm ETD, which is extremely difficult. Figure 17.9 shows the relationship between the optical and mechanical axes and the decentration and angle of deviation in a decentered lens. Table 17.2 demonstrates the relationships between different wedge specifications. These equations provide conversions from one tolerance designation to another. The equations work well for most lenses, but lose accuracy with meniscus lenses as they approach concentricity.

If the tolerance analysis indicates that surfaces must be controlled to a few micrometers, then precision potting of the finished components should be considered. *Precision potting* refers to active alignment of the optic axis to the mechanical axis within the mounting cell. Since it is difficult to center lenses to ETDs of less than 10 μm, assembly techniques have been developed to provide submicrometer alignment. For most optical systems, it is not beneficial to put unusually tight constraints on the lens because the housing in which it will be mounted typically will have more error than the lens.

Clear Aperture

Clear aperture is a specification dimension. It should provide enough aperture for light rays to pass through; however, many problems can result from the clear aperture being specified too close to the outside diameter of the lens. For example, achieving fractional wavelength surface quality will be difficult due to edge roll-off in polishing. In addition, during coating, the lens is held mechanically in a fixture above the coating source. Therefore, it is important to have sufficient clearance

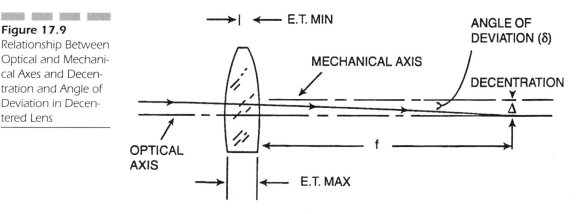

Figure 17.9

Relationship Between Optical and Mechanical Axes and Decentration and Angle of Deviation in Decentered Lens

TABLE 17.2 Centering Tolerance Specifications

	Deviation (Dv)	Edge Runout (ERO)	Surface Runout (ETD)
Deviation (Dv)		$Dv = 1720 \times ERO/f$	$Dv = 3440 \times (n - 1) \times ETD/D$
Edge Runout (ERO)	$ERO = D \times Dv/3440 \times (n - 1)$		$ERO = 2 \times f(n - 1) \times ETD/Dv$
Surface Runout (ETD)	$ETD = D \times Dv/3440 \times (n - 1)$	$ETD = D \times ERO/2 \times f(n - 1)$	

Note: Dv = deviation (in minutes); ERO = edge runout; ETD = edge thickness difference; D = diameter; f = focal length; n = material index of refraction (same value used to calculate focal length)

between physical diameter of the optic and clear aperture. Ideally, a clear aperture-to-diameter difference should be at least 2.0 mm, or 5% of the aperture, whichever measurement is greater.

There is an alternative when the clearance is not adequate for the coating tooling: the lens may be coated before edging. This option is not desirable, however, because the coating will be at risk during the edging process.

Thickness Tolerance

Thickness tolerance is more difficult to achieve on softer glasses that are less resistant to abrasion. When tight thickness tolerance is required along with very stringent cosmetics and fractional wavelength irregularity, the optician must allow the right amount of excess material to accommodate for grinding and polishing of the lens. This causes a wider range of center thickness, which may produce lower production yields. As a result, it may be necessary to start more pieces to account for the expected losses.

Sag Tolerance

Sag tolerance is sometimes specified as the desired clear aperture. This is a difficult feature to measure. A better method is to compute the sag as a function of the clear aperture and the radius. This yields an *axial height*—the on-axis distance from the plane of the sag face to the spherical surface—which can be easily and accurately measured. If the sag face is used as a mounting surface, then the tolerances for the sag and center thickness are cumulative. If the sag is not a mounting surface, then it should be identified as a reference (Ref) surface in order to reduce cost.

Radius Tolerance

Radius tolerance is used to specify the allowable radius measurement deviation from nominal for the test plate (that is, spherical reference tool) that will be used for lens production. For precision optics this measurement is typically 0.1% of the nominal radius and not less than 10 μm for short radii. The optical designer should be aware there are no industry standards for radius measurement and that absolute radius measurement is not possible. If five different optics manufacturers measure a test plate, then there will be five different readings. For radii less than 1000 mm, the variation will be on the order of a few micrometers. For radii over 1000 mm the variation could be several millimeters. Researchers at The National Institute for Standards and Technology are working toward a solution to this problem.

Power Tolerance

Power tolerance is a measure of the deviation from the chosen test plate. This ensures consistency among a group of lenses. In other words, each lens will match the test plate within the power tolerance. From the designer's perspective, the radius tolerance and the power tolerance are cumulative. The original purpose of the power tolerance was to indicate the maximum number of power fringes for which the irregularity fringes could be counted. For example, in order to see two fringes of irregularity, the maximum number of power fringes is 10 fringes; for one fringe irregularity the maximum is five fringes. However, automated interferometric metrology is reducing the need to rely on traditional test plates.

Surface Irregularity

Surface irregularity is a measure of the deviation from a perfect sphere. It is not only a function of the operator's skill and expertise but also a function of the process geometry. As a general rule, multiple blocks with more pieces will have less irregularity than three spots or singles.

The irregularity of lenses processed on multiple blocks will have a tendency to be cylindrical in nature while lenses processed as singles will have a symmetric aspheric profile shape, usually like a sombrero. There are certainly exceptions to this rule, but the general shape of the irregularity will follow these tendencies. Irregularity is defined very well by ISO 10110. See Fig. 17.10 for more information.

Aspheric Lens

Aspheric lens manufacturing technology has progressed rapidly over the past few years. The pace of this progress is limited somewhat by the difficulty in measuring aspheric profiles that include up to sixteenth-order terms. Using bonded diamond-tool generation for brittle materials (for example, glass), convex aspheres are usually easier to fabricate than concave surfaces. The computer-controlled machines can process complex shapes irrespective of best-fit sphere. In contrast, single-point diamond-turning machines can produce convex and concave surfaces on plastic and crystalline materials. However, small departures from best-fit sphere are preferred. The manufacturing cost for aspheres is typically 2 to 5 times that of spherical lenses with short radii. The generally accepted method for metrology is surface profiling to an accuracy of ± 0.1 μm. Aspheric form error on the order of 50 μm may be good enough for a condenser lens, while a precision quality focusing lens would require ± 1 μm tolerance. Greater precision is possible with interferometric testing, which often requires the fabrication of a special null lens.

Figure 17.10
Excerpt from ISO 10110-5 Optics and Optical Instruments. Preparation of Drawings for Optical Elements and Systems—Part 5: Surface Form Tolerances

3.5.2 Irregularity
The *irregularity* of a nominally spherical surface is a measure of its departure from sphericity.
The value of the irregularity of an optical surface is equal to the peak-to-valley difference between the optical surface under test and the approximating spherical surface.

3.5.3 Rotationally Symmetric Irregularity
Surfaces which are rotationally symmetric, but do not have the desired shape, are said to have *rotationally symmetric irregularity*. This error is the rotationally symmetric part of the irregularity function (see subclause 3.5.2).
In order to determine the value of the rotationally symmetric irregularity, d is first necessary to determine the rotationally symmetric aspheric surface which best approximates the surface under test.

Bevels, Chamfers, and Break Edges

Bevels, chamfers, and *break edges* are machining features utilized at the corners of a lens to help prevent edge chipping. Bevels should be specified whenever the included angle of two surfaces on an optic is less than 155°.

Cosmetic Tolerances

Cosmetic tolerances are well defined in MIL-O-13830 and ISO 10110. Most cosmetic inspection of lenses is still done visually by comparing the lenses to scratch-dig reference pieces. Alternatively, defects can be evaluated and categorized using a measuring microscope.

Antireflection Coatings

Antireflection coatings are a significant cost driver and can be reduced with minimal design effort. The most economical solution is to coat all surfaces with a single-layer MgF_2 coating. This enables the lenses to be coated all in one run (depending on size and quantity). Single-layer MgF_2 coating will yield about 1.5% reflection for each low-index surface and less than 1.0% reflection for each high-index surface. For multielement systems, specifying different coatings within the system can minimize coating costs. For example, the high-index glasses may be coated with MgF_2, while a broad band antireflection (BBAR) coating is applied to the low-index material. As a result, coating cost may be reduced because only the low-index glasses are receiving multilayer BBAR coatings, which are more expensive than MgF_2 coating. When using this approach, the designer should consider that BBAR coatings are index dependent. The coater will batch lenses by index—less than 1.60, 1.60 to 1.70, and greater than 1.70. Using glasses within two of these ranges instead of all three will reduce coating costs.

Blocking Quantities

Blocking quantities are a function of the relationship between the radius and the diameter of a lens. The graph in Fig. 7.11 reveals the relationship between radius, diameter, and blocking quantity. For example, lenses with

Figure 17.11
Blocking Graph

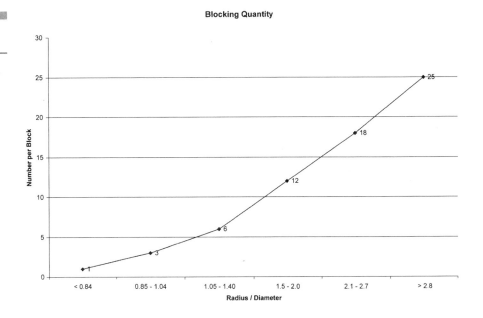

a radius-to-diameter ratio of less than 0.84 will process as a single, a ratio of 0.84 to 1.04 will run three pieces to a block, and so on. The more surfaces per block, the lower the cost per lens to process it. The diameter of the parts and the capabilities of the manufacturer put additional parameters around the number of pieces that can be produced at one time.

Concentric Lenses

Concentric lenses (Fig 17.12) create a problem with centering accuracy. Since the centers of curvature for both surfaces are close to one another, the optician is not able to remove much residual wedge in the centering process. When the concentric lenses have weak curves and the lenses are processed as a multiple block, special care must be taken during blocking and grinding to prevent wedge in the part. In general, it is best to process concentric lenses individually on CNC equipment, where the tolerances can be well controlled.

Hemispheres and Hyperhemispheres

Hemispheres and hyperhemispheres are difficult to process because the polishing tool must rotate beyond the waist of the lens. This requires specialized machines and tooling. Small convex hemispheres are often made by modifying spheres to the desired shape. The use of concave hemispheres should be avoided whenever possible due to manufacturing difficulties associated with these shapes. It is important to note that designing and applying an antireflective coating for all angles of incidence presents another set of challenges such as special coating textures to apply uniform antireflection coatings.

Aspect Ratio

Aspect ratio is the relationship of center thickness to diameter. The higher the ratio, the higher the probability that the glass will distort during processing. The distortion is a function of thermal stress caused by the application of heated pitch and its subsequent cooling. After polishing, the lens is deblocked from the pitch and the stress is relieved. Lenses with extremely thin centers or thin edges are prone to develop surface irregularities during processing. Ideal aspect ratios are less than 6:1 for precision optics with

Figure 17.12
Concentric Lens

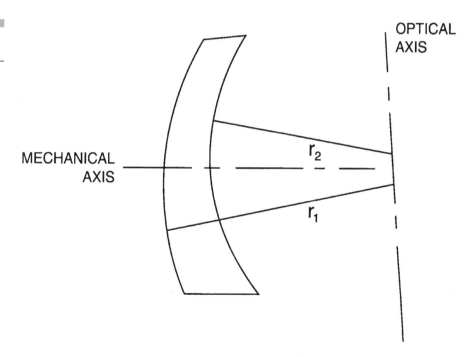

THE TWO AXES CANNOT BE MADE COAXIAL

one half-fringe irregularity. Aspect ratios greater than 10:1 will be more problematic and therefore more costly. There is also a greater likelihood for surface deformation from mounting and the assembly process.

Thin Edges

Thin edges can occur when there is a strong convex surface on at least one side of the lens. When the edge is thin (<1 mm), it is more fragile and prone to chipping, and the optician is able to protect only the edge with a minimal bevel. Thin edges cause flaking out of glass particles during polishing, which leads to difficulty in achieving good cosmetic surfaces. A thin-edge lens is also difficult to hold in place for testing on an interferometer because the slightest amount of pressure causes the lens to distort.

Segmenting

Segmenting refers to special mechanical shaping. It is difficult to polish lenses of noncircular geometries. Therefore, if segmenting is required, the manufacturer will usually perform this step last. Unfortunately, all of the value (material and labor) has already been invested in the lens, which inherently makes this a high-risk process.

Edge Blackening

Edge blackening of the lens helps reduce scattered light and often improves contrast and signal-to-noise ratio. Permanent black ink that is water and alcohol resistant is easy to apply and does not cause mechanical buildup on the surface. Lacquers and epoxies are more opaque; however, they are more difficult to apply and add tens of micrometers to the diameter of the lens. Epoxy is the most durable option, and if factored in during the design of the lens, it will not negatively impact the finished diameter of the lens.

Component Testing

An important consideration before manufacturing begins is *component testing*, which verifies that all parameters of the lens can be measured to the desired accuracy. Inspection data should be provided with all prototype components. In the event that an optical system does not perform as predicted by its design, the system can be computer modeled using the actual test data. In production, it may be helpful to perform inspection in compliance with the military specification MIL-PRF-13830B for a prespecified acceptable quality level (AQL).

Cemented Doublets

Cemented doublets can enhance optical system performance without decreasing light throughput. There are many methods for making doublets and the optimal choice will require some design consideration, including thickness tolerance, surface irregularity, and assembly.

Doublets yield some thickness flexibility for the designer. Most doublets are made from a flint and crown lens and have optical adhesive indices of about 1.5, similar to the crown glass. When tight thickness control is needed on a doublet, rather than give half the tolerance to each half of the doublet, the designer may be able to give the whole tolerance to each half and then have the optician match the thickness of each half before cementing them to make the doublet fall within the tolerance band. This can be a cost-effective solution to controlling doublet thickness.

All optical adhesives have some amount of shrinkage due to curing. This shrinkage can cause deformation of the lens elements and compromise the irregularity of the polished surfaces. Avoiding thin lenses in doublets and selecting a low-shrinkage adhesive help minimize this effect.

The assembly method for a doublet will depend on the wedge tolerance (see Table 17.3). The simplest approach is to center each half of the doublet to the same diameter and use the edges of the lens for alignment. This method is best suited for lenses greater than 15 mm in diameter. Another method is to center one half, the base lens, to a precision diameter and center the other half, the floater, to a smaller and less precise diameter. Then, referencing on the base lens, the optic axis of the floater can be aligned. In special cases the doublet can be built and centered as the final process. This is a very high-risk process and should be avoided when possible.

TABLE 17.3

Centered Doublet
Guidelines

Alignment Method	Mechanical Consideration	Precision (arc min)
V-block aligns the diameters of the two lenses to be cocylindrical.	Precision center the lenses to the same diameter and desired wedge tolerance.	6
Bell clamping aligns the polished surfaces to be coaxial by mechanical positioning.	Precision center the base lens and center the floater to a smaller diameter.	3
Active alignment aligns the polished surfaces to be coaxial by visual interactive positioning.	Precision center the base lens and center the floater to a smaller diameter.	<1

Relative Manufacturing Cost

In addition to the design considerations already described in this chapter, there are several other variables that can significantly impact the relative manufacturing cost of lenses. For example, the tolerances given to manufacturing specifications can lead to additional costs being incurred during manufacture of the lens. Other variables that may influence cost are the aspect ratio and the preferred delivery time.

In the mid-1970s, J. Plummer and W. Lagger wrote an article for *Photonics Spectra*® entitled "Cost Effective Design." The article contrasted the effect of manufacturing tolerance on the cost to make a lens. The chart from that article, represented in Table 17.4, has been updated to detail the cost impact of several variables for a manufacturing process that utilizes the newer deterministic microgrinding technology.

These relative costs are not cumulative, but are clearly interrelated as previous comments in this chapter's discussion have indicated. The total cost impact of several factors would be a complex mathematical function, and would vary from shop to shop, depending on the capabilities and strengths of each shop.

Sourcing Considerations

Every project has specific goals, such as to bring a new product to market before the competition, to develop a new capability, or to reduce

TABLE 17.4

1999 Relative Manufacturing Costs Using Deterministic CNC Processing

Variable	→ More Difficult → →				
Diameter (mm)	±0.10	±0.05	±0.025	±0.0125	>0.0075
	$100	$100	$102	$105	$125
Thickness (mm)	±0.20	±0.10	±0.05	±0.025	±>0.0125
	$100	$103	$115	$140	$200
Stain	<2	2	3	4	5
	$100	$103	$110	$140	$175
Cosmetics (Scr-Dig)	80-50	60-40	40-20	20-10	10-5
	$100	$100	$120	$150	$250
Test (fringes)	5–2	3–1	$2-\frac{1}{2}$	$1-\frac{1}{4}$	$\frac{1}{2}-\frac{1}{8}$
	$100	$105	$125	$175	$250
Wedge (arc min)	3	2	1	$\frac{1}{2}$	$\frac{1}{4}$
	$100	$105	$110	$125	$150
Doublets (arc min)	6	3	2	1	$<\frac{1}{2}$
	$100	$105	$110	$150	$200
Aspect ratio	<10:1	15:1	20:1	30:1	50:1
	$100	$120	$175	$250	$350
Delivery time (weeks)	8	6	4	2	1
	$100	$110	$130	$170	$200

manufacturing cost. In order to be successful, the project manager must determine the priorities among price, quality, and timeliness. The manager must then communicate those priorities to everyone involved with completing the project. The following guidelines are offered for consideration in achieving cost, quality, and delivery goals:

▪ To minimize cost and delivery time, buy from a catalog whenever possible. At the same time, keep in mind that custom lenses are often required in order to achieve a desired optical performance.

▪ To minimize risk on a project (that is, maximize the potential for good quality and on-time delivery), use domestic manufacturers for prototyping and preproduction. Optics manufacturers in the

United States have superior manufacturing capabilities for rapid prototyping, high-precision optics, computer-generated holographic (CGH) and diffractive optical elements (DOEs), laser optics, precision glass aspheres, polarizers, complex optical coatings, and much more.

■ Rapid prototyping can significantly minimize cost and delivery time. Some projects are very time sensitive and optical components become the pacing item. Typical delivery time for rapid prototyping is 8 to 10 weeks. Seek a manufacturer with a proven track record. Several manufacturers have developed the ability to expedite the manufacturing process to achieve shipment of coated optics within a few days. This service may require a premium on the standard price.

■ To reduce cost with relatively low risk, seek a domestic importer with an established offshore facility that has the ability to test and certify product quality. Or for the lowest price, consider working directly with an offshore supplier. However, this is quite risky if

TABLE 17.5

Typical Manufacturing Tolerances

OPTIMAX	Commercial Quality	Precision Quality	Manufacturing Limits
Glass quality (n_d)	±0.001	±0.0005	Melt controlled
Diameter (mm)	+0.00/−0.10	+0.000/−0.025	+0.000/−0.010
Center thickness (mm)	±0.150	±0.050	±0.010
Sag (mm)	±0.050	±0.025	±0.010
Radius (%)	±0.2	±0.1	±0.025
Power - irregularity (fringe)	5–2	3–0.5	1–0.1
Aspheric profile (μm)	±25	±1	±0.5
Wedge lens (TIR, mm)	0.050	0.010	0.005
Prism angles (TIA, arc min)	±3	±0.5	±0.1
Bevels (maximum face width at 45°, mm)	1.0	0.5	No bevel
SCR - DIG	80 - 50	60 - 40	10 - 5
AR coating (average R)	MgF_2 R<1.5%	BBAR, R<0.5%	Custom design

	Radius	Radius tolerance	Power/ Irreg.	Clear aper. diameter.		Edge diameter	Diameter tolerance	CT	Thickness tolerance	Wedge
S1	13.589 CC	TPF	5/3	20.6		31.18	±0.05	4.45	±0.05	0.05 TIR
S2	64.643 CX	TPF	5/3	29.0						

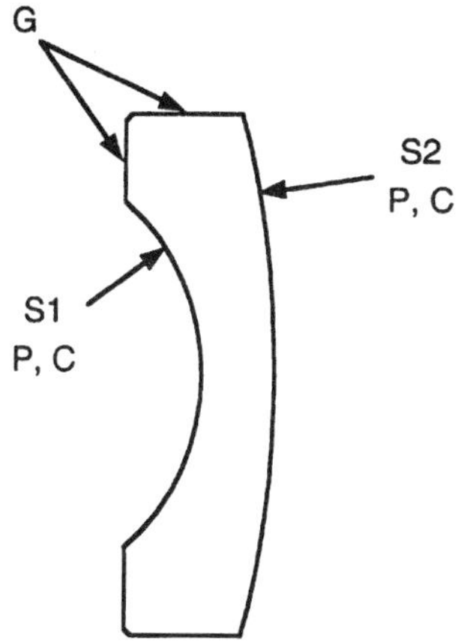

Notes:
1. All dimensions in inches.
2. Material: Optical glass per MIL-G-174 type BK7
   Schott N.o. 517642
   nd = 1.5168 ± 0.0005   V = 64.2 ± 0.8% striae grade
   A, fine anneal
3. Surfaces marked "P" polish to power/irregularity indicated
4. Manufacturer per mil-0-13830
5. Surface quality 60 - 40
6. Surfaces marked "C" coat with high efficiency coating with average reflectance per surface ≤ 0.5% from 420 - 680 nm
7. Surface marked "G" fine grind and blacken with no buildup
8. Bevel edges at 45 degrees to 0.5 max face width
9. Diameter to flat is 19.952 (REF) with surface sag of 5.977 ±0.05 on surface S1

DR	
CHK	
APPD	
SCALE - 2.5:1	

Sample Element Print

Figure 17.13a Conventional Lens Manufacturing Print

you don't have the appropriate metrology to verify the product quality.

Conclusion

This chapter presents a great deal of information to help the designer select attributes and tolerances based on manufacturing considerations. Perhaps the most useful summary is a reference chart that provides a list of reasonable or typical manufacturing tolerances for commercial quality and precision quality lenses (see Table 17.5). This chart is intended as a guideline and assumes a 50-mm-diameter BK-7 lens. The manufacturing limits are not absolute, but represent a pain and/or cost threshold. Job-specific tolerances may vary depending on component size, shape, glass material, and preferred delivery time.

Once a lens has been designed and toleranced, manufacturing drawings are utilized to convey the lens requirements to the optician. Examples of a conventional manufacturing print and a drawing that complies

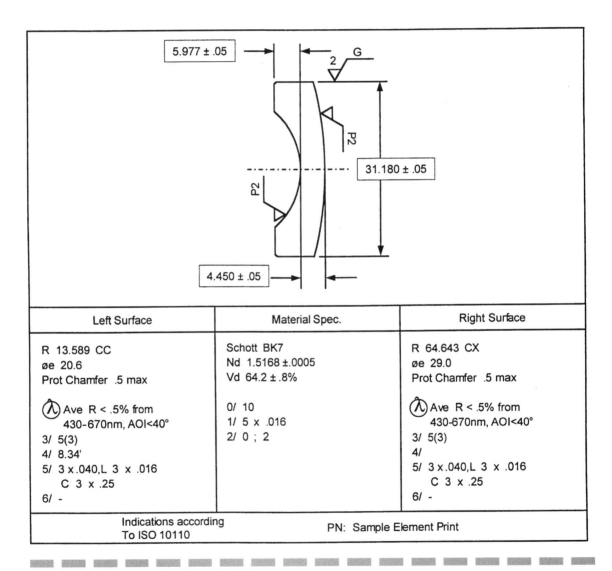

Left Surface	Material Spec.	Right Surface

R 13.589 CC
øe 20.6
Prot Chamfer .5 max
 	
Ⓐ Ave R < .5% from
   430-670nm, AOI<40°
3/ 5(3)
4/ 8.34'
5/ 3 x .040,L 3 x .016
   C 3 x .25
6/ -
	Schott BK7
	Nd 1.5168 ±.0005
	Vd 64.2 ± .8%
	
	0/ 10
	1/ 5 x .016
	2/ 0 ; 2
		R 64.643 CX
		øe 29.0
		Prot Chamfer .5 max
		
		Ⓐ Ave R < .5% from
		   430-670nm, AOI<40°
		3/ 5(3)
		4/
		5/ 3 x .040,L 3 x .016
		   C 3 x .25
		6/ -

Indications according To ISO 10110	PN: Sample Element Print	

Figure 17.13*b* ISO Lens Manufacturing Print

with ISO standards follow (Figs. 17.13*a* and *b*). For more information, see Part 10, "Table Representing Data of a Lens Element," within ISO 10110 ("Optics and Optical Instruments: Preparation of Drawings for Optical Elements and Systems").

## CHAPTER 18

Polarization Issues in Optical Design

Introduction

Optical systems are frequently designed with no consideration given to light's polarization characteristics—and justly so. Many lens designers have successfully designed optics for their entire career without ever writing a Jones matrix. However, this is not true of all lens designers.

The use of polarization in optical systems has become more prevalent for two reasons: (1) Polarized sources are more available. Lasers have strongly influenced the optics industry since they first became commercially available in 1962. The fact that most are polarized means that optical systems utilizing lasers are also polarized. Additionally, sheet polarizers have enabled the easy, inexpensive generation of polarized polychromatic beams. (2) Electronically addressable polarization optics have greatly influenced the use of polarized light. Liquid crystal devices, magnetooptic media, and electrooptic modulators allow information to be electronically encoded into the polarization state of light. Liquid crystal devices have made the largest impact. Projection systems and head-mounted displays are new optical systems which are designed around liquid crystals.

This chapter introduces a variety of polarization topics:

1. It defines polarized light and polarization optics as well as the vocabulary surrounding these subjects.

2. It introduces the mathematics of polarized systems.

3. It discusses polarization aberrations and polarization ray tracing.

4. It discusses general approaches to solving common design challenges involving polarized light.

What Is Polarized Light?

Naturally occurring light, such as from the sun or candles, is often unpolarized—the electric field orientation (and the magnetic field orientation) varies unpredictably in the plane perpendicular to the direction of propagation. *Polarized light* is characterized by an electric field oscillating with a predictable orientation. An example of an electric field propagating through space is represented by Fig. 18.1a. The magnitude of the electric field, indicated by the small arrows, fluctuates sinusoidally as it propagates, indicated by the dashed line. The dots mark the magnitude of the electric field at an instant in time and in space. The projection of the field onto the plane perpendicular to the direction of propagation is a line. This defines *linearly polarized light* and can be seen in Fig. 18.1b. A linearly polarized electric field, associated with a ray propagating along the *z* axis, can have any orientation in the *x-y* plane. For orientations other than horizontal and vertical, the polarization orientation is denoted by the angle from *x* axis to the electric field, measured counterclockwise.

Figure 18.1

Linearly Polarized Electric Field Oscillating As It Propagates: (a) Three-Dimensional Perspective and (b) Looking Down the Propagation Axis Shows Linearly Polarized Light Aligned in the Vertical Direction

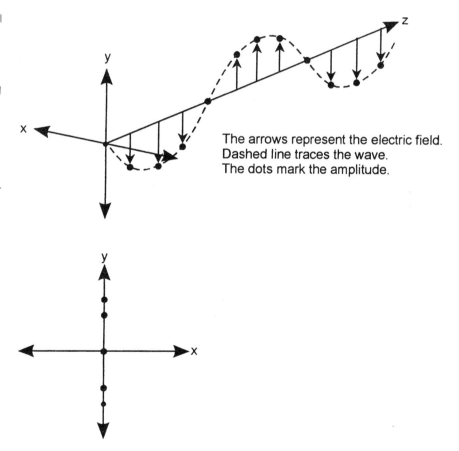

The arrows represent the electric field.
Dashed line traces the wave.
The dots mark the amplitude.

Figure 18.2
(a) Circularly Polarized Electric Field Oscillating As It Propagates and (b) Looking Down the Propagation Axis Shows Circularly Polarized Light

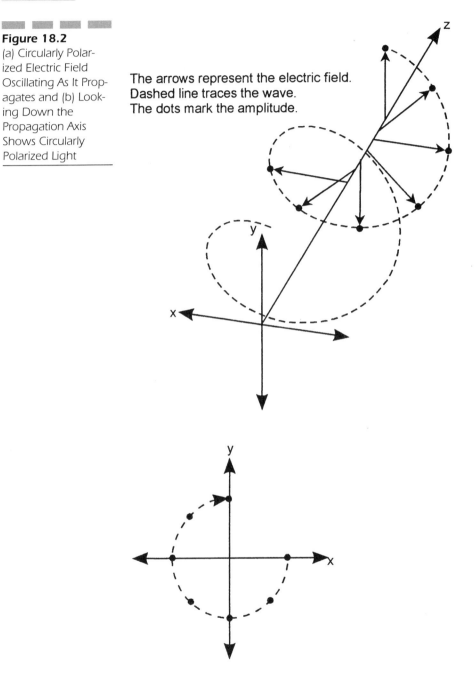

The arrows represent the electric field.
Dashed line traces the wave.
The dots mark the amplitude.

Figure 18.2*a* depicts the electric field of *circularly polarized light.* Figure 18.2*b* shows that the electric field vector traces out a circle. We define this state to be right circularly polarized. The opposite sense of rotation would be left circularly polarized. There is no dominant handedness convention among optics or physics texts.

Circularly and linearly polarized light are two extreme cases of *elliptically polarized light.* Figure 18.3 shows an example of an elliptically polarized electric field. The projection of the electric field vector into the *x-y* plane traces out an ellipse. Elliptical polarization states are specified by the orientation, θ, of the major axis; the *ellipticity*, ε; and the handedness. The ellipticity is defined as $\varepsilon = \tan^{-1}(b/a)$, where *a* and *b* represent the length of the semimajor and semiminor axes of the ellipse, respectively.

Optical systems are often considered to use completely unpolarized or fully polarized light. This proves to be an appropriate approximation for many optical systems. In reality, however, light is always partially polarized—a combination of polarized and unpolarized light. The *degree of polarization* (DOP) is defined as

$$\text{DOP} = \frac{I_p}{I_p + I_u}$$

where I_p is the polarized irradiance and I_u is the unpolarized irradiance.

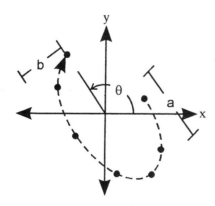

Figure 18.3
Elliptically Polarized
Electric Field

Polarization Elements

Polarization elements are any materials, objects, and media that modify the polarization of light. Polarization elements are fully described by their eigenstates and the magnitudes of their polarization effects. The three polarization effects are retardance, diattenuation, and depolarization.

An *eigenstate* is a polarization state that is unchanged when light interacts with a polarization element. For example, consider a horizontal linear polarizer. Light in one eigenstate, horizontal linear, travels through the polarizer and emerges as horizontal polarized light. The second eigenstate, vertical linear, travels through a horizontal linear polarizer and no light emerges. This is the same as vertical polarized light with zero amplitude.

Two common polarization elements are *linear retarders* and *linear polarizers*. These elements have linearly polarized eigenstates. Other polarization elements include *circular* and *elliptical retarders* and *polarizers*. As their names suggest, their eigenstates are circular or elliptical polarization states. Finally, there are *depolarizers*. Depolarizers do not have eigenstates associated with them.

Retarders

A *retarder* separates one eigenstate temporally from the orthogonal state—in other words, it retards the propagation of one eigenstate with respect to the other. Examples of naturally occurring retarders are quartz and calcite. The crystalline structures of quartz and calcite are such that the molecules are more densely arranged in one direction than in the other, as schematically represented in Fig. 18.4. This characteristic causes *birefringence*, different refractive indices for orthogonal polarization states. (Birefringence can also result from the anisotropy of the individual molecule rather than the arrangement of the molecules.)

In a birefringent medium, the polarization state aligned with the larger index travels through the material more slowly than the orthogonal state. (The slow axis denotes the axis along which a higher index is incurred; the fast axis denotes the axis of lower index.) As the light traverses the medium, a phase delay develops between the two states. The *retardance* is the amount of delay.

Retardance is described in units of waves, degrees (radians), or nanometers. For example, the retardance due to propagation through a birefringent medium might be 200 nm, as calculated by $(n_1 - n_2)L$, where n_1 and n_2 are the indices of refraction and L is the length of propagation in the medium. This quantity is approximately the same for all wavelengths (neglecting dispersion). Alternatively, the element might be said to have 90° of retardance at 800 nm and 180° of retardance at 400 nm. The same retarder, in waves, is a quarter-wave retarder for 800-nm light and a half-wave retarder for 400-nm light. The natural variation of retardation with wavelength is a problem in polychromatic systems.

Retarders have an amount of retardance that is dependent on the direction of the beam through the material. In Fig. 18.4, if the beam were to travel from top to bottom of the crystal instead of from left to right, it would see no birefringence—the molecules are equally spaced in the plane perpendicular to the propagation direction. As the beam direction changes, the retardance of the optic also changes. This is true not only of crystal retarders but also of most retarders.

A retarder converts one polarization state into another without losing energy. For example, linearly polarized light can be converted to circularly polarized light. Figure 18.5 shows a 45° linearly polarized field mathematically decomposed into its horizontal and vertical components. These

Figure 18.4
A Representation of the Molecular Structure of a Birefringent Material

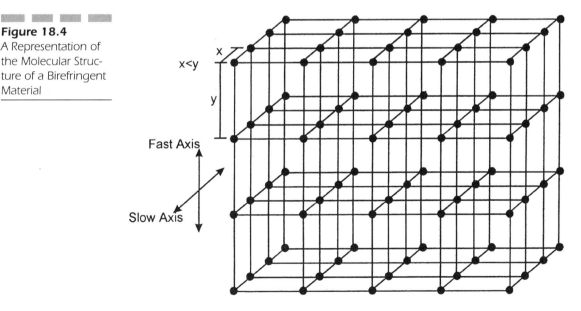

components are of equal magnitude and are in phase. Figure 18.5a to e show the two components oscillating; the sum of the two fields is represented by a dot. As the phase advances, the amplitude of the two fields decreases from the maximum. Figure 18.5f shows all the resulting fields (the dots) from Fig. 18.5a to e. The electric field oscillates along a line oriented 45° from the x axis. When a retarder, aligned such that its eigenstates are horizontal and vertical, delays the phase of one state by a quarter wave, circularly polarized light results. Figure 18.6a shows that the vertical component has zero amplitude initially, when the horizontal component is at its maximum because the retarder has delayed the vertical field by 90°. As the field propagates, the horizontal component decreases and the vertical component increases. Figure 18.6f shows all the resulting fields from Fig. 18.6a to e. The electric field now rotates about the z axis as it propagates, indicating that the resulting field is circularly polarized.

Three common types of linear retarders are crystal retarders, sheet retarders, and liquid crystal devices. Liquid crystal devices can be circular or elliptical retarders.

CRYSTAL RETARDERS   As mentioned earlier, crystals with anisotropic structures can retard one polarization state relative to the orthogonal state. Birefringent crystals occur naturally or are grown in laboratories. The cost and availability of the substrate may limit its use to low-volume applications but its rugged environmental qualities and compactness make it an option for lens designers.

SHEET RETARDERS   Stretching transparent plastic until the molecules become denser in one direction and rarer in the other creates birefringence and thus *sheet retarders*. This phenomenon is demonstrated by taking a piece of plastic wrap between crossed polarizers and stretching it. The colorful display is a result of the chromatic dependence of the retardance.

Sheet retarders are very economical, costing only pennies per square inch. And because of the low cost, it becomes more economically attractive to use multiple layers of retarders to correct for chromatic issues associated with retardance. The principles employed to design such a stack are similar to those in thin-film design.[1]

The disadvantages of this material are mostly environmental. The retarders are made of polymeric materials and are sensitive to high temperatures and excessive ultraviolet flux. Unfortunately, manufacturers frequently supply little data on such constraints. Other disadvantages

Figure 18.5
Two Orthogonal
Fields Summing to
Make 45° Polarized
Light: (a-e) Propagat-
ing Fields As Their
Phases Advance and
(f) Summary of the
Previous Diagrams
Collecting All the
Resultants Together

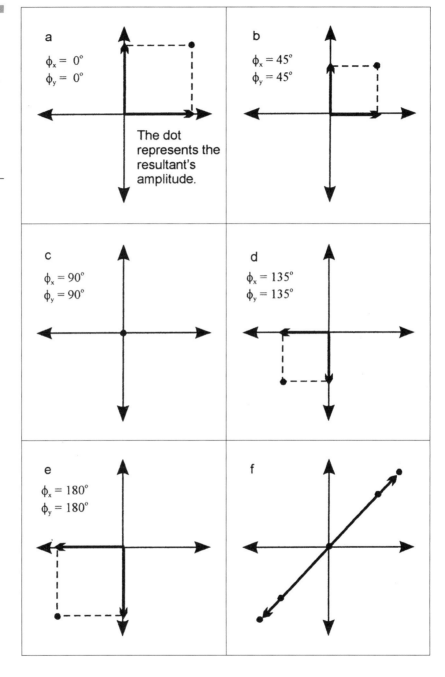

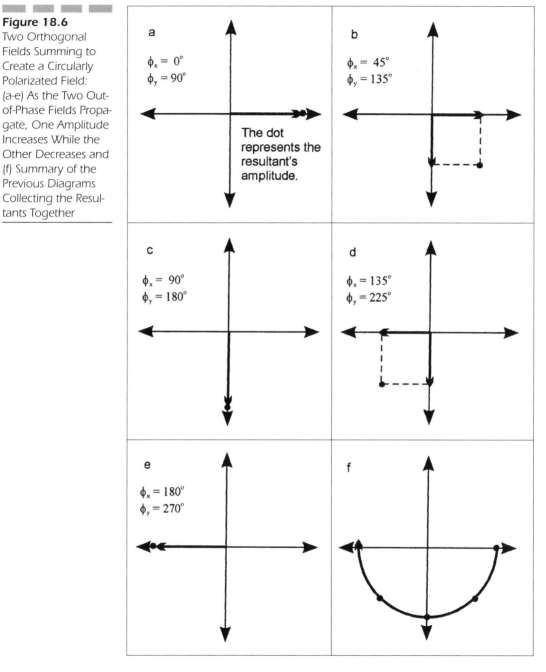

a
$\phi_x = 0°$
$\phi_y = 90°$

The dot represents the resultant's amplitude.

b
$\phi_x = 45°$
$\phi_y = 135°$

c
$\phi_x = 90°$
$\phi_y = 180°$

d
$\phi_x = 135°$
$\phi_y = 225°$

e
$\phi_x = 180°$
$\phi_y = 270°$

f

are distortion and low transmissivity. The irregularity associated with sheet retarders should be considered early in an imaging system design exercise.

LIQUID CRYSTAL DEVICES   *Liquid crystals* (LCs) are elongated polymer molecules that can move freely in a liquid solution. Typically, this solution is contained between two flat substrates of glass. The interior surface of the glass has small grooves rubbed into it. The first layer of liquid crystals aligns itself to the grooves; the second layer then aligns itself to the first layer of liquid crystals; and so on. Since the molecules are elongated, a birefringent condition is established.[2]

The revolutionizing property of LC retarders is the ability to electronically change the retardance. Creating an electric field across a cell gap containing liquid crystals causes the liquid crystal molecules to reorient themselves. To a first-order approximation, the molecules become parallel to the electric field. (There are also some types of liquid crystals that align perpendicular to the electric field.)

For example, a linear LC retarder can result when the rubbing direction on the glass substrates aligns all the molecules along the y axis. The fast axis is along the x axis and the slow axis is the y axis. As a ray propagates through the cell along the z axis, the retardance is proportional to the length of the cell.

Now, if an electric field is applied parallel to the ray propagation, the z axis, the molecules align to the electric field. The slow axis changes to the z axis. Since there is no polarization component along the z axis, the ray sees no birefringence. In other words, the speed of the x polarized light and the y polarized light is the same. No retardance is associated with propagation through the cell when the electric field is applied.

Pixelating this device and placing it between crossed polarizers allows the amount of light transmitted through each pixel to be electronically controlled. Calculators, computers, watches, and numerous other electronic devices use liquid crystals in their displays to selectively transmit light.

Nematic LCs can be used to make circular retarders. Again, the liquid crystals are sandwiched between two plates of glass with grooves. For a circular retarder, however, the rubbing directions of the two plates are perpendicular to each other instead of parallel. The top few layers of liquid crystals are aligned with the x axis, for example, and the bottom few layers of liquid crystals are aligned with the y axis. The layers in between twist until there is a smooth transition in orientation between all the layers of the liquid crystal. This is called a *twisted nematic* (TN) cell.

The physics of this retarder is quite complicated but such a device results in an elliptical retarder. Choosing the thickness of the cell gap carefully can result in a perfect circular retarder for at least one wavelength.[3]

Liquid crystals can have larger retardance per millimeter of thickness than most solid crystals and their thickness can be adjusted by changing the cell gap. LC solutions are expensive but very little is needed usually— an 8-μm gap can create a λ/2 wave of retardance. Circular LC retarders have one additional disadvantage over linear LC retarders. For different wavelengths, the eigenstates of the linear retarder remain constant though the magnitude of retardance varies. In the circular LC retarder configuration, the eigenstates' ellipticity and orientation, as well as the retardance, are functions of wavelength.

Polarizers

Polarizers spatially separate one eigenstate from the orthogonal eigenstate. The rejected eigenstate can be reflected, absorbed, or diverted from the other eigenstate entirely or partially. An unpolarized beam transmitted through (or reflected off) a polarizer can be considered to be two beams. One beam is polarized parallel to the transmission axis (or reflection axis); the other beam is polarized parallel with the extinction axis with a smaller irradiance. The polarizer's performance is described by the extinction ratio

$$R = \frac{I_1}{I_2}$$

where I_1 and I_2 are the irradiances associated with each of these beams, respectively. The extinction ratio is a positive number greater than or equal to 1. For an ideal polarizer, $R = \infty$; for a *partial polarizer,* $1 < R < \infty$. Very good polarizers have extinction ratios of 10^4 or higher.

An equivalent measure of a polarizer's performance is called *diattenuation.* This term refers to the different attenuation coefficients associated with each of the eigenstates. A typical sheet polarizer may transmit 85% of the "aligned" eigenstate and 0.02% of the "crossed" eigenstate. These coefficients, $d_1 = 85\%$ and $d_2 = 0.02\%$, result in a diattenuation calculated by $D = |(d_1 - d_2)/(d_1 + d_2)| = 0.9995$. An ideal polarizer has $D = 1$; a partial polarizer has $0 < D < 1$.

Conducting polarizers and thin-film polarizers are the primary methods of generating linearly polarized light in optical systems. Conducting polarizers utilize a material's anisotropic absorption or reflection to polarize. Thin film polarizers rely on Fresnel coefficients and the geometry of the incoming ray to polarize.

CONDUCTING POLARIZERS   *Iodine* or *dye-sheet polarizers* are by far the most common polarizers in optical systems. These polarizers are composed of iodine or dye molecules suspended in a polymer substrate. Stretching the substrate aligns the elongated molecules very precisely. The polarization component parallel to the molecules is almost completely absorbed and converted into heat. The polarizer substantially passes the orthogonal polarization.

Sheet polarizers are popular because they are inexpensive, costing pennies per square inch. They work very well for most of the visible region, only slightly losing extinction and transmission in the deep blue wavelengths. Also, sheet polarizers work well over large angular cones (see the "Geometrical Issues and the Maltese Cross" section for more information). One disadvantage of sheet polarizers is their susceptibility to damage resulting from heat and exposure to ultraviolet (uv) light. This is not likely to be a problem for irradiances less than 1 W/cm^2 in the visible.

Wire-grid polarizers are conducting polarizers made by depositing thin aluminum wires extremely close together on a substrate. The wires must form a periodic structure with a period much smaller than the wavelength of the light. In contrast to the sheet polarizers, the eigenstate aligned to the wires is reflected rather than absorbed. The other eigenstate is mostly transmitted. Since the rejected state is reflected, the wire grid polarizer can be utilized as a polarization beamsplitter as well as a simple polarizer.

A third type of conductive polarizer, *imbedded silver polarizers*, utilizes aligned silver molecules imbedded in the surface of a glass plate. The silver molecules are analogous to the aluminum wires and the iodine molecules of the previous examples. Presently, wire-grid polarizers and imbedded silver polarizers are common only to infrared and red wavelengths. The performance of these alternatives is highly wavelength dependent. The extinction ratios for long wavelengths are substantially higher than for shorter wavelengths. As a general rule, wavelengths that are 4 times longer than the period of the conductors will have sufficient extinction ratios for most applications.

The reduced absorptivity and more stable materials increase the lifetime of the wire grids and the imbedded silver polarizers substantially over the iodine polarizers. Where environmental concerns are dominant, the cost and performance trade-offs associated with these devices may be worthwhile.

THIN-FILM POLARIZERS A thin-film polarizer's performance is governed by Fresnel's equations. These equations describe the polarization-dependent amplitude coefficients, t and r, associated with the transmitted and reflected rays after interacting with an interface.

Figure 18.7 shows the geometry of this interaction. The plane of incidence (the plane containing the normal to the surface, $\hat{n}$, and the incident ray, $\hat{\imath}$) defines the eigenstates. The polarization state perpendicular to the plane of incidence is the s-eigenstate and the state parallel with this plane is the p-eigenstate. (The letters s- and p- are the initials of two German words, senkrecht and parallel, which translate to perpendicular and parallel, respectively.) An incident ray, $\hat{\imath}$, at angle of incidence θ_0 in a medium of index n_0 divides into two rays at the second medium of index n_1. One ray, $\hat{r}$, will reflect at the angle $-\theta_0$; the other ray, $\hat{t}$, will refract to an angle θ_1 determined by Snell's law. The amplitude of each of these rays is dependent on their polarization states and are described by the following equations:

$$r_s = \frac{n_0 \cos(\theta_0) - n_1 \cos(\theta_1)}{n_0 \cos(\theta_0) + n_1 \cos(\theta_1)} \qquad r_p = \frac{n_1 \cos(\theta_0) - n_0 \cos(\theta_1)}{n_1 \cos(\theta_0) + n_0 \cos(\theta_1)}$$

$$t_s = \frac{2n_0 \cos(\theta_0)}{n_0 \cos(\theta_0) + n_1 \cos(\theta_1)} \qquad t_p = \frac{2n_0 \cos(\theta_0)}{n_1 \cos(\theta_0) + n_0 \cos(\theta_1)}$$

The irradiance coefficients of these rays are t^2 and r^2, respectively.

Brewster's angle, $\theta_B = \tan^{-1}(n_1/n_0)$, is the incident angle for which $r_p = 0$. The reflection coefficient of the p-state is zero, thus the reflected beam is entirely s-polarized. The transmitted beam is also partially polarized since all the p-state is transmitted but also some of the s-state. A "pile of plates" polarizer uses this phenomenon by repeating the diattenuation many times. If N plates of glass are stacked together, the resulting amplitude coefficients are $t_p^{2N} = 1$ (since $t_p = 1$) and $t_s^{2N} \to 0$ (since $t_s < 1$) when N becomes very large.

Thin-film polarizers consist of alternating layers of high-index and low-index dielectric coatings. One eigenstate sees an antireflection-coated interface and is transmitted, while the orthogonal state sees a high-reflectance interface. Essentially this is an optimized pile of plates polarizer.[4] Thin-film polarization beamsplitters can be made in two

Figure 18.7
The Geometry
Associated with
Fresnel's Equations

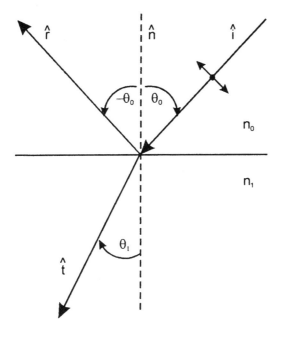

configurations: plates and cubes. Plate beamsplitters have thin dielectric layers on one side of a plate of glass; the other side is often AR coated. The plate is usually tilted at 45° to the incident beam. A cube is constructed of thin dielectric layers sandwiched between two prisms. The prisms form extruded 45°-45°-90° triangles with the dielectric coatings applied to one or both of the hypotenuses. The prisms are cemented together to form a cube, as in Fig. 18.8.

These polarizers work in high-temperature environments and have the same environmental constraints as doublets. They can be used as beamsplitters as well as polarizers. Their performance decreases in systems with increased angular cones. The extinction ratio typically decreases from over 10,000:1 on axis to only around 500:1 at around 10°. (These values are extremely design and wavelength dependent.) Designs optimized for narrower wavelength regions are simpler and generally perform better than designs for large wavelength regions. However, even in a ±10° cone, a sophisticated design can perform well over the entire visible region.

Related to thin-film polarizers, *polymer-film polarizers* are composed of many polymer layers of dissimilar materials. Sheets of birefringent plastic are laminated together to form a multilayer stack. The same principles

Figure 18.8
Thin-Film Polarization
Beamsplitting Cube
(PBSC)

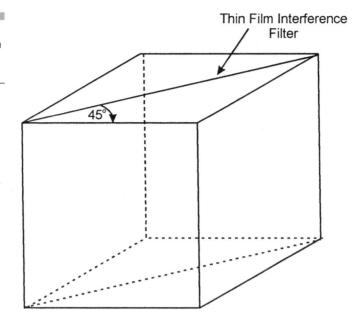

are used as in thin-film polarizing beamsplitters but the asymmetries induced by the birefringence allow the polarizers to be used at much lower angles of incidence.[5]

A polymer-film polarizer has many excellent attributes and a few problems. It can be manufactured in large sheets, measuring many inches on a side. In transmission these polarizers work on axis (extinction ratio on the order of 100s:1) to 50 to 60° (extinction ratio approaching 1000:1). As with sheet polarizers, the largest problem is the environmental reliability due to the polymer absorption. However, the absorption of the polymer-film polarizer is an order of magnitude less than that of the sheet polarizer, so its lifetime is significantly longer.

Depolarizers

The third category of polarization elements includes depolarizers. *Depolarizers* randomize or mix polarization states, resulting in a decreased degree of polarization. Depolarizers in polarized optical systems can reduce the efficiency and contrast, potentially causing substantial problems for the designer. Scattering surfaces producing stray light are the most common sources of depolarized light in polarized optical systems.

Depolarizers can also be desirable. For example, radiometers measure the total irradiance of an object. This measurement should ideally be independent of its polarization state. If the radiometer, as a system, has diattenuation, it can measure different values for the same object depending on the orientation of the object relative to the radiometer. Depolarizing the light before measurement will eliminate this issue.

Depolarizers can be scattering devices such as ground glass or integrating spheres. Scattering devices are, unfortunately, also lossy. A retarder with spatially varying retardance macroscopically depolarizes light, and a retarder with temporally varying retardance depolarizes light observed over time. These devices, known as *pseudodepolarizers,* are more efficient than a scatterer and could be used in imaging systems, with possible loss of image quality.

Other Comments on Polarization Elements

Reflective polarizers such as thin-film, polymer-film, imbedded-silver, and wire-grid polarizers can be used as polarization beamsplitters (PBSs). A PBS splits a beam into two paths depending on the polarization state of the beam. The PBS cube is used such that one beam is reflected at a right angle to the other. Different type PBSs are used at angles which differ from 45° to optimize the polarization characteristics of the polarizer.

Thin-film polarizers always reflect the s-state and transmit the p-state. (The other polarizers work best in this configuration but can be used to transmit the s-state with a reduced extinction ratio.) The extinction ratio for these polarizers is 1000s:1 or higher on transmission, but on reflection the extinction ratio drops to 10s:1. Both polarization states are reflected to a significant degree, resulting in only a partially polarized reflected beam.

The few polarization elements discussed previously in the sections "Polarizers" and "Retarders" are the most common in optical systems. Their size, cost, and availability make them favorable to lens designs. Many more retarding and polarizing elements exist which, in specialized systems, prove more applicable. Some alternative retarders are Fresnel and Mooney rhombs, Babinet and Soleil compensators, Kerr and Pockels cells, Faraday modulators, and optically active crystals and solutions. Some alternative polarizers are Glan-Thompson, Glan-Taylor and Wollaston prisms, cholesteric polarizers, crystalline circular polarizers, and Fresnel composite prisms.[6]

Mathematics of Polarized Light

The understanding of the effects of polarization elements can be enhanced when mathematical modeling of the elements is possible. Two mathematical representations of polarized light, which are helpful for this purpose, are the *Jones calculus* and the *Mueller calculus*. The Jones calculus represents the electric field with a 2 × 1 vector and a polarization element with a 2 × 2 matrix.[7] The electric field represented by has a Jones vector

$$\vec{E} = E_x e^{i\phi_x} e^{i(\vec{k} \cdot \hat{z} - \omega t)} \hat{x} + E_y e^{i\phi_y} e^{i(\vec{k} \cdot \hat{z} - \omega t)} \hat{y}$$

$$\vec{E} = \begin{bmatrix} E_x\, e^{i\phi_x} \\ E_y e^{i\phi_y} \end{bmatrix}$$

where E_x and E_y are the amplitudes of the electric field along the x and y axes and ϕ_x and ϕ_y are the phases of these components. The space- and time-dependent term $e^{i(\vec{k}\,\hat{z} - \omega t)}$ in the phase is not included in the Jones vector. (The Jones vector is dimensionless.) Some examples of Jones vectors are listed in Table 18.1. Jones vectors represent fully polarized fields of monochromatic light, that is, coherent beams.

Table 18.2 contains Jones matrices of common polarization elements. Jones matrices represent nondepolarizing polarization elements. The eigenvectors of the Jones matrix are the polarization eigenstates of the element. The eigenvalues contain the information about the diattenuation and retardance.

The interference of two coherent beams of light is described by the addition of their Jones vectors. For example, a horizontal vector plus vertical vector, which are in phase with one another, results in a 45° vector just as a horizontally polarized beam and a vertically polarized beam coherently makes a 45° polarized beam. The Hermitian inner product of two vectors represents the amplitude of the electric field components which are parallel. If the represented polarization states are orthogonal, the resultant will be zero.

TABLE 18.1

Jones Vectors for Common Polarization States

Linear	Horizontal $\begin{bmatrix} 1 \\ 0 \end{bmatrix}$	Vertical $\begin{bmatrix} 0 \\ 1 \end{bmatrix}$	45° $\dfrac{1}{\sqrt{2}}\begin{bmatrix} 1 \\ 1 \end{bmatrix}$	135° $\dfrac{1}{\sqrt{2}}\begin{bmatrix} 1 \\ -1 \end{bmatrix}$
Circular		Left $\dfrac{1}{\sqrt{2}}\begin{bmatrix} 1 \\ i \end{bmatrix}$		Right $\dfrac{1}{\sqrt{2}}\begin{bmatrix} 1 \\ -i \end{bmatrix}$

TABLE 18.2

Jones Matrices for
General Linear and
Circular Diattenua-
tors, Depolarizers,
and Linear and
Circular Retarders

Linear Diattenuator	$\begin{bmatrix} q\,\cos^2(\theta) + r\,\sin^2(\theta) & q\,\cos(\theta)\sin(\theta) - r\,\cos(\theta)\sin(\theta) \\ q\,\cos(\theta)\sin(\theta) - r\,\cos(\theta)\,sin(\theta) & r\,\cos^2(\theta) + q\,\sin^2(\theta) \end{bmatrix}$
Circular diattenuator	$\dfrac{1}{2}\begin{bmatrix} (q+r) & -i\,(q-r) \\ i\,(q-r) & (q+r) \end{bmatrix}$
Linear retarder	$\begin{bmatrix} \cos^2(\theta) + e^{-i\delta}\sin^2(\theta) & \cos(\theta)\sin(\theta) - e^{-i\delta}\cos(\theta)\sin(\theta) \\ \cos(\theta)\sin(\theta) - e^{-i\delta}\cos(\theta)\sin(\theta) & e^{-i\delta}\cos^2(\theta) + \sin^2(\theta) \end{bmatrix}$
Circular retarder	$\begin{bmatrix} \cos(\delta/2) & \sin(\delta/2) \\ -\sin(\delta/2) & \cos(\delta/2) \end{bmatrix}$
Variables	q, r = diattenuation coefficients of the transmission axis and extinction axis (linear) or right and left circular polarization states (circular). δ = retardance in degrees or radians θ = orientation of the transmission axis (polarizers) or fast axis (retarders)

The interaction of an electric field with a polarization element is mathematically represented by $\vec{E}' = J\vec{E}$, where $\vec{E}$ is the incident Jones vector and J is the Jones matrix. $\vec{E}'$ is the electric field resulting from the interaction. To represent the effects of a sequence of polarization elements, their respective Jones matrices are multiplied. The order is important in multiplication—the Jones matrix of the first element, J_1, should be multiplied on the right of the Jones matrix of the second element, J_2, and so on with additional elements. This procedure results in $\vec{E}' = J_2 J_1 \vec{E}$.

The Mueller calculus, though much more cumbersome, sometimes can offer advantages over the Jones calculus. The Mueller calculus represents incoherent light by a 4×1 vector (called a *Stokes vector*) and a polarization element with a 4×4 real matrix.[8] (The dimensions of the Stokes vector are watts per square meter.) The ability to represent unpolarized light, depolarizing polarization elements, and polychromatic beams can make the Mueller calculus useful in lens design. Stray light analysis, for example, should be modeled with the Mueller calculus since light becomes partially depolarized upon scattering.

Four measurements, shown in Fig. 18.9, fully describe the Stokes vector of any beam of light. The general Stokes vector is

$$\vec{S} = \begin{bmatrix} I \\ H - V \\ 45^\circ - 135^\circ \\ R - L \end{bmatrix}$$

Figure 18.9
Measurement
Technique to Obtain
Stokes Vector

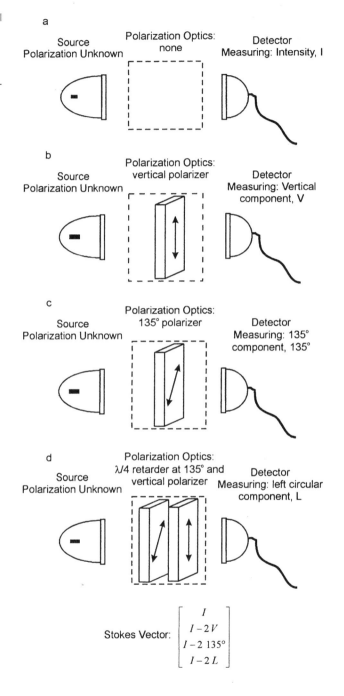

a

Source
Polarization Unknown

Polarization Optics:
none

Detector
Measuring: Intensity, I

b

Source
Polarization Unknown

Polarization Optics:
vertical polarizer

Detector
Measuring: Vertical
component, V

c

Source
Polarization Unknown

Polarization Optics:
135° polarizer

Detector
Measuring: 135°
component, 135°

d

Source
Polarization Unknown

Polarization Optics:
λ/4 retarder at 135° and
vertical polarizer

Detector
Measuring: left circular
component, L

$$\text{Stokes Vector:} \begin{bmatrix} I \\ I-2V \\ I-2\,135° \\ I-2L \end{bmatrix}$$

where I is the total irradiance of the beam and H, V, 45°, 135°, R, and L are the irradiances of the horizontal, vertical, 45°, 135°, right and left polarized components of the beam, respectively. (The 45° and 135° components are the amount of irradiance that is transmitted by a polarizer oriented along the 45° axis or the 135° axis. Likewise, the R and L components are the amount of irradiance transmitted by a right or left circular polarizer.) Stokes vectors of some specific elements are given in Table 18.3. The general Mueller matrices are shown in Table 18.4. The degree of polarization of a Stokes vector is calculated by

$$\text{DOP} = \frac{\sqrt{s_1^2 + s_2^2 + s_3^2}}{s_0}$$

where s_0, s_1, s_2, and s_3 are the first through fourth elements of the Stokes vector.

Incoherent addition of two beams is achieved through addition of the two Stokes vectors representing the beams. Adding horizontal and vertical Stokes vectors results in an unpolarized Stokes vector just as incoherently adding the horizontal and vertically polarized light results in unpolarized light. As in the Jones calculus, the interaction of a beam of polarized light with a polarization element is represented by the multiplication of a Mueller matrix with a Stokes vector. Again, the order of the matrices is important. A beam of light interacting with two elements is represented by $\vec{S}' = \mathbf{M}_2\mathbf{M}_1\vec{S}$, where $\vec{S}$ is the Stokes vector of the resulting beam, $\mathbf{M}_1$ and $\mathbf{M}_2$ represent the elements, and $\vec{S}$ represents the incident beam. $\vec{S}$ is incident upon $\mathbf{M}_1$ first.

Polarization Aberrations and Polarization Ray Tracing

Polarization aberrations are deviations of a polarized field from the ideal polarization. Any rotation of the polarization, change in ellipticity, or change in degree of polarization is an aberration. The ideal polarization varies for different systems. Commonly, polarized systems are linearly polarized at a set orientation. Just as wavefront aberrations generally become more problematic as the lens design utilizes smaller $f/\#$'s or tilted elements, so too is the case with polarization aberrations. Introduction of any optical elements into a beam changes the polarization from the ideal polarization.

TABLE 18.3

Stokes Vectors for Common Polarization States

Linear

Horizontal $\begin{bmatrix} 1 \\ 1 \\ 0 \\ 0 \end{bmatrix}$ Vertical $\begin{bmatrix} 1 \\ -1 \\ 0 \\ 0 \end{bmatrix}$ 45° $\begin{bmatrix} 1 \\ 0 \\ 1 \\ 0 \end{bmatrix}$ 135° $\begin{bmatrix} 1 \\ 0 \\ -1 \\ 0 \end{bmatrix}$

Circular

Right $\begin{bmatrix} 1 \\ 0 \\ 0 \\ 1 \end{bmatrix}$ Left $\begin{bmatrix} 1 \\ 0 \\ 0 \\ -1 \end{bmatrix}$

TABLE 18.4 Mueller Matrices for General Linear and Circular Polarizers, Depolarizers, and Linear and Circular Retarders

Linear diattenuator

$$\frac{1}{2} \begin{bmatrix} (q + r) & (q - r)\cos(2\theta) & (q - r)\sin(2\theta) & 0 \\ (q - r)\cos(2\theta) & (q + r)\cos^2(2\theta) + 2\sqrt{qr}\sin^2(2\theta) & (q + r - 2\sqrt{qr})\cos(2\theta)\sin(2\theta) & 0 \\ (q - r)\sin(2\theta) & (q + r) - 2\sqrt{qr}\cos(2\theta)\sin(2\theta) & 2\sqrt{qr}\cos^2(2\theta) + (q + r)\sin^2(2\theta) & 0 \\ 0 & 0 & 0 & 2\sqrt{qr} \end{bmatrix}$$

Circular diattenuator

$$\frac{1}{2} \begin{bmatrix} (q + r) & 0 & 0 & (q - r) \\ 0 & 2\sqrt{qr} & 0 & 0 \\ 0 & 0 & 2\sqrt{qr} & 0 \\ (q - r) & 0 & 0 & (q + r) \end{bmatrix}$$

Depolarizer

$$\begin{bmatrix} 1 & 0 & 0 & 0 \\ 0 & a & 0 & 0 \\ 0 & 0 & a & 0 \\ 0 & 0 & 0 & a \end{bmatrix}$$

Linear retarder

$$\begin{bmatrix} 1 & 0 & 0 & 0 \\ 0 & \cos^2(2\theta) + \cos(\delta)\sin^2(2\theta) & (1 - \cos(\delta))\cos(2\theta)\sin(2\theta) & -\sin(\delta)\sin(2\theta) \\ 0 & (1 - \cos(\delta))\cos(2\theta)\sin(2\theta) & \cos(\delta)\cos^2(2\theta) + \sin^2(2\theta) & \cos(2\theta)\sin(\delta) \\ 0 & \sin(\delta)\sin(2\theta) & -\cos(2\theta)\sin(\delta) & \cos(\delta) \end{bmatrix}$$

Circular retarder

$$\begin{bmatrix} 1 & 0 & 0 & 0 \\ 0 & \cos(\delta/2) & \sin(\delta/2) & 0 \\ 0 & -\sin(\delta/2) & \cos(\delta/2) & 0 \\ 0 & 0 & 0 & 0 \end{bmatrix}$$

Variables

q, r = diattenuation coefficients of the transmission axis and extinction axis (linear) or right and left circular polarization states (circular)

δ = retardance, in degrees or radians

θ = orientation of the transmission axis (polarizers) or fast axis (retarders)

a = depolarization coefficient, $0 < a < 1$

Lenses, prisms, and mirrors will affect the polarization, as described by Fresnel coefficients. Fortunately, a lens design with good wavefront aberration control will tend to have low angles of incidence onto each surface, often coincidentally minimizing polarization and wavefront aberrations. Tilted surfaces, such as fold mirrors or thin-film filters used in reflection, can be highly polarizing elements as well as variable retarders. Diffracting elements will also reorient a polarization vector. Gratings, CCD arrays, binary optics, obstructions, and irises are examples of common sources of diffraction.

In order to quantify the effects of each of these aberration contributors, a polarization ray trace is needed. *Polarization ray tracing* entails carrying information about the polarization state with each ray and reevaluating the polarization state at every interface the ray sees. Fresnel coefficients, phase shifts due to propagation through birefringent media, diffraction effects, multiple beam interference, and geometrical effects are summed to the resultant polarization vector. Commercial ray tracing programs can model some of these effects and can report results in a variety of formats.

The process for ray tracing is as follows:

1. The eigenstates for a particular element and ray must be determined.

2. The incident beam's polarization must be decomposed into components that are aligned with the eigenstates.

3. The appropriate effects of the element are applied to each component.

4. The components are recombined to achieve the resultant polarization state.[9]

Geometrical Issues and the Maltese Cross

The electric field vector associated with a ray has two degrees of freedom and can be described as such very well by the mathematics of the Jones calculus or the Mueller calculus. For many applications constraining the electric field vector to be in the plane perpendicular to the optical axis, as in the Jones calculus and the Mueller calculus, is an acceptable technique to polarization modeling. However, in polarized optical systems, where the rays cannot be approximated as "paraxial," the third dimension must be included for accurate polarization ray tracing. The following discussion illustrates the necessity.

The plane of polarization for a ray is the plane normal to the direction of propagation. In this plane, any polarization vector is a valid polarization state for that ray. For rays which are parallel, such as in a collimated beam, their planes of polarization are parallel and their polarization states can be described by one set of orthonormal vectors. Here, Jones or Mueller calculus aptly applies.

In a converging (or diverging) beam, however, the rays and corresponding planes of polarization are not parallel. For a strongly converging beam, the planes of polarization can be almost orthogonal. To describe the polarization state of each ray, as required for polarization ray tracing, a different local coordinate system must be defined for each ray.

This three-dimensional aspect of polarized light causes more than just a mathematical modeling challenge—it can also result in problems for an optical system. An experiment demonstrating this phenomenon is sketched in Fig. 18.10a. From a white-light point source, collect an angular cone of approximately ±45° (or larger) into an ideal sheet polarizer. Follow this with another ideal sheet polarizer (called an *analyzer*), which is crossed to the first. From the Jones or Mueller calculus, one might expect that crossed polarizers would result in complete absorption of the beam, and it will on axis. However, in the corners of the illuminated field, white light will leak—there will be around 8% transmission of the beam. (The leakage increases with field angle.) The contours of constant irradiance resemble the shape of a Maltese cross, as shown in Fig. 18.10b and thus this manifestation of the three-dimensional nature of polarized light is referred to as the Maltese cross.

To understand the origin of the leaked light, Fig. 18.11a shows two rays incident on the crossed polarizers. The horizontal and vertical lines represent the absorption axes of each polarizer; the lines with arrows represent the rays. Figure 18.11b and c show the same rays but the diagrams are redrawn in the local coordinate system for each ray. (In the local coordinate system, the plane of polarization is parallel with the plane of the page and the ray propagates normal to the page.) The on-axis ray sees perpendicular absorption axes but the skew ray sees non-perpendicular absorption axes. The dashed line in Fig. 18.11c indicates the orientation of the polarization state passed by the first polarizer. This polarization state is not completely aligned with the absorption axis of the second polarizer and thus light will leak through.

Figure 18.10
(a) Sketch of the
Maltese Cross
Experiment and (b)
The Irradiance Pat-
tern of the Leaked
Beam Resembling a
Maltese Cross

a

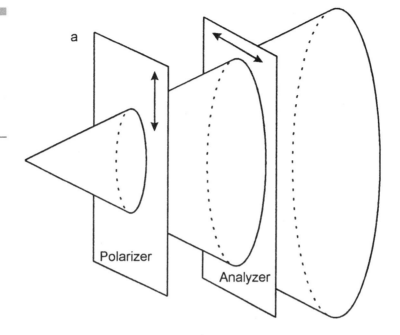

Polarizer

Analyzer

b

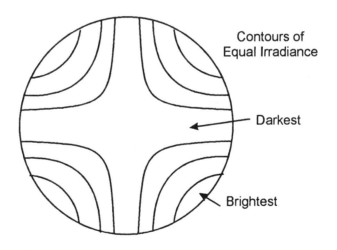

Contours of
Equal Irradiance

Darkest

Brightest

Stress Birefringence

When a material is subject to compression or expansion forces in local regions, a nonuniform refractive index results. This is called *stress birefringence* and is seen as retardance to a polarized beam. Any piece of glass or plastic will have some stress birefringence resulting from two sources: mechanical and thermal stress.

There are numerous opportunities for mechanical stress to be introduced into a lens system. To start, bulk glass quality varies with respect to birefringence, depending on the annealing technique. Then, cutting and polishing glass into sharp corners or aggressively processing the glass can introduces stress. Also, assembly of multiple glass pieces with optical cement interfaces provides a source of stress; the cement can pull

Figure 18.11
(a) Two Rays Incident on the Crossed Polarizers of the Experiment Shown in Fig. 18.10; (b) The Absorption Axes As Seen by the On-Axis Rays; and (c) The Absorption Axes As Seen by the Skew Rays

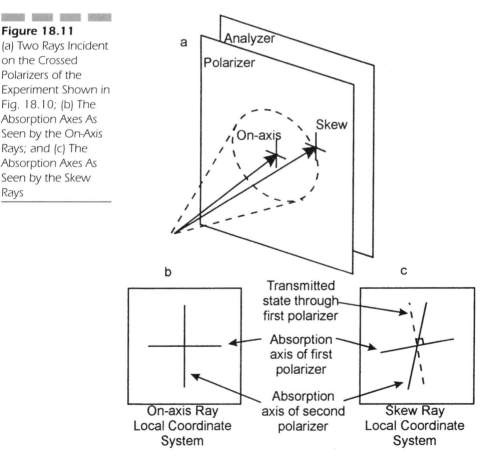

nonuniformly over the surfaces. Finally, mechanical mounting of glass may cause pressure in localized positions.

There are techniques to minimize the stresses in an optical element. Purchasing "special" annealed or "fine" annealed glass ensures that initially the bulk glass will meet a published birefringence specification.[10,11] When grinding and cementing, a good glass shop will have little problem confining the birefringence outside the clear aperture given that the clear aperture is suitably far away from the edge. (For many applications, ~1 mm will be sufficient.)

Thermal birefringence results from thermal gradients within a bulk substrate. The gradient source can be internal (absorption of incident light) or external (ambient environment). An isothermal environment is best for bulk substrates but even then thermal induced birefringence can occur—a uniformly heated (or cooled) substrate cannot expand (or contract) unrestrained at the surface of the material.

All of these birefringence sources can either be avoided or minimized with diligence but the easiest solution starts with an optimal glass type. Glasses vary in their optical stress coefficient, thermal expansion coefficient, and Young's modulus.[12] For example, fused silica, which has essentially no thermal expansion, will allow nonuniform heating to be inconsequential to the polarization. The reduced engineering effort and mechanical simplicity may recover the cost increase associated with using less common glass types.

Plastic lenses, by comparison, are much more troublesome with respect to stress birefringence. Plastic lenses should not be placed in a polarization critical optical space. If the plastic lenses cannot be moved into a more appropriate space, investigation into the lens' polarization characteristics is required.

Polarized Systems and Design Techniques

Why Should a System Use Polarization Optics?

Systems such as some telescopes, sunglasses, and radiometers are illuminated with an external, polarized light source. Since the polarization state of astronomical objects can contain information about the source, polarization optics are sometimes used in telescopes to maximize their utility. Polarized sunglasses selectively remove glare because glare is highly polarized. Some radiometers need to be able to measure the quantity of light independent of the light's polarization state. This can require depolarizing optics to achieve polarization insensitivity.

Magnetooptic storage devices, laptop displays, and projectors are examples of systems that use polarization to encode information. In projectors such as the one shown in Fig. 18.12, pixilated liquid crystal light valves generate spatially varying polarization signatures. Each pixel is individually, electronically addressable. The electric field generated across the pixel selects the amount of retardance associated with the pixel. When polarized light is incident, the pixel can rotate the polarization state or leave it unaffected. A polarizer later in the system (called the *analyzer*) transmits the light from some pixels and rejects light from others (fully or partially), depending on the polarization state generated by the pixel.

Figure 18.12

A Schematic of a Typical LCD-Based Projector System

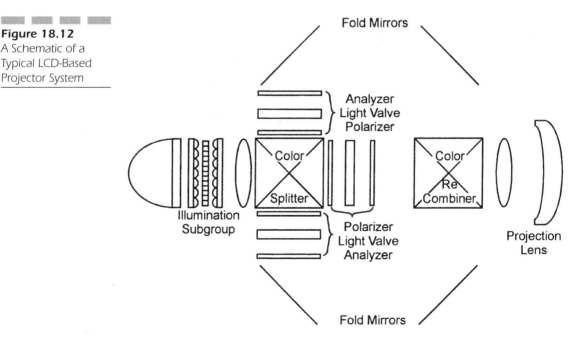

TABLE 18.5

Comparison of
Various Polarizers
in the Visible
Region with
Respect to
Common System
Specifications

	Crystalline Birefringent Polarizers	Polymer Sheet Polarizers	Thin-Film Polarizers	Polymer Polarizers	Wire-Grid Polarizers	Imbedded-Silver Polarizers
Extinction ratio	Very high	Medium to high	Medium to high	Low	Medium	Medium
Angular acceptance	Narrow	Wide	Narrow	Medium to wide	Wide	Wide
Wavelength range	Wide	Medium to wide	Medium to wide	Medium to wide	Only infrared	Red and infrared
Transmissivity	High	Medium	High	Medium	Low	Low
Reflective	No	No	Yes	Yes	Yes	Yes
Cost	High	Low	Medium	Medium	Low to medium	Medium
Optical path length	Long	Short	Short to medium	Short	Short	Short
Environmental ruggedness	High	Medium	High	Medium	High	High

Which Kind of Polarizer Is Best?

For systems that must generate their own light, a polarizer of some
sort is used to polarize the light. The type of polarizer used depends
heavily on the application. Very intense beams should be polarized
with nonabsorbing polarizers. Other determinants to consider include
packaging constraints, environmental specifications, extinction
requirements, transmission requirements, angular acceptance, etc.
Table 18.5 shows a comparison of polarizers with respect to the specifications previously listed.

   One particularly annoying characteristic of polarized systems is the
loss of one-half of the light from the offset. This is not a problem in two
circumstances: (1) systems using a polarized source such as a laser and
(2) systems compatible with polarization recapture optics such as the one
shown in Fig. 18.13.[13] (Polarization recapture optics doubles the etendue
and can be very difficult to match to the rest of an optical system.)

Projectors are excellent examples of systems with a difficult polarizer decision. The projector's two major constraints are environmental requirements and efficiency. The source is extremely bright (sometimes more than 1 W/cm^2 at the object plane) because brightness is a key selling point. Designers of these systems often use the combination of polarization recapture optics (which has very low absorption) followed by a sheet polarizer to polarize the light. The polarization recapture optics does double duty. It recovers some of the lost irradiance while polarizing the beam fairly well. Only a few percent of the beam must be absorbed in the sheet polarizers to obtain a highly polarized beam.

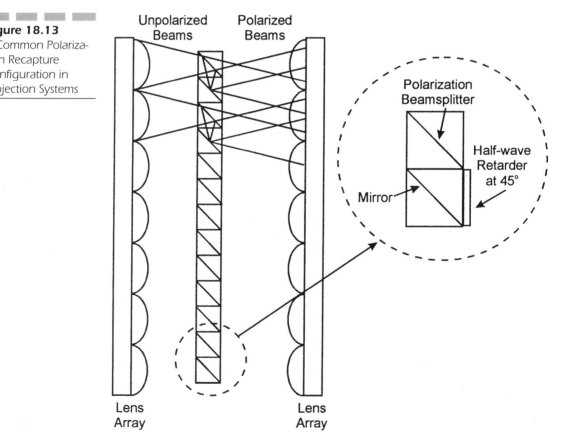

Figure 18.13
A Common Polarization Recapture Configuration in Projection Systems

Where in the Optical Train Should the Beam Be Polarized?

There are places in an optical system which are more friendly to polarization optics than other places. All polarization elements benefit from a high $f/\#$ due to the geometrical issues described in the section "Polarization Aberrations and Polarization Ray Tracing" earlier in this chapter and sometimes because the element is also constrained by its own angular performance. The angular cone in an optical system will vary inversely with the cross section of the beam—it may be worth the cost of a larger polarization optic to have the improved performance of a smaller cone angle.

Telecentric spaces are also good places to put polarization optics. In a nontelecentric space, the chief rays associated with different field points are not parallel. The cone, which converges to an image of any particular field point, will be tilted with respect to the cone of the adjacent field point. Therefore, the polarization state of each field will experience different aberrations than the field point adjacent to it. In telecentric space, each field point is composed of identical, parallel converging cones. The average polarization aberrations experienced are identical for each field point.

When polarization optics are in telecentric space, any angular sensitivity cannot be seen as a spatial nonuniformity. (However, any spatially varying polarization performance will be visible.) For example, the Maltese cross would not be visible if, instead of one point source, there were a continuum of point sources. All the resulting crosses would overlap and be shifted spatially with respect to each other. In such a situation, the cross pattern is lost. The leaked light still exists but it is shared uniformly over the image plane.

Likewise, stops and pupils are comparatively poor places to put polarization optics. In a pupil plane, any angular variation in performance of the part can be seen spatially at an image plane. If a Maltese cross were generated at the stop, then the pattern would be most recognizable at the image plane and less recognizable in any other optical space.

In the circumstance where there are two polarizers or retarders, such as the polarizer and analyzer in projectors, it is beneficial to put the elements conjugate to one another. This will allow the geometric issues to be minimized. For example, if the polarizer of a system must be at the stop, then the optimal place for the analyzer would be at the conjugate to the stop, another pupil plane. This will allow skew rays at the polarizer to be incident on the analyzer with a similar angle. Or, in other words, the orientation of the eigenstates associated with a ray and a polarizer align best with the following set of eigenstates for the same ray but at the second polarizer when the second polarizer is at a conjugate of the first.

The projector system shown in Fig. 18.11 uses many of these ideas at once. The polarizer and analyzer are (1) in telecentric space, (2) in a low working $f/\#$ space, (3) conjugate to each other, and (4) in the image plane, not the pupil plane. This allows the system to be free of the Maltese cross and to have very good contrast. Additionally, the polarizers do not adversely affect the brightness uniformity.

How Else Are Polarization Aberrations Avoided?

There are many techniques for avoiding or correcting for polarization aberrations. For each of these techniques, it is important to keep in mind where the polarization critical optical spaces exist. A polarization critical optical space is where the polarization aberrations must be minimal in order for the system to meet specifications. Once the polarization critical space is determined, decisions on how to influence the aberrations can begin.

In general, the polarization state is critical any place after the first polarizer or diattenuator. The first diattenuator in a system might not always be an obvious polarizer. Mirrors and thin-films stacks can polarize if tilted at a high angle. The polarization effects associated with any tilted element should be well understood to accurately define the polarization critical space. Analyzers, by contrast, can end the polarization critical space—after an analyzer, the polarization state often loses its criticality.

Decreasing the opportunity for aberrations is one option for improving the performance of a system. This is achieved by moving optics from the polarization critical space to another location. Whenever it is possible, fewer elements generally increase the robustness of a design.

Plastic lenses and thick pieces of glass should be placed outside of the polarization critical space. These elements are subject to stress birefringence. Minimizing the polarization aberrations often proves to be more difficult than moving the part. Also, TIR surfaces can influence polarization. Although there is no diattenuation associated with total internal reflection, the phase shift between the eigenstates rotates the polarization. The phase shift, δ, on TIR is

$$\delta = 2\tan^{-1}\left(\frac{\cos(\theta)\sqrt{\sin^2(\theta) - n^2}}{\sin^2(\theta)}\right)$$

where θ is the angle incident and n is the index of refraction of the media.

If moving the part is not an option, sometimes adding a compensating element can negate aberrations. For example, introducing a second tilted element, tilted in the plane perpendicular to the tilt plane of the first, adds polarization aberrations in the orthogonal orientation. (This concept is similar to how an achromatic doublet corrects for chromatic aberration.)

References

1. Jay E. Stockley, et al., "High-Speed Analog Achromatic Intensity Modulator," *Optics Letters,* **19** (10), 1994.
2. P. G. de Gennes, *The Physics of Liquid Crystals,* Oxford: Pergamon, 1977.
3. J.L. Fergason, U.S. Patent 3731986, 1973.
4. H. A. McLeod, *Thin-Film Optical Filters,* New York: American Elsevier, 1969.
5. Polaroid, U.S. Patent 3610729, 1971.
6. Eugene Hecht, *Optics,* Reading, MA: Addison-Wesley, 1974.
7. R. C. Jones, "A New Calculus for the Treatment of Optical Systems: I. Description and Discussion of the Calculus," *J. Opt. Soc. Am.,* **488,** 1941.
8. P. S. Hauge and R. H. Mueller, "Conventions and Formulas for Using the Mueller-Stokes," *Surface Science,* **96,** 1980.
9. Russell Chipman, "Polarization Aberrations," PhD dissertation, Optical Sciences, University of Arizona, 1987.
10. Schott Optical Glass, Schott Glass Technologies Inc.
11. Ohara Optical Glass, Ohara Inc.
12. Schott Optical Glass, Schott Glass Technologies
13. Seiko Epson Corporation, U.S. 5865521, 1997.
14. Max Born and Emil Wolf, *Principles of Optics,* Oxford: Pergamon Press, 1980.

15. William Shurcliff, "Polarized Light," published for the Commission on College Physics, Princeton, N.J., 1964.
16. Russell Chipman, ed., "Polarization Considerations for Optical Systems," *SPIE Proceedings*, **891,** 1988.
17. Russell Chipman, ed., "Polarization Considerations for Optical Systems II," *SPIE Proceedings*, **1166,** 1990.
18. Michael Bass, ed., *Handbook of Optics*, vol. 1, New York: McGraw-Hill, 1995.

For further reading on the fundamental theory of polarized light Born and Wolf's *Principles of Optics* and Shurcliff's *Polarized Light* are excellent resources. The *SPIE Proceedings* entitled "Polarization Considerations for Optical Systems" and "Polarization Considerations for Optical Systems II" contain many discussions on specific polarized system designs. Lastly, the *Handbook of Optics* from the Optical Society of America references many varied topics on polarization.

Optical
Thin Films

Introduction

Optical thin films have become an integral part of almost all optical components and systems manufactured today. Their primary function is to govern the spectral composition and the intensity of the light transmitted or reflected by the optical system. Properly applied to various optical surfaces in a given system, optical coatings can greatly enhance image quality and provide for a convenient way of spectrally manipulating light.

Since light behaves according to the laws of electromagnetic waves, the interaction of light with the media that it travels through, or is reflected from, is directly related to its wave nature, primarily the phenomena of *interference* and *polarization*. Whenever light interacts with a structure of thin films, interference occurs, and a degree of polarization will be a function of the angle of incidence. At normal incidence, no polarization will take place, unless the light is transmitted through a birefringent material (polarizing material, like some crystals and plastics). Besides polarization, at an oblique incidence there is a *spectral shift* of the reflectance or transmittance characteristic toward the shorter wavelength. This is due to the optical path difference between the waves reflected from either side of the film structure. This optical path difference is directly proportional to the cosine of the angle of refraction through the coating.

For an optical designer, besides the fact that the interference and polarization are the most fundamental physical principles in the theory of thin films, an important characteristic is the amount of energy loss, or the *light absorbed* in the coating. In general, for any coating there is a relationship between the transmittance, T, the reflectance, R, and the absorptance, A, in the form of

$$T + R + A = 1 \tag{19.1}$$

where $0 \leq T, R, A \leq 1$.

For materials that are commonly known as *dielectrics*, the coefficient A in Eq. (19.1) is very close to zero, and they basically do not absorb any light. On the other hand, *metals*, besides being highly reflective (90 to 98%), act as light attenuators, and their coefficient of absorption is always greater than zero.

We will refer to Eq. (19.1) later on when we discuss different categories of optical thin films.

Designing Optical Coatings

Without getting into deep analysis of design methods of optical thin films, let us point out that the main building blocks in designing optical coatings are *quarter-wave optical thickness* (QWOT) layers of different materials. The high, medium, and low refractive index QWOT materials are usually denoted as *H, M,* and *L,* respectively. If there are two QWOT layers of the same material next to each other, they form a *half-wave optical thickness* (HWOT) layer. If only a fraction of QWOT appears in a design, say one half of *H,* it is represented either as 0.5*H,* or *H/2.*

The long expressions for some designs can be represented in concise form. For example, a 15-layer longwave-pass filter on BK7 glass given by

$$\text{BK7}\left|\frac{H}{2} \; LHLHLHLHLHLHL \; \frac{H}{2}\right|\text{air}$$

can be written as

$$\text{BK7}\left|\left(\frac{H}{2} \; L \; \frac{H}{2}\right)^{7}\right|air$$

where *H* and *L* refer to high and low index materials, such as TiO_2 and SiO_2.

In principle, the computer programs that assist thin-film engineers in designing optical coatings are very similar to those used by optical designers. Optical design programs are more complex because there are more variables (such as thickness, radius of curvature, refractive index) to simultaneously vary during the optimization. Further, they have a wider spectrum of the target functions to be satisfied at the end of the optimization (either in the form of the aberration functions, wavefront distortion, optical path difference, or the minimum spot size). Thin-film programs, on the other hand, deal with fewer variables (very often only thickness, rarely refractive index), and their target functions are usually in the form of either reflected or transmitted light intensity.

Thin-film computer programs are essential mathematical tools that enable coating engineers to efficiently, and in some cases very quickly, arrive at the best and most economical design once the problem has been formulated. But to successfully apply this math tool to coatings that are manufactured with high reproducibility, it is the engineer's knowledge of the coating materials and processes that determines the coating's final quality and conformity to the spectral and environmental requirements.

Various Categories of Optical Coatings

The most widely applied optical coating is the *antireflection (AR)* coating. Its primary purpose is to reduce the amount of reflected light from the optical surface. Its secondary role is to enhance physical and chemical properties of the surface to which it is applied.

Typically, uncoated glass has between 4 and 8% reflection from the surface. This can be reduced to about 1.0% reflection in the visible by applying a single layer of QWOT low-index material, usually magnesium fluoride (Fig. 19.1). A three-layer design can reduce the reflection in the visible even further (Fig. 19.2). The first layer consists of a QWOT medium-index material (for example, Al_2O_3) next to the glass. The second layer is a HWOT high-index material (for example, Ta_2O_5). The third layer is a QWOT low-index material (for example, MgF_2) as a top layer next to the air. This three-layer design falls in the category of the *broadband* (BB) antireflection coating, often denoted as BBAR coating.

If only one wavelength is considered, a two-layer design of high- and low-index materials will bring the reflection down to virtually zero value. With the layer next to the glass fairly thin (high-index material) and the layer facing the air side (low-index material) somewhat greater than a QWOT, a relatively broad minimum can be obtained (Fig. 19.3). These coatings are usually called *V* coatings.

For much broader antireflective coverage that would include the visible and a near-infrared region, many layers of high- and low-index materials are required. Their thicknesses are computer optimized and monitored throughout the deposition process using either the quartz monitor or the combination of quartz and optical monitoring. For example, the BBAR coating that covers 450 to 1100 nm (Fig. 19.4) would require eight or more layers for the reflection to be less than 1.0% at any wavelength within the region.

Another class of widely used thin-film coatings is the *metallic mirror,* usually consisting of aluminum. Aluminum is a relatively soft metal, so the coating is often protected with silicon dioxide. The reflectance of this coating is about 90%, but can be further increased by adding a few more layer pairs of high- and low-index materials (for example, TiO_2 and SiO_2) to boost reflectance to about 97 to 98% (Fig. 19.5). Since aluminum is a metal, there is a slight light loss associated with its use. This light loss, or absorption, is manifested as heat released within the coating. In

Figure 19.1
Computed
Reflectance at Nor-
mal Incidence of a
Single Surface of
SSK4 Glass ($n = 1.62$)
Coated with a Single
Layer of Magnesium
Fluoride ($n = 1.38$) of
Optical Thickness
One Quarter-Wave at
510 nm. Design:
SSK4| *L* |Air, AOI = 0°

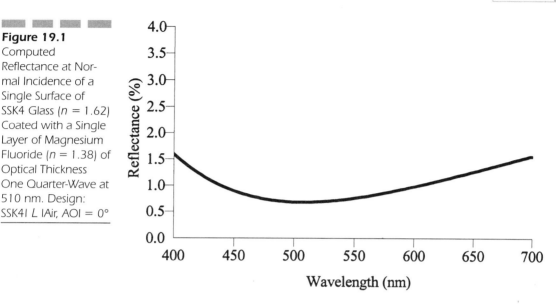

Figure 19.2
Three-Layer Antire-
flection Coating on
BK7 Glass ($n = 1.52$).
(Design: BK7| *MHHL*
|Air at 510 nm, AOI =
0°, $n_H = 2.126$, $n_M = 1.629$, $n_L = 1.384$)

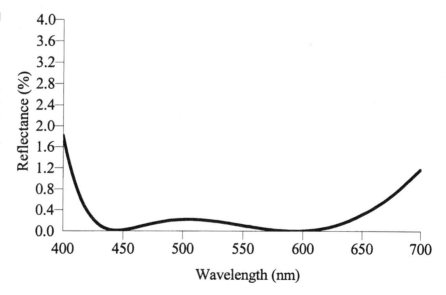

Figure 19.3
The Reflectance of a
Two-Layer Antireflec-
tion Coating on BK7
Glass. (Design: BK7|
0.2681H 1.2702L
|Air at 500 nm,
AOI = 0°, n_H =
2.127, n_L = 1.384)

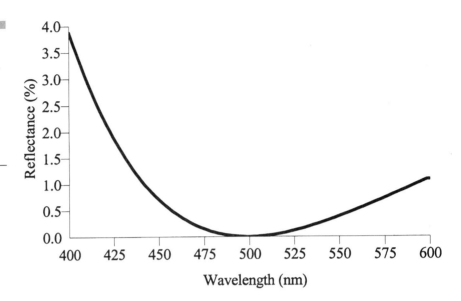

Figure 19.4
Eight-Layer Antireflec-
tion Coating on BK7
Glass (n = 1.52). The
Coating Consists of
Two Materials of
High- and Low-
Refractive Index. This
Design Has Been
Computer Optimized
and Has a Few Thin
Layers (~20 nm) That
Can Only Be Quartz
Monitored. The
Angle of Incidence Is
0°

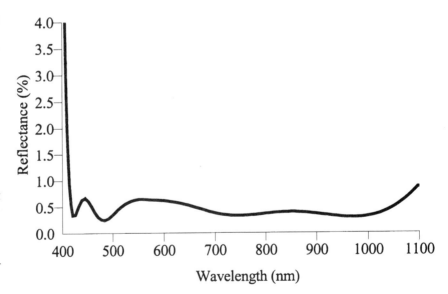

Figure 19.5

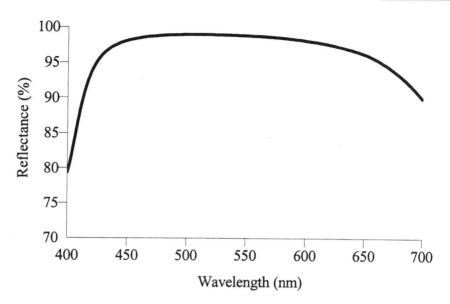

Figure 19.5
The Reflectance of an Enhanced Aluminum Mirror with Four Layers of High- and Low-Index on Top of Aluminum. The Thickness of the Aluminum Layer Is 80 nm. (Design: Al I 0.8$LHLH$ IAir at 520 nm, AOI = 0°, n_H = 2.446, n_L = 1.459)

certain applications, such as high-power lasers, mirrors should be free of absorption to a very high degree. This is achieved through the use of all-dielectric mirrors.

Dielectric mirrors consist of the sequence of the alternating high- and low-index materials (for example, TiO_2 and SiO_2). The more layer pairs in the stack, the higher the reflectance. *Cold mirrors* reflect shorter wavelengths and transmit longer wavelengths (Fig. 19.6). *Hot mirrors* transmit shorter wavelengths and reflect longer wavelengths (Fig. 19.7).

As in the field of electronic circuits, there are many different interference filters in a variety of optical applications. Sometimes the goal is to separate one portion of the spectrum from the other. This separation can be done either at normal incidence or oblique incidence. Whatever the case, the solution will be in the form of an *edge filter* or some kind of *dichroic beamsplitter*.

When there is a need to pass just one narrow bandwidth and reflect a portion of the spectrum on either side of it, use should be made of a *narrowband* interference filter, often called the *Fabry-Perot* filter (Fig. 19.8).

Recently, another class of interference filters has become of great importance in laser and fiber-optic applications: *notch filters*. They reflect one or more narrow bands and transmit the wider regions around the rejection zone (Fig. 19.9). To maintain a narrowband characteristic of the rejection zone, this filter is often designed using low- and medium-index

Figure 19.6

The Transmittance of 52-Layer Cold Mirror at a 45° Angle of Incidence. Design Is Given in *Phase Thicknesses* (Degrees), and P_H and P_L Refer to TiO$_2$ and SiO$_2$, Respectively. [Design: Air|$(100°P_L)(74°P_L$ $74°P_H)^8(90°P_L 90°P_H)^8$ $(108°P_L 108°P_H)^7$ $(105°P_L)(102°P_H)(98°P_L)$ $(90°P_H)(98°P_L)(23°P_H)$ |BK7 at 538 nm]

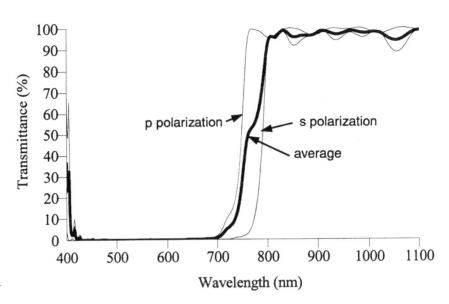

Figure 19.7

Calculated Transmittance of a 44-Layer Computer Optimized Hot Mirror. [Design: Air|$(1.07L(2H2L)^8(2.6H)(2.64L)(2.8H)(2.46L)(2.14H)(2.2L)(2.6H 2.6L)^3(2.6H)(2.74L)(2.9H2.9L)^5(2.74H)(3.08L)(0.4H)$ |BK7 at 415 nm, AOI = 0°, $n_H = 2.239$, $n_L = 1.463$]

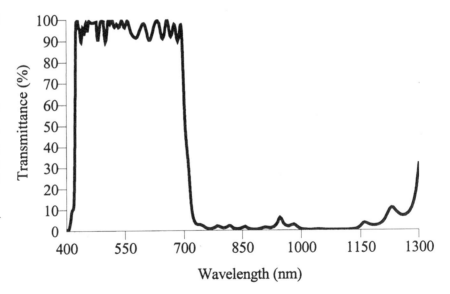

Optical Thin Films

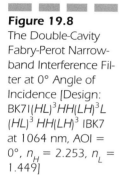

Figure 19.8
The Double-Cavity Fabry-Perot Narrowband Interference Filter at 0° Angle of Incidence [Design: BK7|$(HL)^3HH(LH)^3L$ $(HL)^3HH(LH)^3$ |BK7 at 1064 nm, AOI = 0°, n_H = 2.253, n_L = 1.449]

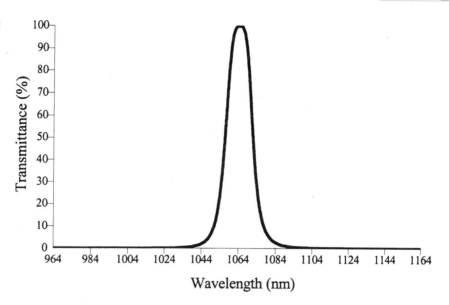

Figure 19.9
Computed Reflectance of a Single-Notch Filter [Design: BK7| $(L3M)^{31}4L$ |Air at 580 nm, AOI = 0°, n_M = 1.626, n_L = 1.457]

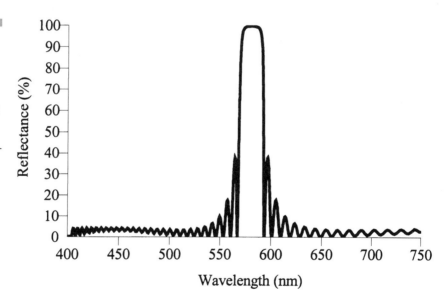

materials. This, in turn, requires many layers to achieve a high reflection. Essentially, their function is just the opposite of the narrowband filters.

With the advent of new polarizing devices in the area of electronic imaging, *polarizing beamsplitters* have become of significant importance. Their role is to maximize the s and minimize the p-reflectance of the unpolarized (randomly polarized) light over the narrow or broadband spectral region. The degree of polarization in transmission is

$$P_T = \frac{T_P - T_S}{T_P + T_S}$$

and in reflection

$$P_R = \frac{R_S - R_P}{R_S + R_P}$$

The *extinction ratio* indicates how well the polarizing beamsplitter discriminates between two planes of polarization. In transmission it is given as a ratio of T_p and $T_{s'}$ and in reflection as a ratio of R_s and R_p. When the degree of polarization is very high, the reflected linearly polarized s-component and the transmitted linearly polarized p-component should each account for 50% of the incoming light intensity. Thus, an ideal polarizing beamsplitter acts as the 50/50 intensity beamsplitter, where each of the two emerging light beams are 100% linearly polarized (Fig. 19.10).

Optical Coating Process

Optical coatings are manufactured in high-vacuum coating chambers. Conventional processes require elevated substrate temperatures (usually around 300°C), whereas more advanced techniques, like ion-assisted deposition (IAD) are utilized at room temperatures. IAD processes not only produce coatings with better physical characteristics compared to conventional ones but also can be applied to substrates made out of plastics. Figure 19.11 shows an operator in front of the optical coating machine. Its main pumping system consists of two cryopumps. Control modules for electron-beam evaporation, IAD deposition, optical monitoring, heater control, pumping control, and automatic process control are in the foreground. Figure 19.12 shows the configuration of the hardware mounted on the base plate of a high-vacuum coating machine. Two electron-beam sources located at each side of the base are surrounded by circular shields

Figure 19.10
The Polarizing Poly-
chromatic Cube
Beamsplitter. The
Computed
Reflectance Repre-
sents a 15-Layer
Design Consisting of
Two Materials of
High and Low Refrac-
tive Index. The Angle
of Incidence is 52°

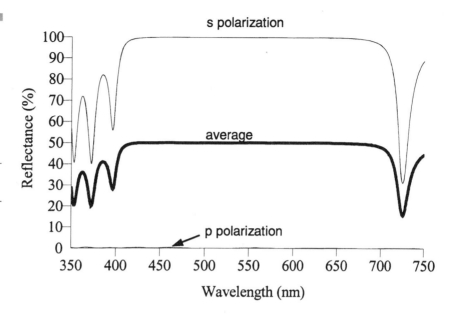

Figure 19.10
The Polarizing Poly-
chromatic Cube
Beamsplitter. The
Computed
Reflectance Repre-
sents a 15-Layer
Design Consisting of
Two Materials of
High and Low Refrac-
tive Index. The Angle
of Incidence is 52°

and covered with shutters. The ion source is located in the middle. The optical monitor windows are in the front of the ion source. Figure 19.13 shows the upper part of the vacuum chamber, which is occupied by the planetary system with six round fixtures. Fixtures are loaded with optics to be coated. A use of a planetary system is a preferred way of maintaining a uniform distribution of the evaporated material across the area of the fixture. Fixtures turn around their common axis and revolve around their own axes. The optical and quartz monitors are in the middle of the planetary drive mechanism, the latter being obstructed by the drive hub. The large opening in the background leads to an additional high-vacuum pump. The substrate heating system consists of four quartz lamps, two at each side of the chamber.

The traditional methods of thin-film deposition have been thermal evaporation either by means of resistance-heated evaporation sources or by electron-beam evaporation. Film properties are determined mostly by the energies of the depositing atoms, which are only around 0.1 eV in conventional evaporation. IAD deposition results in direct deposition of ionized vapor and in adding activation energy to the growing film, typically in the order of 50 eV. Using the ion source, conventional electron-beam evaporation is improved by directing the flux from the ion gun to the surface of the substrate and growing film.

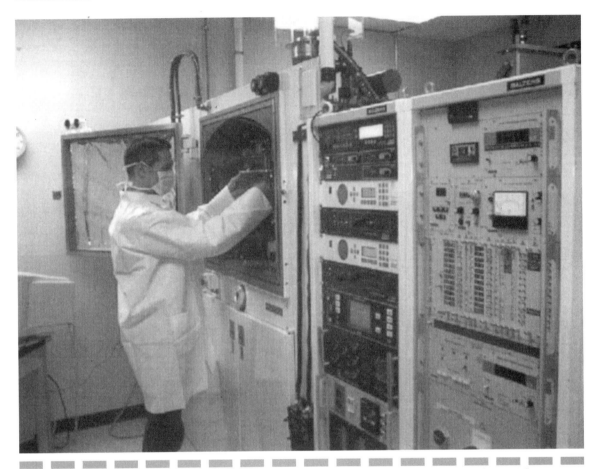

Figure 19.11 An Operator in Front of the Optical Coating Machine (*Courtesy of LaCroix Optical Co., Batesville, Arkansas*)

The optical properties of films, such as refractive index, absorption, and laser-damage threshold, depend largely on the microstructure of the coating. The film material, residual gas pressure, and substrate temperature can all affect the microstructure of the thin films. If the depositing vapor atoms have a low mobility on the substrate surface, the film will contain microvoids, which will be filled subsequently with water when the film is exposed to a humid atmosphere.

We define the *packing density* as the ratio of the volume of solid part of film to the total volume of film (which includes microvoids and pores). For optical thin films, it is usually in the range 0.75 to 1.0, very often 0.85 to 0.95, and rarely as great as 1.0. A packing density that is less

Figure 19.12
The Configuration of the Hardware Mounted on the Base Plate of a High-Vacuum Coating Machine (*Courtesy of LaCroix Optical Co., Batesville, Arkansas*)

than unity reduces the refractive index of evaporated material below the value of its bulk form.

During the deposition, the thickness of each layer is monitored either optically or by using a quartz crystal. Both techniques have advantages and disadvantages that are not discussed here. What they have in common is that they are done in a vacuum while the material is evaporated. Consequently, they represent the refractive index of evaporated material in a vacuum, not the one that the material will acquire after being exposed to humid air. Moisture adsorption in the film results in displacement of air from microvoids and pores, causing an increase in the

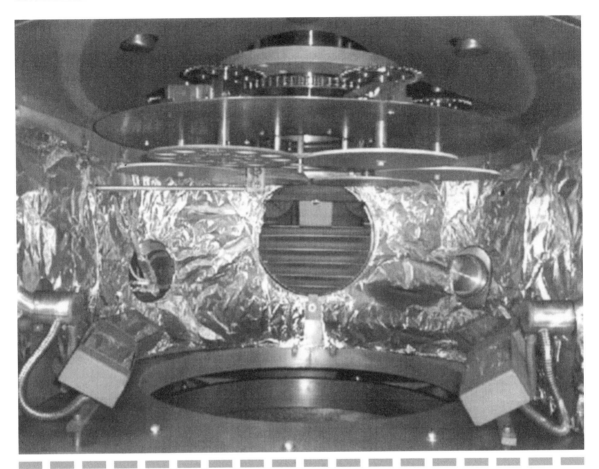

Figure 19.13
The Upper Part of the Vacuum Chamber Is Occupied by the Planetary System with Six Round Fixtures (*Courtesy of LaCroix Optical Co., Batesville, Arkansas*)

refractive index of the film. Since the physical thickness of the film remains constant, this refractive index increase is accompanied by a corresponding increase in optical thickness, which in turn results in the spectral shift of the coating characteristic toward a longer wavelength. To minimize this spectral shift caused by the size and overall population of microvoids throughout the growing film, high-energy ions are employed to convey their momentum to the atoms of evaporating material, thereby largely increasing their mobility during the condensation at the substrate surface.

Coating Performance Versus Number of Layers

We have mentioned earlier that the optical coating materials fall into two groups: *dielectrics* and *metals*. All of the preceding various optical coatings, except some metal mirrors, utilize dielectric materials in their design. Among dielectrics, the most often used are oxides and fluorides. One technological problem associated with the deposition of high-index oxide materials is their tendency to dissociate into oxygen and some lower forms of the original oxide. To avoid absorption in the depleted coating and to keep the coefficient A in Eq. (19.1) as close as possible to zero, it is necessary to *reoxidize* material before it condenses on the substrate, thereby preserving the *stoichiometry* of the bulk material.

One could think that the greater the number of layers, the better the coating performance. However, *given the manufacturing technology, there is a limit to a maximum number of layers that will produce the coating with the best characteristics*. For an optical designer just using an optical design program, it becomes a relatively straightforward conclusion that adding more surfaces and glasses to a certain, already well-corrected lens, for example, a double gauss photographic lens, will cause the image to deteriorate. Thin-film programs do not take into account physical characteristics of the coating microstructure and the atomic and molecular forces that exist between layers of different materials and within each layer. Consequently, thin-film programs cannot predict the physical behavior of the final coating design as much as the optical design programs can predict and characterize the image quality of an optical system. To illustrate this, let us take an example of a high-reflection dielectric mirror. It consists of a sequence of layer pairs of high- and low-refractive index materials (for example, TiO_2 and SiO_2), where each layer is QWOT. Assuming absorbing media, 12 of these layer pairs (implying a coating consisting of 24 layers) would boost reflectance to 99.9% at 530 nm. Adding another eight layers would not result in any considerable improvement. This is shown in Fig. 19.14. This 32-layer coating would have the same reflection of 99.9% at 530 nm, higher absorption, and greater overall thickness. Although with the slightly broader characteristic, it would probably be inferior to the 24-layer design because of a greater possibility of crazing (breaking off the coating because of high-tensile stress) and higher absorption that offsets the gain in reflectance.

Specifying Coating Requirements

Accurate specification of coating requirements assumes an understanding of the coating function, the function of the optical component to which it is applied, and the coating usefulness in a particular application.

For example, to increase the transmittance of an optical glass surface in the visible domain to 99.0% or more would require a broadband antireflection coating (BBAR) from 400 to 700 nm, for which Eq. (19.1) can be written in the following form

$$T = 1 - R - A \geq 0.99$$

or

$$R + A \leq 0.01 \qquad \text{from 400 to 700 nm} \tag{19.2}$$

The last inequality expresses the requirement that the sum of the reflectance and absorptance should not exceed 1.0% for any wavelength in the interval 400 to 700 nm. Very often Eq. (19.2) is written as

$$R \leq 1.0\% \qquad \text{from 400 to 700 nm}$$

Figure 19.14
The Reflectances of Two Dielectric Mirrors at 0° Angle of Incidence. Design Wavelength Is 525 nm, and the Coefficient of Absorption of High-Index Material Is 0.00027. The Upper Curve Represents a 32-Layer Design BK7I $(HL)^{16}$ IAir, and the Lower One Represents a 24-Layer BK7I $(HL)^{12}$ IAir. The Refractive Indices of Two Materials Are 2.336 and 1.461

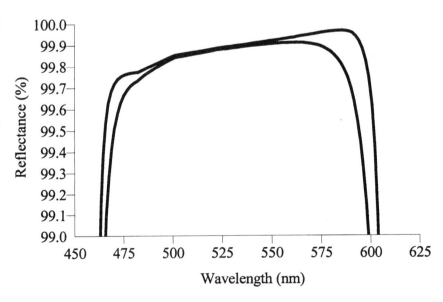

assuming that the absorptance is close to zero ($A \approx 0$). If the glass is BK7 and the angle of incidence (AOI) of the light striking the glass surface is between 0 and 15°, then the fairly complete and accurately formulated requirement would be in the form

BBAR on BK7 glass

R(400 to 700 nm) $\leq$ 1.0% @ AOI = 0 to 15°

$A \approx 0\%$

To avoid some possible misinterpretations of the coefficient A, its maximum value can always be explicitly stated on the coating blueprint.

Relationship Between Production Cost, Tolerances, and Quality

The production cost per run of a particular coating is primarily determined by the size of the coating chamber, the manufacturing technology, and the complexity of the coating. Since the area of the coating chamber that can be used to coat parts is more or less directly proportional to the square of its radius, it follows that the bigger the chamber, the lower the price per coated lens. As an example, if the diameter of one chamber is twice the diameter of the other, then approximately 4 times more lenses can be coated in the first chamber than in the second one.

For some extremely stringent requirements, often found in the production of narrowband filters, it is not always possible to utilize the whole coating area within one chamber but rather one particular segment of it. This is because of the nonuniformity of the coating distribution across the chamber. Therefore, depending on the type of the coating, the capacity of the coating machine can be governed by the tolerances on the spectral characteristics of the coating.

For well-designed coating machines, the distribution of the spectral characteristic of evaporated material stays within ±1% of the nominal value. For example, the coating represented by Fig. 19.3 would have the range of reflectance minima from 495 to 505 nm. The inconsistency between different runs could further increase this range, say from 490 to 510 nm.

Besides the spectral conformity of the coated lens to the prescribed value, its quality is further governed by the least amount of coating voids, good adhesion and hardness, environmental stability, and the high packing density.

Different deposition techniques have been invented over the past 20 years in order to increase the packing density of evaporated material to the value close to unity. The most important ones are ion-assisted deposition (IAD), ion-beam deposition (IBD), and ion plating. We could finally say, the closer the packing density to unity, the more expensive the coating.

Bibliography

Holland, L. (1956) *Vacuum Deposition of Thin Films*. London: Chapman & Hall.

Jacobson, M. (1986) *Deposition and Characterization of Optical Thin Films*. New York: Macmillan.

Macleod, H. A. (1986) *Thin-Film Optical Filters*. New York: Macmillan.

Pulker, H. K. (1984) *Coatings on Glass*. Amsterdam: Elsevier.

Rancourt, James D. (1987) *Optical Thin Films Users' Handbook*. New York: Macmillan.

Thelen, Alfred (1989) *Design of Optical Interference Coatings*. New York: McGraw-Hill.

20

Hardware Design Issues

There are many optical system design issues which relate directly to the ultimate hardware implementation, yet are different from the subjects we have covered thus far. It is important that the designer be reasonably fluent in these areas. They include the use of off-the-shelf optics, baffling and stray light control, and optomechanics.

Off-the-Shelf Optics

Off-the-shelf optics is, in effect, catalog optics. One of the significant advantages of off-the-shelf optics is that if what you need is in stock, you can have nearly immediate delivery. Unfortunately, the converse is also true: if what you need is *not* in stock, you may be faced with a long delivery time, perhaps in the order of 12 to 16 weeks.

The forms of off-the-shelf optics follow.

Precision Lens Assemblies

This first class of off-the-shelf optics includes relatively precision lenses such as camera lenses, relay lenses, enlarging lenses, and other multielement lens assemblies of reasonable quality. These lenses are most often mounted in nice-looking anodized housings, and may have adjustable f/numbers and focusing capability.

The optical and mechanical quality of these lenses may or may not be good. Just because the lenses are mounted in a beautiful black anodized housing with red, blue, green, and yellow engraving and just because the lenses are coated with a nice deep blue antireflection coating, there is no assurance whatsoever that the optical performance is any good. In addition, the focal length and f/number may or may not be per the specification. Moreover, the image quality may or may not be good. This is not to say that the specifications are not as advertised, nor is the performance necessarily poor, we only bring this up as a caution so that you are not misled by the external appearance of the lens assembly.

As with most off-the-shelf optics, these types of lenses are available almost immediately. If your lens is out of stock, delivery could take as long as 3 to 4 months, or longer, if indeed the same lens is ever again available. When you are dealing in a commercial commodity-like product line arena such as with 35-mm camera lenses, there is a rapid changeover in products, making future availability of a given lens a real questionable issue.

The cost of off-the-shelf optics in the form of completed lens assemblies can range from under $100 to over $500, or more, depending on manufacturing costs, volume, and, of course, quality. A good example of such a lens is a name brand 50-mm focal length $f/2.0$ 35-mm camera lens, which we used recently for a laboratory test. It cost less than $200 and performed extremely well for the intended purpose. The lens had six elements, and while its housing was partially plastic, it seemed robust enough for most applications.

Single Elements and Achromatic Doublets

The major catalog companies have several hundred different single-element lenses and achromatic doublets available. They typically range from approximately 1.5- to 2000-mm focal length, in diameters from approximately 1.5 to 150 mm. Their optical quality and level of tolerances are generally reasonable for many nondemanding applications and they are generally available uncoated as well as antireflection coated. Do not, however, expect to find extremely high-precision optics in this commodity area.

The cost of catalog single elements and doublets of small diameters up to approximately 50 mm are in the order of $60 to $150 each. The cost of *custom* single elements and achromatic doublets can be approximately $350 per element in low quantities, and delivery can be 6 to 8 weeks. Delivery in 1 week is available from several vendors, naturally at a premium price. As soon as the quantities increase to between 50 and 100, the price of custom lenses drops to prices close to catalog levels.

Other Forms of Off-the-Shelf Optics

There are many other forms of off-the-shelf optics available, including prisms, windows, mirrors, beamsplitters, polarization components, filters, and more. Also, there is the relatively new class of microoptics available off the shelf such as laser diode collimators and focusing optics.

How to Effectively Work with Off-the-Shelf Optics

If you are careful in use of off-the-shelf optics, you can be highly successful. On the other hand, if you are too casual and don't pay attention to details, your project could easily end up in trouble. Some guidelines gleaned over the years follow.

Complete lens assemblies are the most difficult to deal with. Manufacturers, such as the major camera companies, simply will not share with anyone the lens design prescription. Your ability, therefore, to input the design into one of the lens design and analysis software programs and interface it with other off-the-shelf, or even custom optics, becomes difficult, if not impossible.

If not given by the manufacturer, you could certainly have some of the basic parameters measured such as the focal length, f/number, and entrance and exit pupil locations. While not trivial to characterize, these parameters can indeed be measured. However, what you cannot do easily is measure the residual lens aberrations in order to factor them into a more complex system model to be used with other lens groups. Thus, incorporating off-the-shelf lenses with custom lenses can lead to serious problems.

There is one very important matter that must be considered, and that is that a lens designed for one set of specifications or parameters may or may not perform well under different conditions. For example, if we procure a 35-mm focal length f/2.8 double gauss camera lens from a well-known manufacturer, and we then proceed to use it at a near-unit magnification to image postage stamps or integrated circuit chips onto a CCD sensor, we will likely be very disappointed in its performance. The lens was most likely designed for an infinite object distance or a distant object, and at a unity magnification, it will most likely perform very poorly. In addition to spherical aberration, the lens will suffer seriously from astigmatism and other off-axis aberrations.

The same holds for other specifications. Again, using this camera lens as an example, if we use it over a wider field of view or a larger spectral band than it was designed for, we will likely have poor performance. In addition, there may be distortion, which is not an image quality issue but rather a mapping error. If your application requires a precision machine vision lens with low distortion, then this must be measured for the proposed off-the-shelf lens.

Working with Off-the-Shelf Singlets and Doublets

This task is far more straightforward than working with complex lens assemblies due to two factors:

1. Many of the lens design software packages have included the design prescriptions of singlets and achromatic doublets from most of the major catalog suppliers. For example, Zemax has resident lens prescriptions from Edmund Scientific, Melles Griot, Opto-Sigma, Rolyn, Newport Corporation, Coherent, Spectra Physics, and Spindler and Hoyer. Fortunately, the suppliers of these lenses realize that they can serve the technical community far better by providing this information rather than being secretive.

2. The lenses are fundamentally simple lenses with little to be concerned about with respect to pupils for example. In addition, even if the prescription is not available, you could generate a candidate design and have a moderate level of confidence that the real lens will be close to the catalog lens.

For example, let's assume that you find in some new catalog an achromatic doublet which has a focal length of 78 mm and a diameter of 10 mm, but the catalog is not resident in your design package. If you are confident that the lens was designed for an infinite object distance, you could in a matter of a few minutes emulate it with reasonable confidence of the design being at least sufficiently close to the actual design to be useful in your computer modeling. You might, for example, select BK7 glass for the crown element and SF2 glass for the flint and optimize it. The results should be reasonably close to the real lens.

In developing a lens design for which you intend to explore the potential of using off-the-shelf singlets and/or doublets, a good procedure to follow is to first perform the design yourself so as to meet the system's first-order and performance specifications. You may want to begin with a first-order design using so-called paraxial lenses, and later convert it to a real design. Once you feel comfortable with your design, then you need to evaluate the focal length of the singlets and/or doublets which you intend to match to off-the-shelf components. If you intend to use a planoconvex singlet, then in the design you should also use a planoconvex element; the same holds true for planoconcave lenses and equiconvex or equiconcave lenses. Now you need to look in one or more catalogs for lenses that match closely the parameters of your lenses (in particular, the focal length and diameter) and replace your lenses with the catalog lenses. Most of the software packages allow you to simply insert any off-the-shelf lens into an otherwise custom design. Make sure you pay attention to the lens orientation, or which way the crown and flint elements are oriented.

At this point, you may find that your performance and other specifications are adequately met, in which case you can freeze the design and procure the lenses. On the other hand, you may find that for one reason or another the design requires further optimization, in which case you need to comply. This may require customization of one or more lens groups for example. Often your final design might include a mix of off-the-shelf components as well as custom components. You will likely find that as you incorporate more off-the-shelf components into a given design, its performance will degrade from optimum. However, the important question to be answered is whether the performance is *good enough.*

Example of Lens Used at Conjugates Different from What It Was Designed

To illustrate some of the preceding issues, Fig. 20.1 shows the layout and performance for a 35-mm focal length $f/2.8$ double Gauss lens designed for an infinite object distance. Figure 20.2 shows the performance of the same lens with an object distance of 0.5 m. Note that the plot scales are maintained and are identical in all of the figures in this analysis, and the MTF data are plotted to 50 line pairs/mm. At first glance, the transverse ray aberrations look similar to the previous nominal design data, and indeed there is not a significant degradation. However, note that the MTF has suffered a significant drop, especially off axis. In Fig. 20.3, we show the same lens at a 100-mm object distance (this results in a demagnification of $3\times$). The performance is significantly degraded from the nominal lens performance. Figure 20.4 shows the performance if the lens is used at 1:1, or unity, magnification. In this case, the performance is extremely poor. At the edge of the field there is over an order of magnitude increase in spot diameter when the lens is used at 1:1 magnification!

Figure 20.5 shows parametrically how the performance degrades as a function of object distance for the previous lens design example. Note that the nominal design gives an rms blur diameter of approximately 9 μm over most of the field of view. This doubles for the 0.5-m object distance, and for a 100-mm object distance the rms blur diameter increases to about 50 μm over the central region of the field of view. At the unit magnification position, due to the extreme aberrations introduced, the spot diameter ranges from 70 to about 340 μm at the edge of the field. If you were using a CCD chip with a 12-μm pixel pitch, an object distance of no more than 0.5 to 0.75 m would be viable in order to maintain a reasonable modulation at the Nyquist frequency.

Pupil Matching

In addition to the basic specifications, performance, and aberrations, the extremely important issue of entrance and exit pupils must be considered when working with off-the-shelf optics. Clearly, if you were simply using an off-the-shelf lens to image an object onto a CCD array, the location of the pupils is of little concern or interest. However, if you had a multiple-stage relay system, then the exit pupil of one stage must be coincident or nearly coincident with the entrance pupil of the next stage, and so on. As

Figure 20.1
A 35-mm Focal Length ƒ/2.8 Lens at Infinity

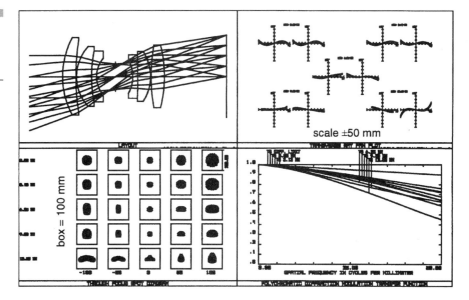

Figure 20.2
A 35-mm Focal
Length *f*/2.8 Lens at
500-mm Object
Distance

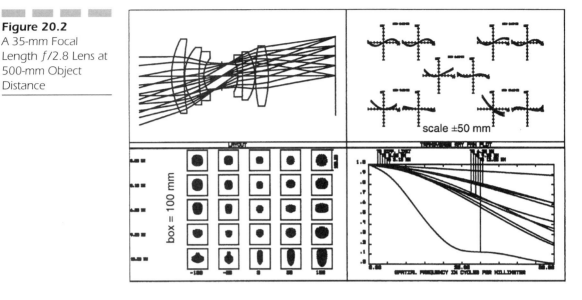

Figure 20.3
A 35-mm Focal
Length *f*/2.8 Lens at
100-mm Object
Distance

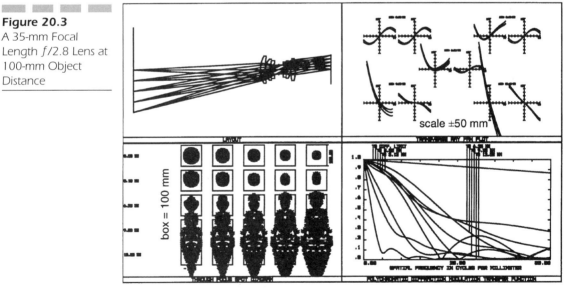

Figure 20.4
A 35-mm Focal Length ƒ/2.8 Lens at Unit Magnification (32.41-mm Object Distance)

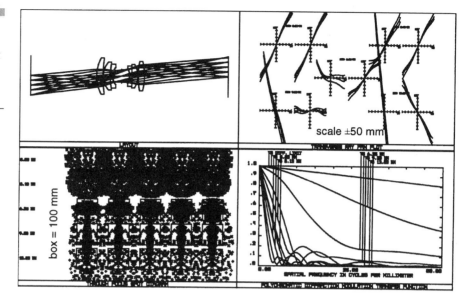

Figure 20.5
RMS Blur Diameter for a 35-mm Focal Length ƒ/2.8 Lens As a Function of Object Distance

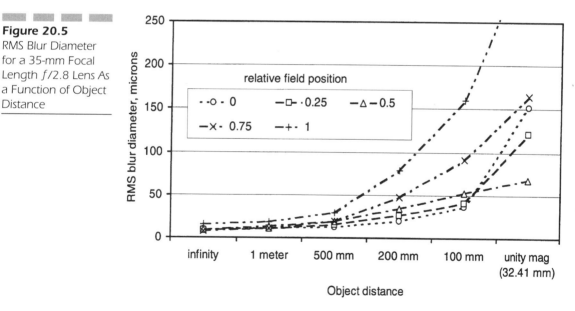

we learned earlier, field lenses are indispensable in this task, as one of their primary roles is to reimage the exit pupil of one lens group into the entrance pupil of the next lens group. This issue can be of major concern when one or more of the off-the-shelf lenses are zoom lenses, since the pupils can translate or move over great distances as the lenses zoom, and having a mismatch in pupils is very likely, if not inevitable.

Development of a Lab Mockup Using Off-the-Shelf Optics

There are many situations where determining the level of performance of your system quickly is to your advantage. Situations where this approach is helpful are when validating important aspects of your specifications. For example, assume you are designing a new visual telescope. Parameters, which are important, include field of view, magnification, eye relief (clearance from the last element to the eye), and of course image quality. You could manufacture a prototype of your custom production design, which would likely take several months and cost many dollars. An alternate approach is to build up a unit using off-the-shelf optics. It should be straightforward to nearly meet the magnification and field-of-view specifications, and likely the eye relief too. The image quality may be degraded from your custom production design; however, the overall ability to assess the general nature of the system specifications and performance is often quite valuable. This is especially true for some of the specifications, such as field of view and magnification, which may have been based on a judgment or best-effort basis. You can take your mockup system outside and use it in a near-real functional environment. There is always an anticipated level of performance associated with every lens design, and your system performance can often be demonstrated using off-the-shelf optics.

Stray Light Control

The suppression of stray light is often ignored until it is too late, and then it becomes costly and time consuming to fix the problem. Good engineering in this area is imperative. The best way to learn the subject is through the following two examples: (1) a machine vision system and (2) a reflective Cassegrain telescope.

Machine Vision System

We will first relate a true story regarding a potentially serious stray light problem:

- We were called in to visit a colleague who said that he had just installed a new vision system and the contrast was badly degraded from prior systems. The system was very basic and consisted of a microscope objective and a CCD camera. The contrast was indeed poor on the video monitor.

- We first removed the camera from the tube assembly and looked in with our eye at a location similar to where the CCD chip was. It was immediately evident that there was a lot of stray light reflected from the interior of the tube assembly. While the tube interior was black anodized, at near-grazing angles of incidence black anodizing is quite reflective. Figure 20.6a shows the situation.

- We then asked our host if he had any flat black paper, and he did. We rolled the paper into a tube shape and put it into the anodized

Figure 20.6
Example of Stray-Light Control in Machine Vision System

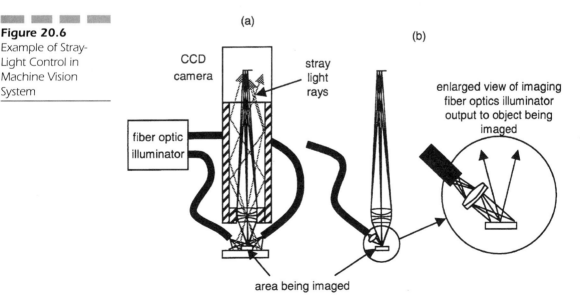

(a)

(b)

CCD camera

fiber optic illuminator

stray light rays

enlarged view of imaging fiber optics illuminator output to object being imaged

area being imaged

tube assembly. Visually, with our eye again looking into the tube, we found that the situation was indeed improved but not perfect.

- After reinstalling the camera, our host said, "wow, that's a lot better...but it isn't as good as it used to be." This was consistent with our observation.

- We then took a careful look at the overall system layout and we realized that the fiber-optic illuminators were illuminating an area far exceeding the object being imaged, as shown in Fig. 20.6a. We asked our host if he had a small positively powered singlet or doublet lens and he did. We then cut a small aperture in a piece of black paper and fastened it to the end of the fiber bundle. The lens was now used to reimage the aperture onto the object being imaged, as shown in Fig. 20.6b. A little experimentation with the magnification resulted in a situation where we were just overfilling the object area of interest.

- We now turned the system on and our host said, "wow, that's better than it has ever been!"

If you were to design the microscope tube assembly for optimum stray-light attenuation, it would be best to incorporate baffle features on the interior of the tube, as shown in Fig. 20.7. Figure 20.7a shows a series of baffle structures similar to washers. This is one of the most efficient baffle forms; however, it is somewhat costly to machine or otherwise implement. Figure 20.7b shows a coarse thread with which we can derive good results. Note that we show the multiple bounce path of several representative rays, and in the case shown none of the rays reaches the CCD sensor until after three bounces, which is a good guideline. Do keep in mind that there will inevitably be scattering and diffraction coming from the tops of the threads, no matter how perfectly they are machined, and you will be better off with a coarse thread rather than a fine thread with more thread tops. One final tip: it will help if you make the inner diameter of the tube and associated baffles as large as possible.

Cassegrain Telescope

One system that always requires efficient stray-light baffling is the Cassegrain telescope, which was discussed in Chap. 8. Without baffling,

Figure 20.7

Use of Threads and
Baffles for Stray-Light
Attenuation in
Machine Vision
System

(a)

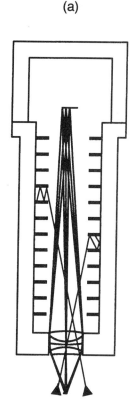

(b)

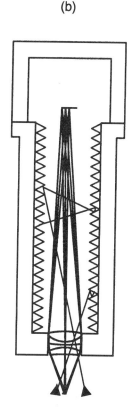

there is generally a direct stray-light path from the object space to the
sensor, and that could be a serious problem. We generally use two basic
baffles, one a conical baffle extending aft from the edge of the secondary mirror along the limiting imaging ray bundle, and the other a
tubular baffle extending forward from the hole in the primary mirror.

In order to show how to baffle a Cassegrain, we first generated a candidate design. We selected an $f/8$ system with a 100-mm entrance pupil
diameter covering a full 1° field of view. The goal for our baffles is that
a limiting ray that just passes by the two baffles described earlier shall
not directly strike the image plane. In order to quickly and efficiently
reach a solution, we added a central obscuration to the computer model
and traced 500 rays into the entrance pupil at each field of view. Figure
20.8a shows the model. Areas in black are fully populated with rays, and
the clear regions extending aft from the secondary mirror and forward

from the primary mirror are available for baffles. We show also the limiting ray which just clears the two baffle ends and reaches the image plane.

Figure 20.8*b* shows an implementation of this baffle. Note that we have added vane-type baffle segments as presented earlier in this chapter. We have also added an outer-tube assembly with interior baffling.

It is important to realize that good common sense and a little dedicated work will generally provide you with efficient stray-light baffling. If you need a specific attenuation factor, then you will need to use one

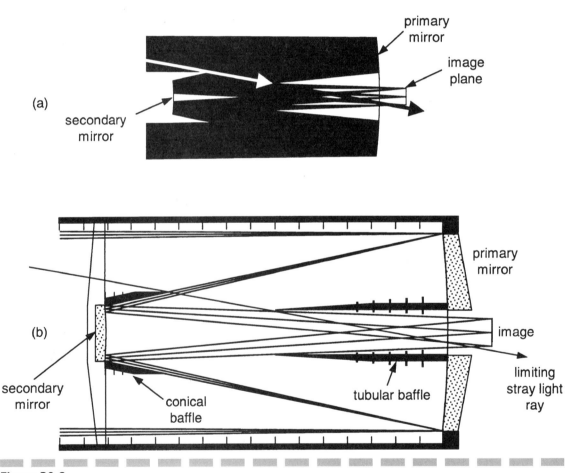

Figure 20.8
Stray-Light Baffling of Cassegrain Telescope

of the stray-light software packages. For example, in space applications, where a system may be observing a black sky to within a few degrees of the Sun, stray-light attenuation in the order of 10^{-15} or more is often required. We showed earlier in Fig. 8.7 a configuration especially well suited for efficient stray-light attenuation. This system is a three-mirror configuration consisting of primary, secondary, and tertiary mirrors. Let us assume that we are in space looking within a few degrees of the Sun into a black sky. There will be a large amount of light scattered and diffracted from the edge of the primary mirror since it is receiving direct solar radiation. If we now locate a stop further aft in the system at an image of the primary and slightly reduced in size from the image of the primary mirror, we will effectively block this light from proceeding further through the system. This is known as a *Lyot stop* after the French astronomer Lyot. While Fig. 8.7 is not to scale, it does illustrate the principle involved.

Optomechanical Design

The design of the mechanics to support your imaging optics is extremely important. Design issues relating to the optomechanics are the following:

- The mechanics supports the lenses and/or mirrors in the system. In order to keep the image quality within the specification, every optical component must be held to its nominal position within the required tolerances, as derived from your tolerance analysis and system performance error budget.

- The mechanics, along with the optics, must perform over the required thermal environment. The designer must allow for thermal expansion and contraction of the optical components to prevent any catastrophic problems.

- Maintaining focus through temperature is very dependent on the optics as well as the mechanics, and athermalization may be required.

- The mechanics must fit within the desired packaging space and be within its weight goal.

■ The mechanics must aid in attenuation of stray light. This is often accomplished by blackening the housing interior as well as threading and providing stray light baffles at strategic locations.

Figure 20.9 shows a typical housing for a projection lens. In use, a reflective display device is located to the left of the light-injection prism. The image generated on the display is then projected to a screen to the right of the lens system.

We have pointed out some of the important mechanical design features to incorporate into the design. These are

■ In this design the aperture stop is between the two smaller elements, and we use a spacer as a physical aperture stop.

■ The elements are held in place by a front-threaded retainer and a rear retainer.

■ A thread is applied to the conical spacer between the two left-most powered lens elements. This threading is to attenuate any stray light which may be incident on the housing.

We show for reference a different lens housing in Fig. 20.10. Note in this lens the two left elements are bonded into the housing as evidenced

Figure 20.9
Typical Lens Housing

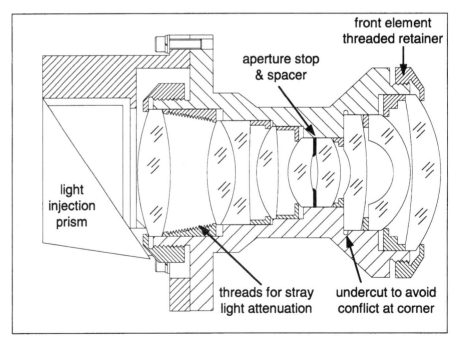

Figure 20.10
Typical Lens Housing

Injection holes for
element bonding

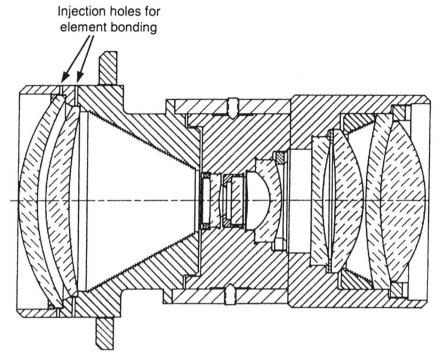

by the two material-bond injection holes. The bond material is typically
a semicompliant epoxy or RTV. This is done in situations where shock
and vibration may be a problem. The elements are centered using shims
or by rotating the housing on a precision air bearing and assuring that
the runout of the housing and the elements are per the tolerance
callout.

Lens Design
Optimization
Case Studies

In this chapter we will guide you through several representative design studies and parametric analyses in order to demonstrate the design process. We will begin with the design, from basic principles, of an achromatic doublet. Included will be the detailed computer input and output using the Zemax software package, one of the industry's standards. Following a successful design effort on the doublet, we will show the design of a low f/number double Gauss lens similar to a high-quality 35-mm camera lens. And then we will work through a case study for a digital camera lens. Following this, we will show the design for a 7×50 binocular. And, finally, we will show a parametric analysis of single-element and achromatic doublets using various manufacturing technologies, including aspherics, diffractive surfaces, and others.

Error Function Construction

Prior to embarking on several design examples, we need to discuss how the measure of performance, or the *error function,* is computed in a lens design program. Since this error function must be computed a very large number of times during the optimization process, it must be kept as simple as possible so it computes quickly. The construction of an error function was discussed in Chap. 9.

We could use the third-, fifth-, and seventh-order aberrations for our error function. These are very fast to compute; however, with today's complex systems, these aberrations rarely represent sufficiently well the real performance. We could alternatively combine these third-, fifth-, and seventh-order aberrations with specific ray aberrations or optical path differences at selected fields of view and entrance pupil coordinates. This approach can solve the problems of the higher-order aberration residuals; however, there is a lot of user interaction involved, which makes this a user-intensive methodology.

Perhaps the best and easiest to use error function is the rms blur diameter at the image formed by a grid of rays traced into the entrance pupil. The error function could also take the form of the rms wavefront error, or other similar criteria. Regardless of which method is used, it is specified at each wavelength and at each field position along with appropriate weightings.

The grid of rays in the entrance pupil is shown in Fig. 21.1, where we show on the left the default grid of three rings and six arms. Specifically, the rays traced are at the intersection of the rings and arms. On the right we show a denser grid formed by six rings and twelve arms. The default grid represents 18 rays in the entrance pupil per color per field position, and the denser grid represents 72 rays in the entrance pupil per color per field position. Denser grids are used when higher-order aberrations are present so as to better sample the aberrations. This is often the case when aspheric surfaces are used, for example. Overall, you must consider constantly whether you are sampling the rays or OPDs sufficiently well in the entrance pupil, the fields of view, and the wavelengths. If not, more rays, more fields, and/or more wavelengths are required. The computer really doesn't care what grid density is used; it will minimize the ray or OPD aberrations specifically at the grid points you specify, and *only* at those points.

The merit function is a numerical representation of how closely an optical system meets a specified set of goals. The *operands* in the merit function represent not only the image quality but also focal length, magnification, size constraints, etc.

Achromatic Doublet Lens Design

The specifications for our doublet are shown in Table 21.1.

Figure 21.1
Ray Grid of Three Rings and Six Arms (Left) and Six Rings and Twelve Arms (Right)

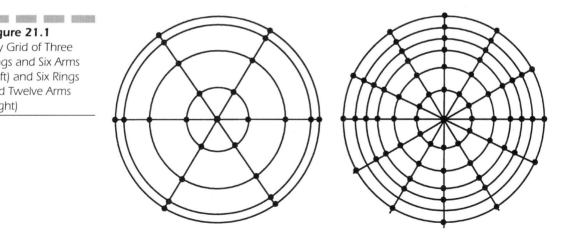

3 rings, 6 arms, 18 rays/color   6 rings, 12 arms, 72 rays/color

TABLE 21.1

Achromatic
Doublet Design
Example

Parameter	Specification
Entrance pupil diameter (mm)	50.8
Focal length (mm)	254
f/number	f/5
Full field of view (degrees)	±2
Spectral range (μm)	Visual (C, d, F) (0.6563, 0.5876, 0.4861)

To begin, we will derive a simple achromatic doublet so as to have a decent starting point for the computer optimization. The V number, or Abbe number, of optical glass is

$$V\# = \text{Abbe}\# = \frac{n_d - 1}{n_f - n_c}$$

where n_f is the refractive index at 0.486μm or shorter wavelength and n_c is the refractive index at 0.6563μm or longer wavelength. Further, we showed in Chap. 6 for lens elements a and b that

$$\phi = \phi_a + \phi_b \qquad \phi_a = \frac{\phi V_a}{V_a - V_b} \qquad \phi_b = -\frac{\phi V_b}{V_a - V_b}$$

We will assume that the positive crown element is BK7 glass with a refractive index $n_a = 1.517$ and a dispersion $V_a = 64.5$, and for the negative flint element we assume SF2 glass with a refractive index $n_b = 1.620$ and a dispersion $V_b = 36.3$. Based on these glass assumptions, we find that $\phi_a = 0.009$ (focal length = 111 mm), and $\phi_b = -0.0051$ (focal length = −197.1 mm). Further, let us assume that the positive crown element is equiconvex and the negative element is planoconcave, and the elements are cemented. For a thin lens we have shown that

$$\phi = \frac{1}{f} = (n - 1)\left(\frac{1}{r_1} - \frac{1}{r_2}\right)$$

For the crown which is to be equiconvex, $r_2 = -r_1$ and we derive the radius to be $r = 114.681$ mm. For the negatively powered planoconcave flint element we find that r_2 = infinity and $r_1 = -122.806$. We will cement the two elements, as shown in Fig. 21.2.

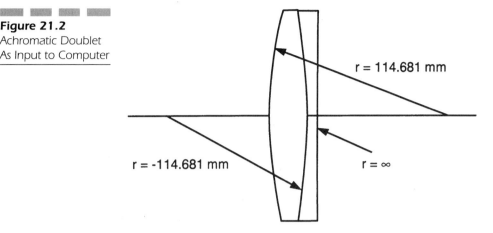

Figure 21.2
Achromatic Doublet
As Input to Computer

r = 114.681 mm

r = -114.681 mm

r = ∞

We now show in Fig. 21.3 the lens prescription data on the top and the optimization data on the bottom. The following points relate to the lens data input:

▪ The data are input in spreadsheet format.

▪ The radii of 114.681 and −114.681 are input in the radius column.

▪ These two radii, as well as the rear flat surface, are assigned the letter "V" which indicates that the radii are variable in the optimization.

▪ The thicknesses are assigned reasonable values. Surface 3 (surface numbers are on the far left) is designated an "M" to the right of the thickness. This means that the thickness is to be the distance to where the paraxial ray height equals zero (287.3968 mm).

▪ The aperture stop is on surface 1. Surface 1 is also variable in thickness in order to allow a reasonable edge thickness for the element. Note that if you do control the edge thickness of an element, you *must* vary its thickness.

▪ The glasses are listed as appropriate on surfaces 1 and 2.

▪ Separately in what are known as the "General," "field," and "wavelength" editors in Fig. 21.4, we input the entrance pupil diameter, the fields of view, and the wavelengths with their associated weights.

▪ We add a surface (number 4) which we vary independently in thickness. This is the refocusing from the paraxial focus to the best

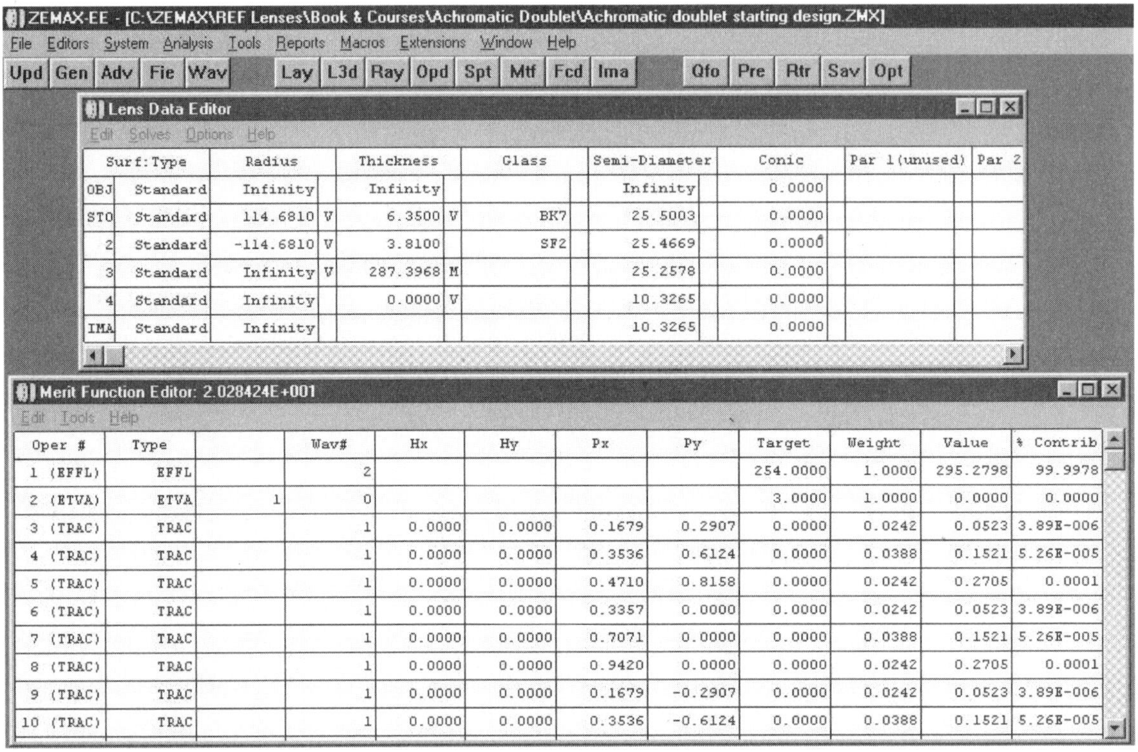

	Surf:Type	Radius	Thickness	Glass	Semi-Diameter	Conic	Par 1(unused)	Par 2
OBJ	Standard	Infinity	Infinity		Infinity	0.0000		
STO	Standard	114.6810 V	6.3500 V	BK7	25.5003	0.0000		
2	Standard	-114.6810 V	3.8100	SF2	25.4669	0.0000		
3	Standard	Infinity V	287.3968 M		25.2578	0.0000		
4	Standard	Infinity	0.0000 V		10.3265	0.0000		
IMA	Standard	Infinity			10.3265	0.0000		

Oper #	Type		Wav#	Hx	Hy	Px	Py	Target	Weight	Value	% Contrib
1 (EFFL)	EFFL		2					254.0000	1.0000	295.2798	99.9978
2 (ETVA)	ETVA	1	0					3.0000	1.0000	0.0000	0.0000
3 (TRAC)	TRAC		1	0.0000	0.0000	0.1679	0.2907	0.0000	0.0242	0.0523	3.89E-006
4 (TRAC)	TRAC		1	0.0000	0.0000	0.3536	0.6124	0.0000	0.0388	0.1521	5.26E-005
5 (TRAC)	TRAC		1	0.0000	0.0000	0.4710	0.8158	0.0000	0.0242	0.2705	0.0001
6 (TRAC)	TRAC		1	0.0000	0.0000	0.3357	0.0000	0.0000	0.0242	0.0523	3.89E-006
7 (TRAC)	TRAC		1	0.0000	0.0000	0.7071	0.0000	0.0000	0.0388	0.1521	5.26E-005
8 (TRAC)	TRAC		1	0.0000	0.0000	0.9420	0.0000	0.0000	0.0242	0.2705	0.0001
9 (TRAC)	TRAC		1	0.0000	0.0000	0.1679	-0.2907	0.0000	0.0242	0.0523	3.89E-006
10 (TRAC)	TRAC		1	0.0000	0.0000	0.3536	-0.6124	0.0000	0.0388	0.1521	5.26E-005

Figure 21.3 Achromatic Doublet As Input to Computer

focus position. Using this technique can sometimes result in a better-controlled optimization process.

In the lower part of Fig. 21.3 we show the optimization window, and we note here the following:

- The first line is labeled "EFFL" which means the *effective focal length.* Our target is 254 mm and its current value is 295.2798. We assign this a weight of unity in the optimization.
- The second line is "ETVA" which is the *edge thickness value* on surface 1, the positive first element. Our goal here is 3 mm. From line 3 on, we have as labeled "TRAC" the transverse shift from the ideal image locations in the image plane. There is a very *important* side note here: Recall that we have varied the thickness of the first element (surface 1). If we had not varied the element thickness but did require the

Figure 21.4

Means for Entering
Basic System
Parameters, Including
Entrance Pupil
Diameter, Field
Angles, and
Wavelengths

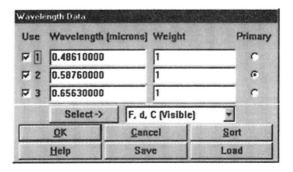

constraint of the edge thickness of the element to be 3 mm we would be overconstraining the lens. The program would reach a solution; however, we would, in effect, be constraining the power of the first element by this edge thickness constraint, and the net result would be a poorer level of optical performance.

Figure 21.5 shows the initial lens design and its performance. The transverse ray aberrations are plotted on a 200-μm scale and the spot diagrams are on a 500-μm scale.

We now execute the optimization, and only a few seconds later a local minimum in the error function is reached. Figure 21.6 shows the design data as well as the error function construction, only here for the optimized doublet. Figure 21.7 shows the layout and the performance of the design. The transverse ray aberrations are now plotted on a scale of 50 μm which is 25% of the initial scale factor, and the spot diagrams are on a scale of 100 μm, which is 20% of the initial scale. The lens is clearly better now than the starting design.

If we required a further performance improvement, we would likely introduce a small airspace between the two elements and possibly change the glass types. The airspace would allow for a balancing of third- and fifth-order spherical aberration which can often make a

Figure 21.5
Performance of
Achromatic Doublet
As Input to Computer

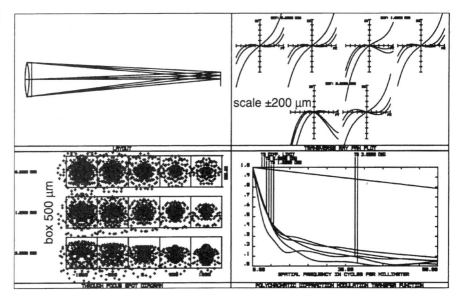

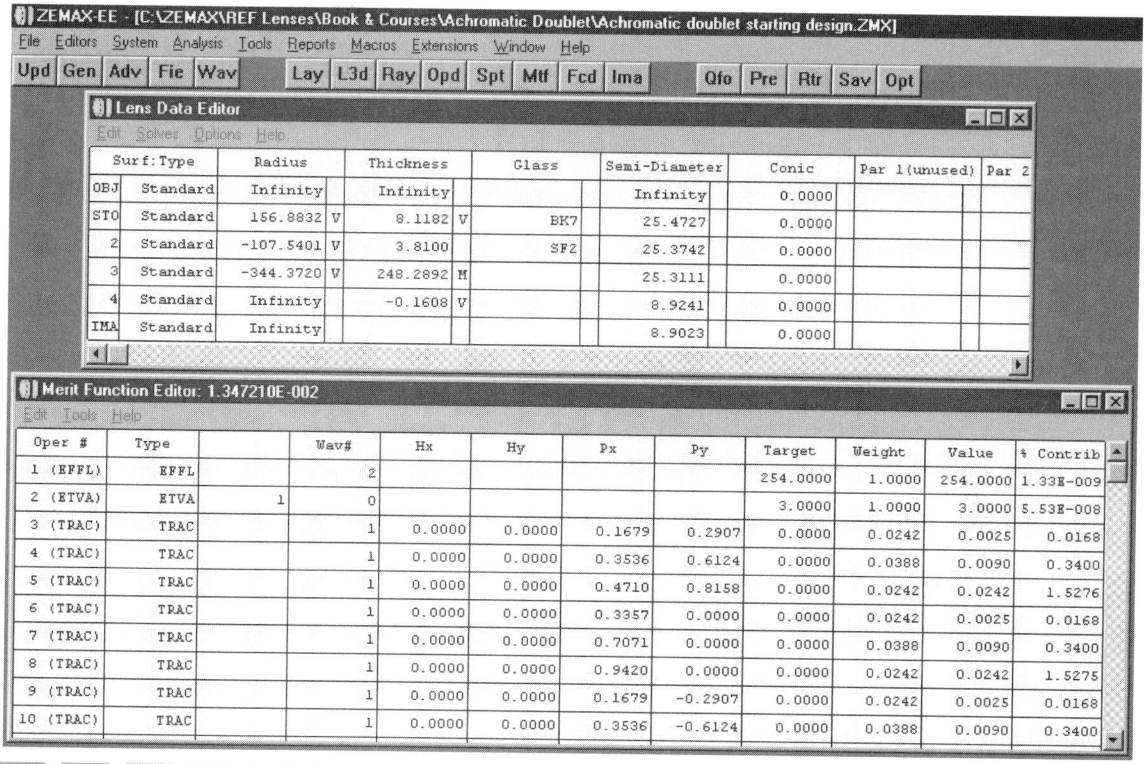

Surf:Type		Radius		Thickness		Glass		Semi-Diameter		Conic	Par 1(unused)	Par 2
OBJ	Standard	Infinity		Infinity				Infinity		0.0000		
STO	Standard	156.8832	V	8.1182	V	BK7		25.4727		0.0000		
2	Standard	-107.5401	V	3.8100		SF2		25.3742		0.0000		
3	Standard	-344.3720	V	248.2892	M			25.3111		0.0000		
4	Standard	Infinity		-0.1608	V			8.9241		0.0000		
IMA	Standard	Infinity						8.9023		0.0000		

Oper #	Type		Wav#	Hx	Hy	Px	Py	Target	Weight	Value	% Contrib
1 (EFFL)	EFFL		2					254.0000	1.0000	254.0000	1.33E-009
2 (ETVA)	ETVA	1	0					3.0000	1.0000	3.0000	5.53E-008
3 (TRAC)	TRAC		1	0.0000	0.0000	0.1679	0.2907	0.0000	0.0242	0.0025	0.0168
4 (TRAC)	TRAC		1	0.0000	0.0000	0.3536	0.6124	0.0000	0.0388	0.0090	0.3400
5 (TRAC)	TRAC		1	0.0000	0.0000	0.4710	0.8158	0.0000	0.0242	0.0242	1.5276
6 (TRAC)	TRAC		1	0.0000	0.0000	0.3357	0.0000	0.0000	0.0242	0.0025	0.0168
7 (TRAC)	TRAC		1	0.0000	0.0000	0.7071	0.0000	0.0000	0.0388	0.0090	0.3400
8 (TRAC)	TRAC		1	0.0000	0.0000	0.9420	0.0000	0.0000	0.0242	0.0242	1.5275
9 (TRAC)	TRAC		1	0.0000	0.0000	0.1679	-0.2907	0.0000	0.0242	0.0025	0.0168
10 (TRAC)	TRAC		1	0.0000	0.0000	0.3536	-0.6124	0.0000	0.0388	0.0090	0.3400

Figure 21.6 *Design Data and Error Function for Achromatic Doublet Final Design*

significant difference, and the glass change would allow for improved chromatic aberration correction. A further improvement would be realized by adding a third element. If this new element were near the focal plane, it would be able to minimize both the field curvature as well as the astigmatism. We leave these exercises to the reader.

Double Gauss Lens Design

This example is the design and tolerancing of a 50-mm focal length double Gauss lens for a 35-mm camera application. The basic lens specifications are shown in Table 21.2.

Figure 21.7
Performance of
Achromatic Doublet
Final Design

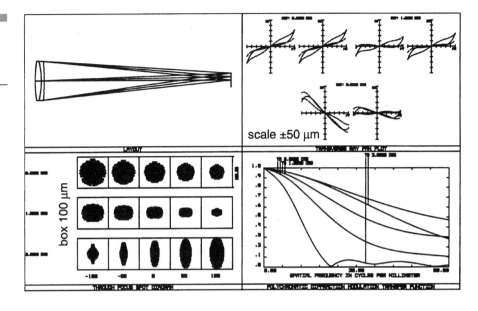

scale ±50 μm

box 100 μm

TABLE 21.2

Double Gauss
Design Example

Parameter	Specification
Entrance pupil diameter (mm)	25.4
Focal length (mm)	50.8
f/number	f/2.0
Full field of view (degrees)	±16
Spectral range	Visual (C, d, F)
Distortion (%)	<2.5
Vignetting (%)	<50 at edge of field
Back focus (mm)	>25.4

Relative to the optical performance, we will derive our own specification based fully on the functional performance requirements of our 35-mm camera lens. We will then show how this derivation comes extremely close to what is often used in the industry.

Let us assume that the goal for lens resolution is that the image blur from a point object is barely discernable by the eye when viewed on a 200- × 254-mm enlargement at a distance of 254 mm. This is a pretty reasonable specification, and the beauty is that we did not need to know any complex aberration theory to come up with it. It is a functional specification based solely on the application of the lens, which is, of course, to give the user a good quality photograph.

Figure 21.8 shows the situation. A person with good visual acuity can resolve about 2 min of arc per line pair, which equates to 1 min of arc per line, which is approximately 0.0003 rad. At a distance of 254 mm, this equates to an image blur diameter of 0.076 mm. (For reference, 1 min of arc is a spot 1 mm in diameter at a distance of 3 m.) A 35-mm negative which measures 24 × 36 mm is 7.06 times smaller than the enlargement, which means that we are looking for an image blur diameter of 0.0107 mm on the negative.

Prior to getting into our design example, consider the common rule of thumb that a 35-mm camera lens should have an MTF of >0.3 at 50 line pairs/mm and an MTF of >0.5 at 30 line pairs/mm. We can rederive this guideline as follows: Our image blur diameter goal is 0.0107 mm at the lens focal plane. This is approximately equivalent to a line $\frac{1}{100}$ mm

Figure 21.8
Performance Derivation for Double Gauss Lens

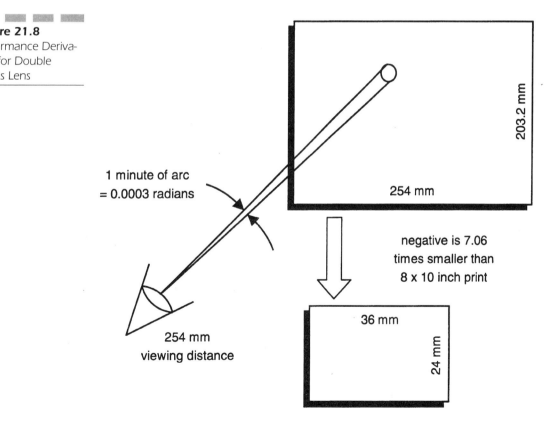

wide. A line pair, which is a dark line and an adjacent bright line, is therefore $\frac{1}{50}$ mm wide, which equates to 50 line pairs/mm. This very interestingly matches our rule of thumb perfectly. We could then conclude that a reasonable contrast level for such a lens would be an MTF of about 0.3 at 50 line pairs/mm, exactly what our rule of thumb calls for!

Let us now select as a starting point a double Gauss lens design from a 1938 patent. after setting up the prescription on the computer, we have the results shown in Fig. 21.9. These data include a lens drawing or layout, a plot of the transverse ray aberrations, a through-focus geometrical spot diagram, and a plot of the MTF out to 50 line pairs/mm. The error function is a combination of different constraints, the ray aberrations, and other performance criteria. Our constraints include the focal length, the edge thickness of the positive elements, the minimum back focus distance, and the distortion. The pure lens quality portion of the error function is 0.018, and it is this metric that we will be following as we optimize the lens.

Step 1. We now take the initial patent lens prescription and establish our variables. Variable are all of the lens radii, the airspaces surrounding the aperture stop, the thicknesses of the positive elements, and the back focus distance from the rear lens vertex to the image plane. Figure 21.10 shows the result of this initial optimization. We have a spherical aberration residual,

Figure 21.9
Performance of Double Gauss Starting Design from Patent

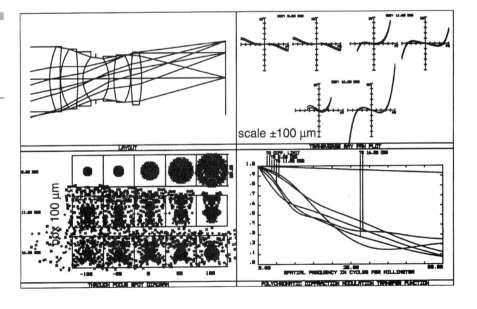

Figure 21.10
Performance of Initial
Optimization

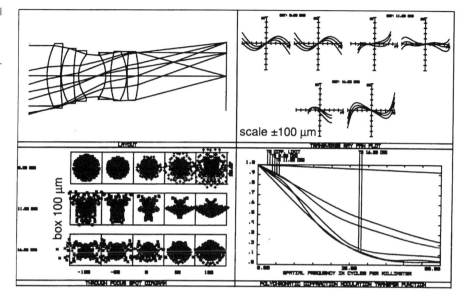

primary axial color, and field curvature, among other residual aberrations. The error function has reduced from 0.018 for the starting design to 0.0106.

Step 2. We now will vary the inner doublet glasses using a routine called "Hammer" optimization in Zemax. The Hammer optimization uses a random search algorithm in the solution space surrounding the starting design. The program will, by definition, end up with real glasses, so the user does not need to be concerned with fictitious glasses after a long optimization. We allowed the Hammer optimization to run approximately 30 min, enough time to realize a moderate improvement in the lens performance. Figure 21.11 shows the results. The spherical aberration is somewhat reduced and the chromatic aberration is nearly eliminated. The error function has reduced from 0.0106 to 0.0060, another reasonable improvement.

Step 3. We now allow all of the glasses to vary, including those of the outer elements. Figure 21.12 shows the results, and the error function reduces from 0.0060 to 0.0059, only a slight improvement. Note here that the MTF is about 0.15 at the corner of the field of view in the tangential target direction

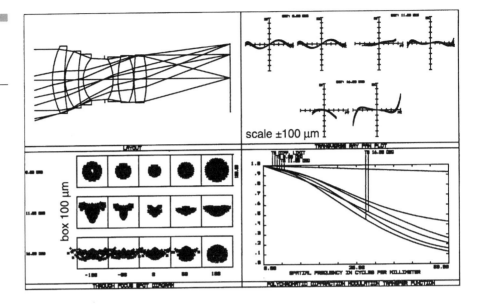

Figure 21.11
Vary Inner Doublet Glasses, 30-min Hammer

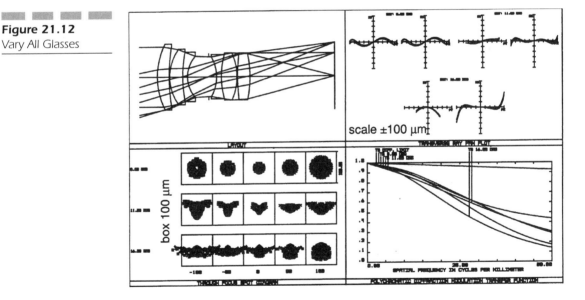

Figure 21.12
Vary All Glasses

at 50 line pairs/mm. Our goal is 0.3 minimum at the edge of the field of view, so we have a way to go.

Step 4. Up to this point, we have been optimizing and analyzing our design at the center of the field of view, 0.7 of the semidiagonal of the field of view, and at the corner which is the maximum field of view. This is often an adequate sampling of the fields of view, especially when there are not significant changes in performance with field. But we have a reasonably low f/number lens with a reasonably wide field of view, so we elected to increase the number of fields of view to 5 in equal increments. Figure 21.13 shows the performance. Something extremely noteworthy has happened, and that is the inner fields of view really do not perform well! Note that at 25 and 50% of the field of view our rms blur diameter is 25% larger than at the corner of the field of view! This is evident in the ray trace plots as well as the spot diagrams. We actually did two other things in this model: We defined the aperture diameters on each surface, and we then adjusted the vignetting factors so that realistic vignetting would result. Unfortunately, if the user does not go through this exercise, the vignetting may be fictitious and not representative of what will happen in hardware. The net result of this was that the higher-order flare in the sagittal ray fan at the corner of the field of view was eliminated which improved the off-axis performance. This was partially responsible for the error function coming down to from 0.0059 to 0.0047, even though there was no reoptimization of the lens in this step.

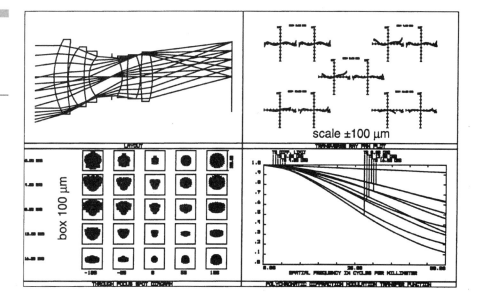

Figure 21.13
Five Fields, Set
Apertures Then
Vignetting, No
Reoptimization

scale ±100 µm

box 100 µm

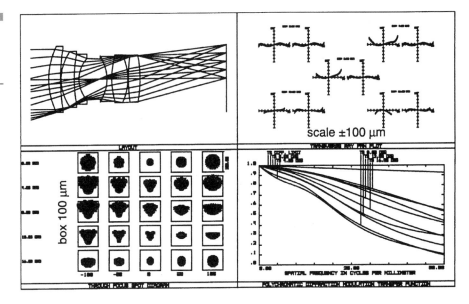

Step 5. In the next step we applied the basic optimization
algorithm and Fig. 21.14 shows the result. The error function
comes down from 0.0047 to 0.0042. The inner fields of view
still show more degradation in performance than the center
or corner of the field, and we will need to do something
about this problem. We could increase the field weights at
these positions; however, as you will see, this was not
necessary.

Step 6. In our next iteration we continued to use the increased field
sampling of five fields of view, and further we increased the
ray sampling in the pupil to six rings and 12 arms in order to
assure an adequate sampling. While this was not mandatory,
we do have some higher-order aberrations, and it makes good
sense to increase the sampling at about this stage in the design.
Figure 21.15 shows the result. While the error function only
reduced from 0.0042 to 0.0038, the performance at the inner
fields of view are clearly improved. We should note here that
any time you change the ray sampling via the number of
rings and arms, the field and/or spectral weights, or other
similar parameters, you need to recompute your error

function. Note also that the new error function will likely be different from what it was before due to the different sampling, changes in field and/or wavelength weights, or other factors, which have changed. And this is true even if the lens itself is unchanged.

Step 7. In the next iteration we allowed all glasses to vary one final time and executed a 30-min Hammer optimization. The error function reduced slightly from 0.0038 to 0.0034, and the MTF actually seemed to degrade somewhat from the prior design, so we are not showing the results.

Step 8. In the final iteration we allowed the Hammer optimization to run for a full 12 h, and a much improved design resulted, as shown in Fig. 21.16. The error function has come down from 0.0034 to 0.0023, approximately a 30% reduction. The ray trace curves, spot diagrams, and MTF all show a notable improvement in both basic performance as well as uniformity of performance over the field of view. Note that the lowest MTF is now 0.6 at 50 line pairs/mm! The rms blur diameters range from 6 to 9 μm over the full field of view.

In the previous optimization sequence we took eight separate and independent steps in the optimization of the double Gauss lens. This is

Figure 21.15
Five Fields, Tighter Ray Grid (Six Rings, 12 Arms), Basic Optimization

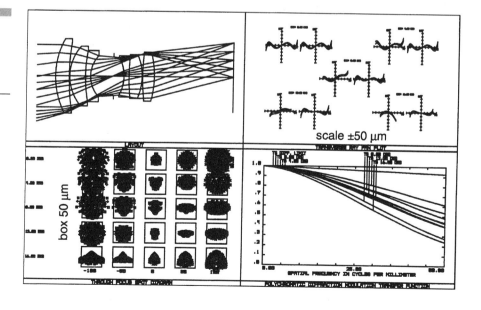

Figure 21.16
Final 12-h Hammer
Optimization

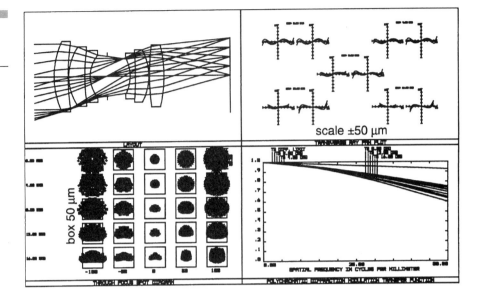

summarized graphically in Fig. 21.17, where we plot the steps taken in the abscissa and the error function in the ordinate. At the conclusion of each one of the individual steps, we reached a local minimum in the error function and we had to apply an outside influence prior to taking the next step. Notice that we raise the question "can we reach zero" in the error function? In order to reduce the error function more, we would need to add additional elements in order to further minimize the residual aberrations. A good way to think about the answer to this question is to recall the microlithography lens of Glatzel in Chap. 5. This lens has about 17 elements, and while its performance is not perfect, it is clearly diffraction limited. Of course the specifications are not the same, but we can conclude that a lens with a similar level of complexity to the Glatzel lens may be required to bring the error function much closer to zero.

We now will stop the lens down to $f/4.0$ as in Fig. 21.18. It is important to evaluate the lens over the functional range over which it will be used, and in 35-mm photography, lenses are often used stopped down in f/number (at higher f/numbers).

We will now proceed to tolerance the lens. Figure 21.19 shows the Zemax input table where initial tolerances are input. Due to the reasonably tight level of the tolerances expected for our low f/number double Gauss lens, we will select a starting mix of tolerances which are representative

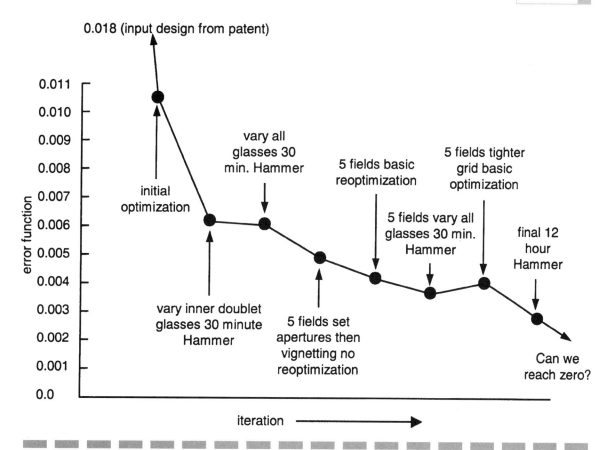

Figure 21.17
Progression of Error Function (Image Quality Portion Only) During Double Gauss Lens Design Example

of somewhat tight, yet achievable values. The following are the tolerances and the rationale:

■ We will assume that prior to manufacture, all radii will be matched to existing testplates, and thus for the surface radii we will input power fit to testplate as four fringes. If we had to custom manufacture one or more testplates, then a specific radius tolerance would be necessary, indicating the accuracy to which the testplate were manufactured. As noted earlier, fitting 100% of the radii of a given lens design to existing testplates is generally done.

■ All element thicknesses and airspaces are assumed to be ±0.05 mm, a reasonable assumption.

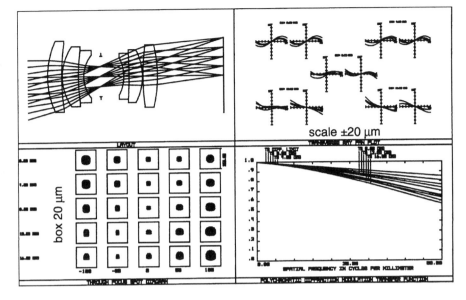

- We have elected not to use surface decentrations as mathematically element wedge takes this into account.

- Surface tilts of 0.025-mm total indicator runout (TIR) on each surface is assumed for the element wedge. We have intentionally split the wedge between each surface of each element, so in effect the net total wedge of any given element is 0.05 mm TIR. We may be able to further refine the accuracy of our wedge model once we know the specific manufacturing methods.

- Surface irregularity is assumed in Zemax to be a mix of spherical aberration due to a fourth-order OPD contribution and astigmatism from cylinder. We have assumed one fringe of irregularity per surface.

- For refractive index and Abbe number we are using the default values of ±0.0002 and ±0.01, respectively.

- The previous specifications are for surfaces. For elements we have decentration in *x* and *y* of ±0.05 mm.

- For element tilt we are using 0.114°, which equates to approximately 0.05 mm TIR.

Note at the bottom that we are using focus compensation as a compensator. What this means is that for each and every tolerance perturbation we

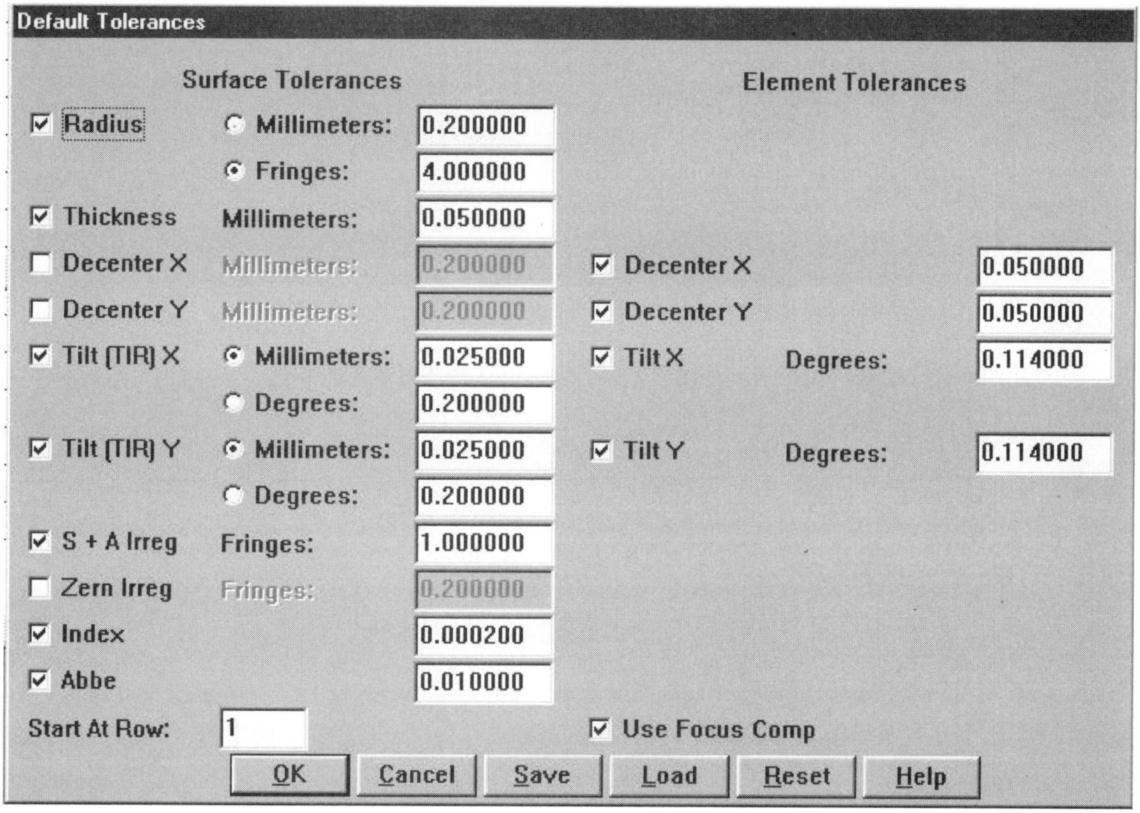

Figure 21.19 Input of Tolerance Values

will assume that the lens can be refocused. Since the lens is definitely refocused during final testing and assembly, this is a fair assumption.

We now show the initial output page, where the various assumptions are listed for the analysis. Note that the so-called merit function is the average of sagittal and tangential target orientation MTF at 30 line pairs/mm. Averaged over the field of view the overall nominal MTF is 0.836.

```
Analysis of Tolerances
Title: Final Final 12 Hour Hammer
Units are Millimeters.
Fast tolerancing mode is on. In this mode, all
compensators are ignored, except back focus error.

Merit: Diffraction MTF average S&T at 30.0000 lp/mm
Nominal Merit Function (MF) is 0.83564909
Test wavelength: 0.6328
```

```
Fields: User Defined Angle in degrees
 #     X-Field      Y-Field      Weight      VDX     VDY     VCX     VCY
 1   0.000E+000   0.000E+000   1.000E+000   0.000   0.000   0.000   0.000
 2   0.000E+000   4.000E+000   1.000E+000   0.000   0.006   0.009   0.091
 3   0.000E+000   8.000E+000   1.000E+000   0.000   0.008   0.047   0.205
 4   0.000E+000   1.200E+001   1.000E+000   0.000   0.000   0.100   0.337
 5   0.000E+000   1.600E+001   1.000E+000   0.000  -0.004   0.248   0.493
```

The results are shown as follows.

Fringes of Power Fit to Testplate

We show here the drop in MTF for each tolerance, both averaged over the field of view (labeled "All" under the field column) as well as at each specific field of view. The first tolerance listed is "TFRN" which is fringes of power fit to testplate. For brevity, we show only surfaces 2 through 6 which is the front half of the lens (surface 1 is a dummy surface forward of the lens). The greatest degradation here is on surface 6 which is the strong concave surface prior to the stop, and at fields 1 and 2 the MTF drops approximately 0.012. This is not a large MTF drop at all, and, in fact, most of the power fit to testplate tolerances can likely be increased from four fringes to five or more fringes with little effect.

```
Sensitivity Analysis:
                       !———————— Minimum ————————! !———————— Maximum ————————!
Type Sf1 Sf2 Field    Value      MF     Change     Value      MF     Change
TFRN      2  All  -4.000000  0.837422  0.001773   4.000000  0.832193 -0.003456
             1              0.806590  0.002522              0.798613 -0.005454
             2              0.828502  0.002910              0.820480 -0.005112
             3              0.854749  0.002028              0.849222 -0.003499
             4              0.850717  0.001319              0.847165 -0.002233
             5              0.851824 -0.000270              0.851948 -0.000145
TFRN      3  All  -4.000000  0.833350 -0.002299   4.000000  0.836925  0.001276
             1              0.801921 -0.002147              0.805537  0.001469
             2              0.823507 -0.002085              0.827025  0.001433
             3              0.851510 -0.001211              0.853042  0.000321
             4              0.847407 -0.001991              0.850012  0.000614
             5              0.847876 -0.004217              0.854610  0.002517
TFRN      4  All  -4.000000  0.837272  0.001623   4.000000  0.831326 -0.004323
             1              0.806835  0.002767              0.796489 -0.007579
             2              0.829831  0.004239              0.817930 -0.007662
             3              0.856489  0.003768              0.847042 -0.005679
             4              0.850653  0.001255              0.846977 -0.002421
             5              0.847577 -0.004516              0.855512  0.003419
TFRN      5  All  -4.000000  0.835808  0.000158   4.000000  0.835468 -0.000181
             1              0.804200  0.000132              0.803907 -0.000161
             2              0.825782  0.000190              0.825377 -0.000215
             3              0.852948  0.000226              0.852473 -0.000248
```

```
            4              0.849651  0.000253        0.849126 -0.000272
            5              0.852092 -0.000001        0.852078 -0.000015
TFRN    6  All -4.000000   0.828345 -0.007304  4.000000  0.837541  0.001892
            1              0.792139 -0.011928        0.806558  0.002491
            2              0.813809 -0.011783        0.830302  0.004710
            3              0.844253 -0.008468        0.856661  0.003940
            4              0.845195 -0.004203        0.850895  0.001497
            5              0.854287  0.002194        0.848446 -0.003647
```

Thickness: Both Element Thicknesses and Airspaces

We show next thickness tolerances (TTHI) of ±0.05 mm. The largest MTF drop is approximately 0.083 on axis for thicknesses 4 and 5, which is the inner doublet prior to the stop. This is of some significance and we should revisit these tolerances after we complete the analysis. Note that the most sensitive tolerances are highlighted in **bold** text and via an asterisk.

				─── Minimum ───			─── Maximum ───		
Type	Sf1	Sf2	Field	Value	MF	Change	Value	MF	Change
TTHI	2	3	All	-0.050000	0.835880	0.000231	0.050000	0.835116	-0.000534
			1		0.804020	-0.000048		0.804120	0.000053
			2		0.825708	0.000116		0.825453	-0.000139
			3		0.851833	-0.000888		0.853440	0.000718
			4		0.848441	-0.000957		0.849847	0.000449
			5		0.855198	0.003105		0.848101	-0.003992
TTHI	3	6	All	-0.050000	0.835371	-0.000279	0.050000	0.829620	-0.006029
			1		0.806188	0.002121		0.792026	-0.012041
			2		0.829775	0.004184		0.815661	-0.009931
			3		0.858979	0.006258		0.844557	-0.008165
			4		0.851557	0.002159		0.844201	-0.005197
			5		0.835523	-0.016570		0.860443	0.008350
TTHI	4	6	All	-0.050000	0.789088	-0.046561	0.050000	0.756450	-0.079199
			1		0.720801	-0.083266		0.660283	-0.143785*
			2		0.783468	-0.042124		0.715721	-0.109871*
			3		0.850413	-0.002309		0.788121	-0.064600
			4		0.840267	-0.009130		0.816846	-0.032552
			5		0.777086	-0.075007		0.851939	-0.000154
TTHI	5	6	All	-0.050000	0.789928	-0.045721	0.050000	0.757571	-0.078078
			1		0.722150	-0.081918		0.661822	-0.142246*
			2		0.784430	-0.041162		0.717043	-0.108549*
			3		0.850703	-0.002018		0.789293	-0.063428
			4		0.840283	-0.009115		0.817884	-0.031514
			5		0.778232	-0.073861		0.852130	0.000037
TTHI	6	7	All	-0.050000	0.835714	0.000065	0.050000	0.835576	-0.000073
			1		0.804068	0.000000		0.804067	-0.000000
			2		0.825633	0.000041		0.825550	-0.000042
			3		0.852748	0.000027		0.852690	-0.000031
			4		0.849347	-0.000050		0.849433	0.000035
			5		0.852432	0.000339		0.851731	-0.000362

Element Decentration

We show next several element decentrations (TEDX and TEDY) as well as element tilts (TETX and TETY). Most of the MTF drops here are in the order of 0.02 to 0.05, and we should revisit these later.

Type	Sf1	Sf2	Field	Value	MF (Min)	Change (Min)	Value	MF (Max)	Change (Max)
			5		0.851583	-0.000510	0.852224	0.000131	
				── Minimum ──			**── Maximum ──**		
Type	**Sf1**	**Sf2**	**Field**	**Value**	**MF**	**Change**	**Value**	**MF**	**Change**
TEDX	2	3	All	-0.050000	0.817067	-0.018582	0.050000	0.817067	-0.018582
			1		0.781078	-0.022990		0.781078	-0.022990
			2		0.803667	-0.021925		0.803667	-0.021925
			3		0.832935	-0.019786		0.832935	-0.019786
			4		0.832676	-0.016722		0.832676	-0.016722
			5		0.842073	-0.010020		0.842073	-0.010020
TEDY	2	3	All	-0.050000	0.808191	-0.027458	0.050000	0.817327	-0.018322
			1		0.781078	-0.022990		0.781078	-0.022990
			2		0.795995	-0.029596		0.814195	-0.011397
			3		0.835070	-0.017651		0.823512	-0.029209
			4		0.826343	-0.023055		0.821939	-0.027459
			5		0.807515	-0.044578		0.853223	0.001130
TETX	2	3	All	-0.114000	0.809352	-0.026297	0.114000	0.808853	-0.026796
			1		0.790527	-0.013541		0.790527	-0.013541
			2		0.812954	-0.012638		0.802454	-0.023138
			3		0.805097	-0.047624		0.838547	-0.014174
			4		0.798082	-0.051315		0.824232	-0.025166
			5		0.844726	-0.007367		0.793068	-0.059025
TETY	2	3	All	-0.114000	0.813819	-0.021830	0.114000	0.813819	-0.021830
			1		0.790527	-0.013541		0.790527	-0.013541
			2		0.809240	-0.016352		0.809240	-0.016352
			3		0.829636	-0.023086		0.829636	-0.023086
			4		0.819439	-0.029959		0.819439	-0.029959
			5		0.822740	-0.029354		0.822740	-0.029354
TEDX	4	6	All	-0.050000	0.813285	-0.022364	0.050000	0.813285	-0.022364
			1		0.783273	-0.020795		0.783273	-0.020795
			2		0.803831	-0.021761		0.803831	-0.021761
			3		0.828743	-0.023978		0.828743	-0.023978
			4		0.823886	-0.025512		0.823886	-0.025512
			5		0.831138	-0.020955		0.831138	-0.020955
TEDY	4	6	All	-0.050000	0.801162	-0.034487	0.050000	0.788111	-0.047538
			1		0.783273	-0.020795		0.783273	-0.020795
			2		0.805776	-0.019816		0.790366	-0.035225
			3		0.796606	-0.056116		0.815828	-0.036893
			4		0.790714	-0.058684		0.790983	-0.058415
			5		0.833211	-0.018882		0.763443	-0.088651
TETX	4	6	All	-0.114000	0.785336	-0.050313	0.114000	0.801904	-0.033745
			1		0.789832	-0.014236		0.789832	-0.014236
			2		0.793959	-0.031633		0.808478	-0.017114
			3		0.811466	-0.041255		0.797176	-0.055545
			4		0.779829	-0.069569		0.789780	-0.059618
			5		0.755544	-0.096549		0.826712	-0.025381
TETY	4	6	All	-0.114000	0.818623	-0.017027	0.114000	0.818623	-0.017027
			1		0.789832	-0.014236		0.789832	-0.014236
			2		0.810120	-0.015472		0.810120	-0.015472

3	0.834414 -0.018307	0.834414 -0.018307
4	0.828629 -0.020769	0.828629 -0.020769
5	0.834238 -0.017856	0.834238 -0.017856

Element Wedge

We show the tolerances for element wedge as the total indicator runout (TIRX and TIRY). We had assigned 0.025 mm for each surface, which is, in effect, 0.05 mm for the element. Most of the surfaces listed are not too sensitive, except for surface 4 which is the front of the forward doublet where the largest MTF drop is approximately 0.07 at one of the outer field positions.

Type	Sf1	Sf2	Field	─ Minimum ─ Value	MF	Change	─ Maximum ─ Value	MF	Change
TIRX		2	All	-0.025000	0.826102	-0.009547	0.025000	0.826102	-0.009547
			1		0.791277	-0.012791		0.791277	-0.012791
			2		0.813671	-0.011921		0.813671	-0.011921
			3		0.842599	-0.010122		0.842599	-0.010122
			4		0.841593	-0.007805		0.841593	-0.007805
			5		0.848177	-0.003917		0.848177	-0.003917
TIRY		2	All	-0.025000	0.828801	-0.006848	0.025000	0.820179	-0.015470
			1		0.791277	-0.012791		0.791277	-0.012791
			2		0.822472	-0.003120		0.807437	-0.018155
			3		0.840980	-0.011741		0.842928	-0.009793
			4		0.839645	-0.009753		0.836891	-0.012507
			5		0.856950	0.004856		0.827498	-0.024595
TIRX		3	All	-0.025000	0.830928	-0.004721	0.025000	0.830928	-0.004721
			1		0.800116	-0.003952		0.800116	-0.003952
			2		0.821408	-0.004184		0.821408	-0.004184
			3		0.847847	-0.004875		0.847847	-0.004875
			4		0.843688	-0.005709		0.843688	-0.005709
			5		0.846723	-0.005370		0.846723	-0.005370
TIRY		3	All	-0.025000	0.827219	-0.008430	0.025000	0.836174	0.000524
			1		0.800116	-0.003952		0.800116	-0.003952
			2		0.817082	-0.008510		0.827514	0.001922
			3		0.841542	-0.011179		0.856938	0.004216
			4		0.836525	-0.012873		0.853683	0.004285
			5		0.845017	-0.007076		0.849618	-0.002475
TIRX		4	All	-0.025000	0.830680	-0.004969	0.025000	0.830680	-0.004969
			1		0.800850	-0.003217		0.800850	-0.003217
			2		0.821652	-0.003940		0.821652	-0.003940
			3		0.847230	-0.005491		0.847230	-0.005491
			4		0.842622	-0.006776		0.842622	-0.006776
			5		0.845819	-0.006274		0.845819	-0.006274
TIRY		4	All	-0.025000	0.794785	-0.040864	0.025000	0.765899	-0.069750
			1		0.800850	-0.003217		0.800850	-0.003217
			2		0.806401	-0.019191		0.802323	-0.023269
			3		0.789915	-0.062806		0.795301	-0.057420
			4		0.775756	-0.073642		0.736589	-0.112809*
			5		0.802509	-0.049584		0.710186	-0.141907*

Type	Sf1	Field						
TIRX	5	All	-0.025000	0.835219	-0.000430	0.025000	0.835219	-0.000430
		1		0.803645	-0.000423		0.803645	-0.000423
		2		0.825162	-0.000430		0.825162	-0.000430
		3		0.852279	-0.000443		0.852279	-0.000443
		4		0.848959	-0.000439		0.848959	-0.000439
		5		0.851658	-0.000435		0.851658	-0.000435
TIRY	5	All	-0.025000	0.834858	-0.000791	0.025000	0.835580	-0.000069
		1		0.803645	-0.000423		0.803645	-0.000423
		2		0.824740	-0.000852		0.825634	0.000042
		3		0.851027	-0.001694		0.853591	0.000870
		4		0.847527	-0.001871		0.850388	0.000990
		5		0.852874	0.000780		0.850362	-0.001731
TIRX	6	All	-0.025000	0.820268	-0.015381	0.025000	0.820268	-0.015381
		1		0.791865	-0.012202		0.791865	-0.012202
		2		0.811931	-0.013661		0.811931	-0.013661
		3		0.835974	-0.016747		0.835974	-0.016747
		4		0.830315	-0.019082		0.830315	-0.019082
		5		0.835288	-0.016805		0.835288	-0.016805
TIRY	6	All	-0.025000	0.734318	-0.101331	0.025000	0.765926	-0.069723
		1		0.791865	-0.012202		0.791865	-0.012202
		2		0.784413	-0.041179		0.791466	-0.034126
		3		0.767164	-0.085557		0.752842	-0.099880
		4		0.695016	-0.154382		0.730450	-0.118948
		5		0.659546	-0.192547		0.768909	-0.083184

Surface Irregularity

We show the sensitivities for surface irregularity (TIRR) where one fringe is assumed. Most of the sensitivities are reasonable, and we may be able to loosen some of these tolerances to perhaps two to three fringes of irregularity.

					Minimum			Maximum	
Type	Sf1	Sf2	Field	Value	MF	Change	Value	MF	Change
TIRR	2		All	-1.000000	0.833455	-0.002194	1.000000	0.828633	-0.007017
			1		0.803273	-0.000794		0.792029	-0.012039
			2		0.822489	-0.003103		0.818111	-0.007481
			3		0.845813	-0.006908		0.850890	-0.001831
			4		0.845717	-0.003680		0.845692	-0.003706
			5		0.855439	0.003345		0.843624	-0.008470
TIRR	3		All	-1.000000	0.826018	-0.009632	1.000000	0.832446	-0.003203
			1		0.788850	-0.015218		0.802526	-0.001542
			2		0.815685	-0.009907		0.821513	-0.004079
			3		0.849174	-0.003547		0.844043	-0.008678
			4		0.843523	-0.005875		0.844021	-0.005377
			5		0.840126	-0.011967		0.855549	0.003456
TIRR	4		All	-1.000000	0.833276	-0.002373	1.000000	0.828988	-0.006661
			1		0.802227	-0.001840		0.791542	-0.012526
			2		0.822124	-0.003468		0.817340	-0.008252
			3		0.846462	-0.006259		0.850571	-0.002151
			4		0.846556	-0.002842		0.846653	-0.002745
			5		0.854702	0.002609		0.846504	-0.005589

					Minimum		Maximum	
TIRR	5	All	-1.000000	0.835714	0.000065	1.000000	0.835579	-0.000070
		1		0.804195	0.000127		0.803933	-0.000134
		2		0.825669	0.000077		0.825509	-0.000083
		3		0.852715	-0.000006		0.852723	0.000001
		4		0.849428	0.000030		0.849364	-0.000034
		5		0.852171	0.000077		0.852014	-0.000080
TIRR	6	All	-1.000000	0.828973	-0.006676	1.000000	0.833132	-0.002517
		1		0.791582	-0.012486		0.801799	-0.002269
		2		0.816892	-0.008700		0.822012	-0.003580
		3		0.850059	-0.002662		0.846626	-0.006095
		4		0.846769	-0.002629		0.846613	-0.002784
		5		0.847286	-0.004807		0.854362	0.002269

Refractive Index and Abbe Number

Here we show the sensitivities to refractive index and Abbe number (TIND and TABB). These are quite insensitive and could be loosened if there is a reason to do so.

					Minimum			Maximum	
Type	Sf1	Sf2	Field	Value	MF	Change	Value	MF	Change
TIND		2	All	-0.000200	0.836125	0.000476	0.000200	0.835101	-0.000548
			1		0.804687	0.000619		0.803357	-0.000711
			2		0.826309	0.000718		0.824797	-0.000795
			3		0.853150	0.000429		0.852222	-0.000500
			4		0.849676	0.000279		0.849054	-0.000344
			5		0.852362	0.000269		0.851773	-0.000320
TIND		4	All	-0.000200	0.836882	0.001233	0.000200	0.833834	-0.001815
			1		0.805930	0.001862		0.801194	-0.002874
			2		0.827573	0.001981		0.822850	-0.002742
			3		0.853965	0.001244		0.850964	-0.001758
			4		0.850101	0.000703		0.848374	-0.001024
			5		0.852190	0.000097		0.851854	-0.000239
TIND		5	All	-0.000200	0.833863	-0.001786	0.000200	0.836876	0.001227
			1		0.801434	-0.002633		0.805796	0.001728
			2		0.823050	-0.002542		0.827435	0.001844
			3		0.851014	-0.001708		0.853923	0.001201
			4		0.848190	-0.001208		0.850245	0.000847
			5		0.851592	-0.000501		0.852392	0.000299
TIND		8	All	-0.000200	0.834542	-0.001107	0.000200	0.836503	0.000853
			1		0.802123	-0.001945		0.805510	0.001442
			2		0.823845	-0.001747		0.826991	0.001399
			3		0.851480	-0.001242		0.853779	0.001058
			4		0.848730	-0.000668		0.849976	0.000578
			5		0.852504	0.000411		0.851640	-0.000454
TIND		9	All	-0.000200	0.836471	0.000822	0.000200	0.834587	-0.001062
			1		0.805510	0.001442		0.802141	-0.001927
			2		0.827001	0.001409		0.823849	-0.001743
			3		0.853712	0.000991		0.851557	-0.001164
			4		0.849825	0.000427		0.848893	-0.000505
			5		0.851674	-0.000419		0.852484	0.000391

Type		Sf								
TIND	11	All	-0.000200	0.836082	0.000433	0.000200	0.835162	-0.000487
		1		0.804780	0.000712		0.803254	-0.000814
		2		0.826269	0.000677		0.824842	-0.000750
		3		0.853184	0.000463		0.852216	-0.000505
		4		0.849615	0.000218		0.849156	-0.000241
		5		0.852076	-0.000018		0.852099	0.000006
TABB	2	All	-0.010000	0.835612	-0.000037	0.010000	0.835684	0.000035
		1		0.804060	-0.000007		0.804072	0.000005
		2		0.825548	-0.000044		0.825633	0.000041
		3		0.852631	-0.000091		0.852810	0.000088
		4		0.849292	-0.000106		0.849501	0.000103
		5		0.852145	0.000051		0.852039	-0.000054
TABB	4	All	-0.010000	0.835571	-0.000079	0.010000	0.835717	0.000068
		1		0.804047	-0.000020		0.804075	0.000008
		2		0.825499	-0.000093		0.825674	0.000082
		3		0.852548	-0.000174		0.852885	0.000164
		4		0.849205	-0.000193		0.849581	0.000183
		5		0.852163	0.000070		0.852015	-0.000078
TABB	5	All	-0.010000	0.835784	0.000135	0.010000	0.835460	-0.000189
		1		0.804059	-0.000009		0.804007	-0.000060
		2		0.825751	0.000159		0.825371	-0.000221
		3		0.853054	0.000332		0.852336	-0.000386
		4		0.849772	0.000374		0.848978	-0.000420
		5		0.851954	-0.000139		0.852194	0.000101

We will now list the 10 worst offenders, in other words, the parameters and their associated tolerances giving the biggest drop in MTF. These data are for the average MTF computed over the entire field of view. Note that most of the more sensitive tolerances are element wedges. We also see here the nominal MTF (averaged over the field) to be 0.84, the estimated change in MTF of −0.23, and the estimated MTF of 0.61. There is additional statistical information including the expected compensator amount.

```
Worst offenders:
Type Sf1 Sf2    Value       MF      Change
TIRY      6   -0.025000  0.734318  -0.101331
TTHI  4   6    0.050000  0.756450  -0.079199
TTHI  5   6    0.050000  0.757571  -0.078078
TIRY      4    0.025000  0.765899  -0.069750
TIRY      6    0.025000  0.765926  -0.069723
TIRY      8   -0.025000  0.778044  -0.057605
TIRX      8   -0.025000  0.778506  -0.057143
TIRX      8    0.025000  0.778506  -0.057143
TIRY      8    0.025000  0.779299  -0.056350
TETX  4   6   -0.114000  0.785336  -0.050313
```

The net prediction for the expected MTF at 30 line pairs/mm is shown here. This is for the average MTF over the field of view and we see that the prediction is for an MTF of 0.64 at 30 line pairs/mm. Note also that the expected total range in back focus adjustment is approximately ±180 μm, with a standard deviation of 31 μm.

```
Nominal MTF           :          0.84
Estimated change      :         -0.19
Estimated MTF         :          0.64

Merit Statistics      :
Mean                             0.824034
Standard Deviation    :          0.018316

Compensator Statistics:
Change in back focus:
Minimum               :         -0.183763
Maximum               :          0.184228
Mean                  :          0.000007
Standard Deviation    :          0.031307
```

Perhaps the most important way to assess the overall tolerance situation is via a *Monte Carlo analysis*. In this analysis, every parameter is perturbed between its plus and minus tolerance extremes according to a normal probability distribution. The MTF is then computer averaged

```
Monte Carlo Analysis:
Number of trials: 20

Statistics: Normal Distribution
```

Trial	Merit	Change	0.0, 0.0 Field 1	0.0, 0.3 Field 2	0.0, 0.5 Field 3	0.0, 0.8 Field 4	0.0, 1.0 Field 5
1	0.671008	-0.164641	0.598434	0.598588	0.674171	0.743139	0.784013
2	0.739904	-0.095745	0.654500	0.751300	0.815063	0.781986	0.725598
3	0.683551	-0.152098	0.636928	0.662700	0.710189	0.703255	0.711801
4	0.671622	-0.164027	0.586082	0.620614	0.682513	0.719141	0.789717
5	0.752621	-0.083028	0.727852	0.714027	0.743741	0.778635	0.811677
6	0.815043	-0.126764	0.663265	0.709160	0.728300	0.712060	0.737265
7	0.708885	-0.108246	0.678229	0.718370	0.741885	0.738249	0.768576
8	0.727403	-0.092848	0.729252	0.767542	0.745248	0.719039	0.755914
9	0.742801	-0.108210	0.654835	0.687834	0.761382	0.784023	0.773559
10	0.678215	-0.157435	0.603121	0.639159	0.684315	0.718385	0.774066
11	0.691634	-0.144015	0.654557	0.666323	0.719306	0.721229	0.702888
12	0.691634	-0.094413	0.688700	0.713367	0.762045	0.765506	0.789961
13	0.741237	-0.091746	0.647524	0.705033	0.773729	0.700739	0.840690
14	0.641301	-0.194348	0.634102	0.701845	0.719443	0.628549	0.548493
15	0.653456	-0.182194	0.591334	0.604764	0.638324	0.687568	0.779041
16	0.805344	-0.030305	0.766817	0.809427	0.839262	0.821381	0.797463
17	0.754233	-0.081416	0.776285	0.766568	0.766268	0.737186	0.728399
18	0.681817	-0.153832	0.600345	0.637742	0.701103	0.731612	0.767892
19	0.764597	-0.071052	0.734169	0.747044	0.781917	0.777415	0.787125
20	0.558923	-0.276726	0.570030	0.628170	0.579650	0.494046	0.534242
Nominal	0.835649		0.804068	0.825592	0.852721	0.849398	0.852093
Best	0.805344		0.776285	0.809427	0.839262	0.821381	0.840690
Worst	0.558923		0.570030	0.598588	0.579650	0.494046	0.534242
Mean	0.706995		0.659818	0.692479	0.728393	0.728107	0.745419
Std Dev	0.053359		0.059708	0.057695	0.058114	0.068625	0.075600

```
Compensator Statistics:
Change in back focus:
Minimum               :         -0.159362
Maximum               :          0.277543
Mean                  :         -0.031688
Standard Deviation    :          0.116491
```

over the field of view and at each of the five separate fields. The resulting statistics shows for the 20 Monte Carlo samples the nominal, best, worst, mean, and standard deviation in the MTF. Recall that our MTF goal is 0.5 at 30 line pairs/mm. Field 4 (75% of the way to the corner) shows a worst MTF encountered of 0.494. The best MTF at field 4 was 0.821. Here we show the results of 20 Monte Carlo samples.

The final results of the Monte Carlo analysis are shown next, where we see that 90% of the lenses have an MTF of greater than or equal to 0.703. The standard deviation in the compensator motion (refocusing) is 0.116 mm.

```
90% of Monte Carlo lenses have an MTF above 0.703
50% of Monte Carlo lenses have an MTF above 0.774
10% of Monte Carlo lenses have an MTF above 0.812
```

The final result of the tolerance analysis is that the lens will, with the input tolerances, meet our MTF performance goal of 0.5 minimum at 30 line pairs/mm. If we were to take the analysis further, we would discuss the more sensitive parameters with the optical shop to see if we can tighten them. Then we would tighten these tolerances and simultaneously loosen many of the less sensitive tolerances so that the net result is meeting the MTF performance requirement while making the lens more producible at a lower cost and higher confidence.

The previous tolerances were computer for one side of the field of view. Due to the asymmetrical nature of some tolerances, both sides of the field should be modeled. When this is done, the 90% confidence level MTF at 30 line pairs/mm reduces to 0.634, still well above out goal of 0.5.

Digital Camera Lens

This case study is based on a VGA digital camera, which has some rather unique specifications. Specifically, some time ago we bought a digital camera which has a $1/3$-in CCD chip. The lens is $f/2.0$, and the camera manual states that objects from 533 mm (21 in) to infinity will be in focus. Many of us who have worked a lot in 35-mm camera photography will remember that if we focus on the front of someone's nose at $f/2.0$, that person's earlobes will be out of focus. The depth of field is incredibly small at such a low f/number. What is it that allows for such a large depth of field in our digital camera? The answer will be given later. First, we will summarize in Table 21.3 the specifications for our camera lens.

The Airy disk diameter is 2.8 μm, or about one-third of a pixel. Let us use Newton's equation, which relates object distance to the amount of defocus. Newton's equation states that

$$-xx' = (\text{focal length})^2 \qquad x' = \frac{f^2}{-x} = \frac{f^2}{\text{object distance}} = \text{defocus}$$

where x' is the amount of refocus required for an object at distance x. If we compute the defocus required for a given object distance, we can easily determine the blur diameter by multiplying the defocus by $1/(f/\text{number}) = 0.5$. Figure 21.20 shows the situation.

Table 21.4 shows the defocus along with the associated blur diameter.

What this means is that if we have an otherwise perfect lens at $f/2.0$ focused for an object at infinity, the image distance will change by the amount "δ image distance" as a function of the object distance. If our sensor were to remain fixed at the infinity focus position, then the image would blur to the diameter indicated in the third column. Thus, an object at 0.5 m will blur to a diameter of 23 μm. This is approximately three pixels of image blur, which seems excessive. However, what if we select an intermediate object distance at which to focus our lens nominally, so that the blur is equalized at infinity and at 0.5 m. This distance is approximately 1 m, and the residual image blur with the object at infinity and at 0.5 m is approximately 11 μm which is in the order of 1.5 pixels. Thus, the bottom line is that at the factory the lens will be focused for an object distance between 1 and 2 m, and in use the

TABLE 21.3

Digital Camera Lens Design Example

Parameter	Specification
Sensor type	CCD
Sensor size	$\frac{1}{3}$ in (3.6 × 4.8 mm, 6-mm diagonal)
Number of pixels	640 × 480
Pixel pitch	7.5 μm
Lens f/number	$f/2.0$
Lens focal length	4.8 mm
Comparable focal length in 35-mm camera	35-mm focal length
Stated depth of field	533-mm to infinity

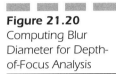

Figure 21.20
Computing Blur
Diameter for Depth-
of-Focus Analysis

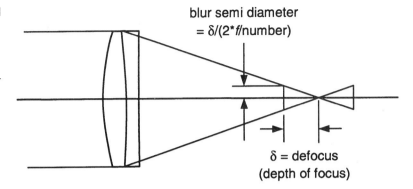

TABLE 21.4 Depth-of-Focus Calculation for $f/2$ Digital Camera Lens

Object Distance	δ Image Distance (μm) (Sensor Focused for Infinity)	Blur Diameter (μm) (Sensor Focused for Infinity)	δ Image Distance (μm) (Sensor Focused for 1-m Distance)	Blur Diameter (μm) (Sensor Focused for 1-m Distance)
Infinity	0	0	−23	11.5
3	7.68	3.84	−15.3	7.7
2	11.5	5.75	−11.5	5.75
1	23.0	11.54	0	0
0.5	46.1	23	23	11.5

maximum image blur diameter from a point object everywhere from 0.5 m to infinity will be about 10 μm, or about 1.5 pixels. This large depth of field achieved with an $f/2.0$ lens explains how a digital camera can have a fixed focus.

Why is it then that the $f/2$ 35-mm camera lens cannot have a fixed focus, and even with the adjustable focus, some objects in the field of view are less sharp than the others? To answer this question, we will compare two lenses: a digital camera lens and a 35-mm camera lens. Let us choose the 35-mm camera lens to have the same field of view as the digital camera. This determines the focal length of the camera lens to be 35 mm. If both cameras are focused to infinity, we can compare the depth of field allowing the same angular blur in both cameras. This can be expressed as the equal relative linear blur in the image plane:

$$\left[\frac{\text{linear blur}}{\text{Linear field of view}}\right]_{\text{35-mm camera}} = \left[\frac{\text{linear blur}}{\text{linear field of view}}\right]_{\text{digital camera}}$$

$$\frac{f_1^2}{D_1\, f\#\, \text{fov}_1} = \frac{f_2^2}{D_2\, f\#\, \text{fov}_2}$$

where f_1 and f_2 are, respectively, focal lengths of two cameras, D_1 and D_2 are distances at which the angular image blur is acceptable and equal in both cases, and fov_1 and fov_2 are the respective image sizes.

$$D_1 = D_2\left(\frac{f_1}{f_2}\right)^2 \frac{\text{fov}_2}{\text{fov}_1} = D_2\left(\frac{f_1}{f_2}\right)^2 \frac{f_2}{f_1} = D_2 \frac{f_1}{f_2}$$

$$D_1 = 1500\,\frac{35}{4.8} = 10937$$

We see that if with the digital camera we can go from infinity to 1.5 m, for the same allowable image blur, with the 35-mm camera lens we can go from infinity to only about 11 m. Depth of field is inversely proportional to the focal length of the lens.

Prior to designing our lenses, we need to determine at what spatial frequency the lens should be evaluated. Consider Fig. 21.21, where we show the representation of a pixelated sensor such as a CCD. For our VGA CCD sensor the pixel pitch is 7.5 μm. The maximum spatial frequency of an image, which can effectively be resolved by a pixelated sensor without aliasing, is the spatial frequency where the bright and dark bars line up with adjacent rows or columns of the sensor as shown in Fig. 21.21. This frequency is called the *Nyquist frequency*. At higher-image spatial frequencies we will get so-called aliasing, where the image is undersampled by the sensor. Angled lines look like staircases due to the undersampling. For our camera case study the 7.5-μm pixel pitch results in a Nyquist frequency of 66.6 line pairs/mm. We will thus evaluate our lens performance at this value.

In order to determine the quality of the lens that we bought, we decided to measure two basic characteristics: camera resolution and distortion. This may be useful and serve as a reference during the design of the lens. To determine the resolution, we took the picture of a resolution chart, shown in Fig. 21.22 from a distance of 1 m. We measured the smallest diameter in the photo of the chart where the radial lines were resolved, and found a corresponding width of a line pair, which was 52 line pairs/mm at the CCD chip. This corresponds to a modulation of only a few percent. Taking into account normal manufacturing errors, we can conclude to a close approximation that the nominal design of

Figure 21.21
Imagery onto
Pixelated Sensor
and the Nyquist
Frequency

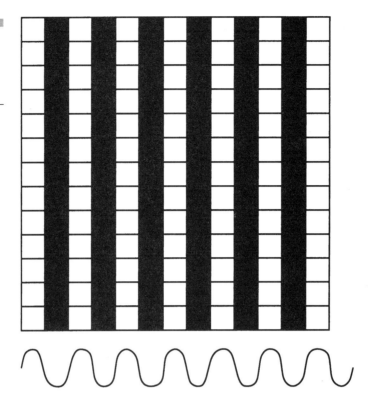

Figure 21.21
Imagery onto
Pixelated Sensor
and the Nyquist
Frequency

this camera lens has an MTF in the order of 0.3 to 0.4 at 52 line pairs/mm. Distortion was measured from the picture of an object with a straight edge whose geometry is shown in Fig. 21.23. From the measured sag of the bowed image of the straight line, we calculated the distortion to be less than 3%.

Note that the previous assessment of image quality and distortion was done very quickly with extremely rudimentary equipment and without removing the lens from the camera. These forms of tests can often be extremely useful, even though they are not highly quantitative.

We will now proceed to look at several candidate design solutions for the lens. The design parameters are as follows:

- Focal length of 4.8 mm
- $f/2$ lens
- Field of view (diagonal) of 64°
- Nyquist frequency of 66 line pairs/mm

▬▬ ▬▬ ▬▬ ▬▬

Figure 21.22
Image of Resolution Chart Taken with Digital Camera

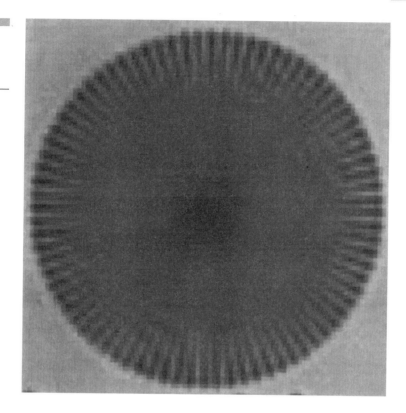

Design with Three Glass Elements

Figure 21.24 shows a design for an all-glass lens with spherical surfaces. The field of view is quite large, and we would not expect to have a nicely behaved lens with small angles of incidence as the rays proceed through the lens. In this lens, rays enter the second component at very large incident angles, and also reach the detector at very large angles at the corner of the CCD. After optimizing different configurations, one criterion that is used to determine which lens is better than the other is how strongly rays refract on each surface throughout the lens. The manufacturing tolerances have to be tighter in the locations of strong ray bending, so that the lens with the smoother ray travel through the lens is preferable.

We will analyze a few configurations comparing their performance shown in four diagrams. The first is the lens layout. The second is the MTF curve for four field angles shown to the Nyquist frequency of 66 line pairs/mm. The third diagram is the field curvature and the distortion curve, given on the scale of 10%. The fourth diagram is the rms wavefront

Figure 21.23
Measurement of
Lens Distortion

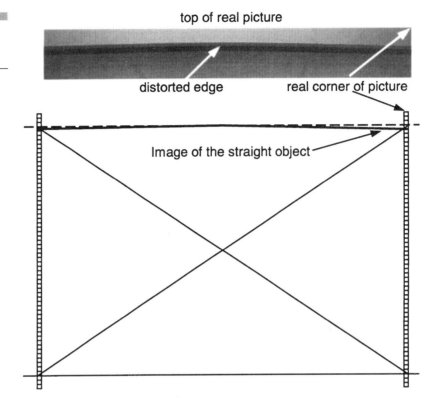

error plotted on a two-wave scale, as a function of field of view. The MTF of this lens should be higher, although the measured resolution of the digital camera suggests that its MTF is probably a little lower than the MTF of our design shown here. The distortion is definitely unacceptable, and it will have to be more tightly controlled. In the next step, we increased the weights on the chief ray heights in the merit function, in order to reduce the distortion below 3%. We also allowed both surfaces of the third component to be aspheric. The result was a reduced distortion down to 4%, but the MTF and the lens shape remained the same.

Design with Four Glass Elements, Two of Them Aspheric

All lenses (Fig. 21.25) are made of high-index glass. Ray bending is smoother than in the previous configuration. Distortion is very low, but unfortunately, MTF is somewhat lower.

Figure 21.24
Three All-Spherical-
Glass Elements for
Digital Camera Lens

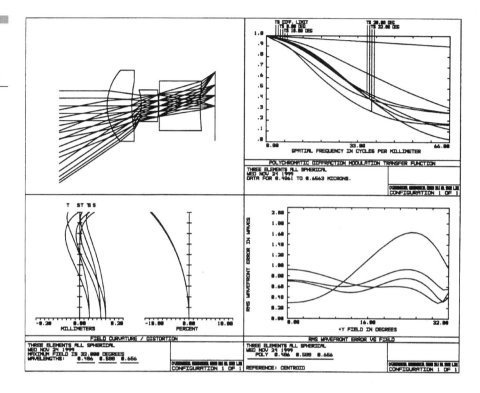

Design with Four Glass Elements, Three of Them Aspheric

Figure 21.26 shows this configuration. This lens has very good performance. The first element may be difficult to manufacture cost effectively, and note also that the angles of incidence on the sensor are high at the edge of the field.

Design with Three Elements, Two of Them Glass Spherical and One Plastic Aspheric

The last configuration (Fig. 21.27) that we are going to show is a three-component lens. It has one plastic component, which is generally cheaper than a glass one. The performance is satisfactory, although the MTF is lower than in the previous four-component case. One parameter that should be controlled and the lenses compared to is the total track, which is the distance from the lens front surface to the CCD chip. We did not control this

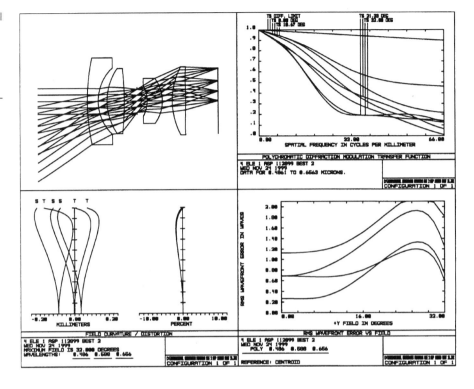

parameter, simply because we could not measure it in our camera. The last three-component lens shown has a shorter total track than the previous four-component lens, and it is preferable. It would be useful to investigate design forms with one diffractive surface, but we will stop at this point. The last two configurations shown could be good candidates for the final design. In the next step, a tolerance analysis, manufacturability, and cost analysis should be performed, and the final design chosen.

In our short exercise, we attempted to derive the design of a digital camera lens closest to the one in our camera. However, without destroying our camera, we may never know precisely what lens design form was used.

Binocular Design

This example is the design of a reasonable quality 7 × 50 binocular. The binocular should be low cost and as compact as possible. A binocular system is more compact with a Pechan than with Porro erecting prisms. However, a Pechan prism is more expensive, since it is a roof prism with

Figure 21.26

Four Glass Lenses for
Digital Camera Lens,
Three of Them
Aspherics

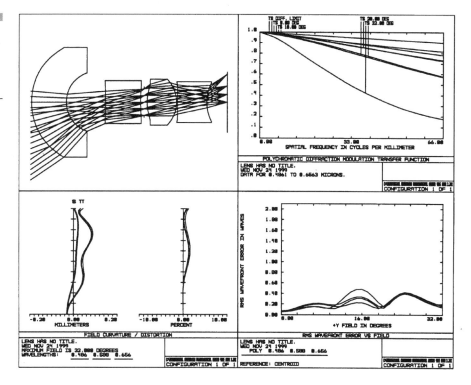

a tight tolerance on a roof angle in the order of 2 to 3 arc s. That is the
reason why we will design the system with a Porro erecting prism
rather than a Pechan prism. The simplest objective is a cemented achromatic doublet. An $f/4$ achromatic doublet gives a reasonable quality
image. This equates to a 200-mm focal length for the objective. The basic
binocular specifications are listed in Table 21.5.

Good optical performance would require the system to have resolution
in the exit pupil similar to the resolution of the human eye. If the eye can
resolve 2 arc min/line pair, a good system should have, at the center of the
field of view, not more than 2 min of spherical aberration and 2 min of
chromatic aberration over the exit pupil size of 3 mm. For our system
here, the optical performance can tolerate somewhat larger aberrations.
The magnification of $7\times$ and the entrance pupil diameter of 50 mm give
an exit pupil diameter of 7 mm. The design and analysis will be done for
a 7-mm exit pupil diameter, and only in the final analysis, we will look at
the system performance having an exit pupil diameter of 3 mm.

The design of the system starts with the design of the objective. First,
we enter the doublet into the design program, and optimize it to

Figure 21.27
Two Glass and One
Plastic Lens for Digital
Camera

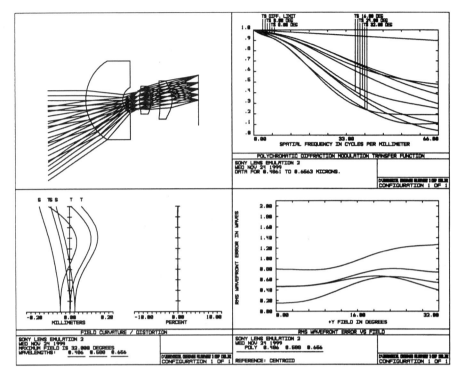

TABLE 21.5

7 × 50 Binocular
Design Example

Parameter	Specification
Entrance pupil diameter (mm)	50
Magnification	7×
Objective focal length (mm)	200
Objective ƒ/number	ƒ/4.0
Eyepiece focal length (mm)	28.6
Full field of view (degrees)	6
Spectral range	Visual (C, d, F)
Distortion (%)	<12
Vignetting (%)	<30 at edge of field
Diameter of eyepiece assembly (mm)	<38
Eye relief (mm)	>23

minimize the aberrations and achieve the focal length 200 mm. In the next step we add two blocks of SK5 glass of the right thickness to simulate two right-angle prisms of the Porro system. In each right-angle prism there are two internal reflections, which can be simulated with tilted surfaces inside the glass block. The location of the prisms was chosen to leave minimum 40 mm from the exit surface of the second prism to the image plane. This is necessary to have enough room for the nose and comfortable resting of the binocular on the face. At this point, we decided to introduce a small amount of vignetting. The layout of the objective with the Porro prism and its performance is shown in Fig. 21.28. The second graph in Fig. 21.28 shows the rms wavefront error on a five-wave scale as a function of the field. The third graph shows the transverse ray aberration curves on a ±100-μm scale. There is some field curvature and astigmatism, as well as some lateral color aberration. The last graph shows a distortion grid, with practically no distortion.

In the following step, we fold the prisms in four places where the reflections take place, check to see if the image plane is behind the plane in which the corner of the first right-angle prism is, and make adjustments to the location and size of the prisms, if needed. Now we can freeze the objective and add the eyepiece.

Figure 21.28
7 × 50 Binocular Objective with Porro Erecting System

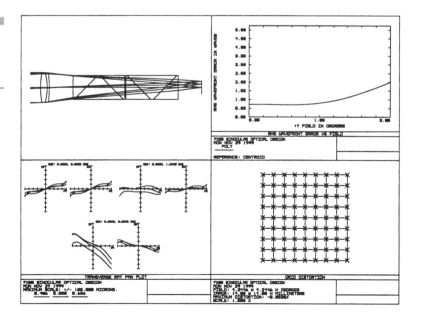

The full field of view of 6° in object space gives an apparent field of view to the user of 6 × 7 = 42°, where 7 is the system magnification. A field of view of 42° is a little larger field than a symmetrical form of eyepiece is designed for, but it is a low-cost eyepiece which performs reasonably well, and we will design our system with it. We now add the symmetrical eyepiece and a paraxial lens in the exit pupil of the system. The optimized system with the symmetrical eyepiece is shown in Fig. 21.29. The eye relief is 25 mm. The second graph in Fig. 21.29 shows the rms wavefront error on a five-wave scale as a function of field of view. We notice some degradation in image quality toward the outer periphery of the field. The third graph shows the transverse ray aberration curves on the ±2000-μm scale. These transverse ray aberrations are at the image of our paraxial lens, which was used to evaluate our afocal system. We used a paraxial lens of 1000-mm focal length. This means that the angular blur of 1 mrad coming into this paraxial lens corresponds to 1000-μm blur in the image plane, or that our scale shows ±2 mrad in the exit pupil. There is quite a lot of astigmatism at the edge of the field, otherwise performance is not bad. The last graph shows the distortion grid, with a maximum of 8.2% distortion.

We will now try a different form of eyepiece. If we start with a cemented doublet and two singlets, varying glasses and allowing the

Figure 21.29
7 × 50 Binocular Design—Objective with Folded Porro Erecting System and Symmetrical Eyepiece

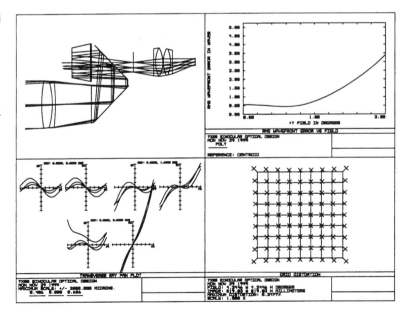

Figure 21.30
7 × 50 Binocular Design—Objective with Folded Porro Erecting System and Four-Element Eyepiece

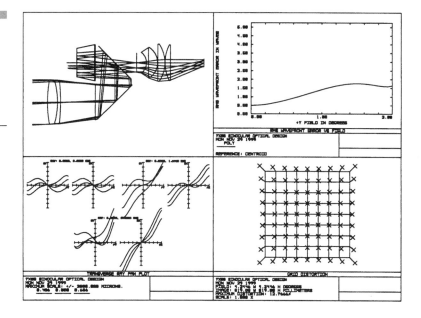

doublet to acquire a meniscus form, the resulting design has reduced astigmatism. This system is shown in Fig. 21.30. However, this design has 12.7% maximum distortion, the diameter of the eyepiece assembly is larger, and the eyepiece is more expensive.

Let us go back and analyze the binocular design with the symmetrical eyepiece, with the 3-mm exit pupil diameter. The performance is shown in Fig. 21.31. We can see that the total blur on axis, including all colors, is smaller than 1 mrad. This is good, indeed. Some degradation in the image can be noticed in the last 25% of the field of view, which may be acceptable in our case.

Parametric Design Study of Simple Lenses Using Advanced Manufacturing Methods

In order to illustrate the relative benefits of conventional as well as advanced manufacturing methods, it is often valuable to compare these methods parametrically. For lenses of the following specifications we

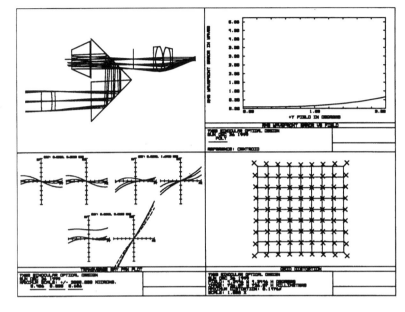

optimized the performance of the design forms listed in Table 21.6. Also shown in Table 21.6 are the figure numbers:

- Clear aperture diameter of 12.5 mm
- Field of view of ±2°
- Spectral band visible
- Materials BK7 and SF2 for doublets and BK7 for singlet, unless otherwise noted

The figures show for each design the following:

- Lens layout
- Transverse ray aberrations on a scale of ±100 μm
- Spot diagrams with a box width scale of 200 × 200 μm
- MTF plotted to a spatial frequency of 50 line pairs/mm

Note that the scales for the optical performance are identical for all of the designs. While some of the data are off the scale, you will get a better understanding of the relative performance of each lens as it relates to the other approaches by using the same scale for the data.

The limiting aberrations in the *singlet designs* is spherical aberration. This is especially prevalent at the low *f*/# of *f*/2. At *f*/4 the spherical

TABLE 21.6

Parametric Lens
Design Examples

Lens Form	First f/Number	Figure Number	Second f/Number	Figure Number
Single element	$f/2$	21.32	$f/4$	21.33
Achromatic doublet	$f/2$	21.34	$f/4$	21.35
Achromatic doublet with aspheric surface	$f/2$	21.36	$f/4$	21.37
Single element with y^2 DOE	$f/2$	21.38	$f/4$	21.39
Single element with y^2 and y^4 DOE	$f/2$	21.40	$f/4$	21.41
Spherical acrylic Fresnel lens	$f/2$	21.42	$f/4$	21.43
Aspheric acrylic Fresnel lens	$f/2$	21.44	$f/4$	21.45
Gradium G1SFN gradient index glass	—	—	$f/4$	21.46
Planoconvex simple lens	—	—	$f/4$	21.47

aberration is reduced to a level commensurate with the residual primary axial color. In both singlet designs the MTF is poor at 50 line pairs/mm, but this is not at all surprising.

For the *achromatic doublet* the chromatic aberrations are reasonably well corrected with higher-order spherical aberration the limiting aberration for the $f/2$ lens. At $f/4$ the design is approaching being diffraction limited, except for the astigmatism residual at the edge of the field.

An *aspheric surface on the achromatic doublet* allows for the correction of the residual spherical aberration with the astigmatism evident off axis. The chromatic aberrations are well corrected.

The *diffractive lens with a y^2 kinoform* period surface still has a residual of spherical aberration, and this is evident at both $f/2$ and $f/4$.

If we also allow, in addition to the y^2, a y^4 *kinoform* period we can achieve a nearly complete control over the spherical aberration.

We have included *Fresnel lenses* for completeness. It is apparent that the residual spherical aberration of the spherical surface emulation of the Fresnel lens produces significant spherical aberration, in fact, far more than the equivalent conventional spherical single element.

The *aspheric Fresnel lens* is well corrected for spherical aberration; however, the primary residuals are primary axial color and coma off axis.

The $f/4$ planoconvex lens is shown for comparison with the $f/4$ Gradium lens.

The *Gradium axial refractive index* gradient material is a material which permits most of the spherical aberration to be corrected while using spherical surfaces.

Figure 21.48 shows a summary of the rms blur diameters for each of the designs presented in this parametric study.

Design Data for Double Gauss

For reference we include here the optical design prescription data for the double gauss design example of the section "Double Gauss Lens Design" earlier in this chapter.

```
Title: Double Gauss Starting Design From Patent
System Aperture  : Entrance Pupil Diameter = 25
Eff. Focal Len.  :          50 (in air)
Image Space F/#  :           2
Entr. Pup. Dia. :           25
Field Type: Angle in degrees
#       X-Value        Y-Value        Weight
1       0.000000       0.000000       1.000000
2       0.000000      11.000000       1.000000
3       0.000000      16.000000       1.000000
Vignetting Factors
#      VDX        VDY        VCX        VCY
1   0.000000   0.000000   0.000000   0.000000
2   0.000000   0.000000   0.000000   0.300000
3   0.000000   0.000000   0.000000   0.500000
Wavelengths    : 3  Units: Microns
#         Value        Weight
1       0.486100       1.000000
2       0.587600       1.000000
3       0.656300       1.000000
SURFACE DATA SUMMARY:
Surf    Type        Radius      Thickness       Glass       Diameter
 OBJ STANDARD      Infinity     Infinity                           0
   1 STANDARD      Infinity          7.5                           0
   2 STANDARD        32.715         4.06       SSK51        25.34117
   3 STANDARD       122.987         0.25                    24.33643
   4 STANDARD        20.218          7.3       SK10         23.12816
   5 STANDARD       -112.78         2.03       F8           20.79916
   6 STANDARD        12.548         5.08                    16.57081
 STO STANDARD      Infinity         5.08                    16.09063
   8 STANDARD       -14.681         2.03       F15          15.37717
   9 STANDARD        40.335          6.6       SSK2         16.93081
  10 STANDARD       -19.406         0.25                    17.96581
  11 STANDARD       82.8556         4.11       SK10         19.73168
  12 STANDARD      51.96156     32.24573                    20.49336
 IMA STANDARD      Infinity                                 28.6088
```

```
Title: Double Gauss Final Design 12 Hour "Hammer" Optimization
System Aperture    : Entrance Pupil Diameter = 25
Eff. Focal Len.    :          50 (in air)
Image Space F/#    :          2
Entr. Pup. Dia.    :          25
Field Type:  Angle in degrees
#      X-Value      Y-Value      Weight
1     0.000000     0.000000     1.000000
2     0.000000     4.000000     1.000000
3     0.000000     8.000000     1.000000
4     0.000000    12.000000     1.000000
5     0.000000    16.000000     1.000000
Vignetting Factors
#       VDX         VDY          VCX         VCY
1   0.000000    0.000000    0.000000    0.000000
2   0.000000    0.006457    0.008781    0.090794
3   0.000000    0.008019    0.047112    0.204823
4   0.000000    0.000120    0.100113    0.337036
5   0.000000   -0.004495    0.248497    0.493301
Wavelengths       : 3   Units: Microns
#         Value          Weight
1       0.486100       1.000000
2       0.587600       1.000000
3       0.656300       1.000000
SURFACE DATA SUMMARY:
Surf   Type      Radius      Thickness     Glass     Diameter
OBJ STANDARD    Infinity     Infinity                       0
  1 STANDARD    Infinity         7.5                        0
  2 STANDARD    32.32399     5.345731     LAFN28    26.82534
  3 STANDARD    84.92015         0.25               24.72447
  4 STANDARD    19.13959     6.304279     LAFN10    22.61449
  5 STANDARD    75.0351          2.03     LAF9      19.66355
  6 STANDARD    12.66662     6.520931               16.11785
STO STANDARD    Infinity     8.724626               15.09638
  8 STANDARD   -15.0799          2.03     SF9       13.99697
  9 STANDARD  -140.6069     5.060231     LAK33     17.28576
 10 STANDARD   -21.26407        0.25               19.84303
 11 STANDARD    91.13499     5.477296     LAK8      22.46901
 12 STANDARD   -49.21186        25.4               23.51296
IMA STANDARD    Infinity                            28.43362
```

Figure 21.32
f/2 BK7 Single
Element

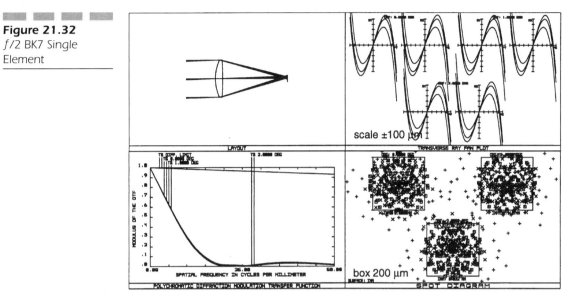

Figure 21.33
f/4 BK7 Single
Element

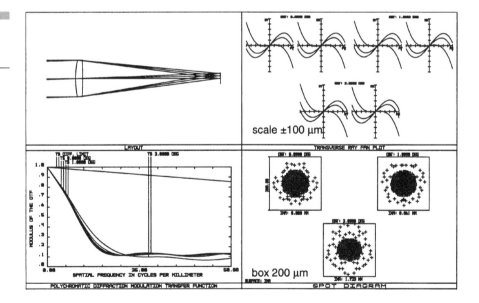

Figure 21.34
f/2 Achromatic
Doublet

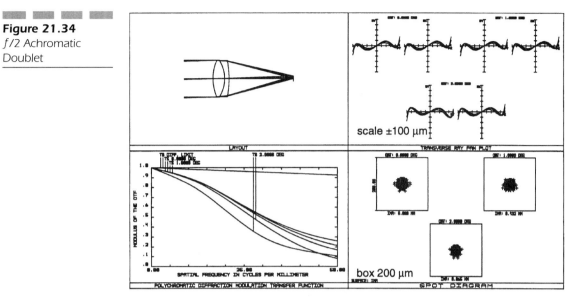

scale ±100 μm

box 200 μm

Figure 21.35
f/4 Achromatic
Doublet

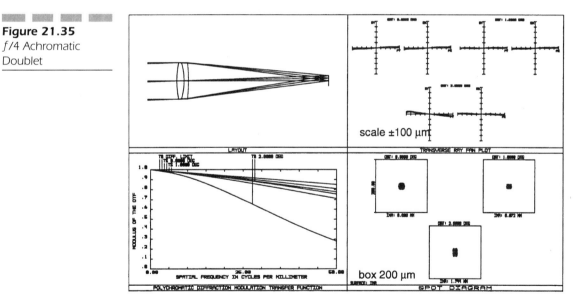

scale ±100 μm

box 200 μm

Figure 21.36
$f/2$ Achromatic
Doublet with
Aspheric Surface

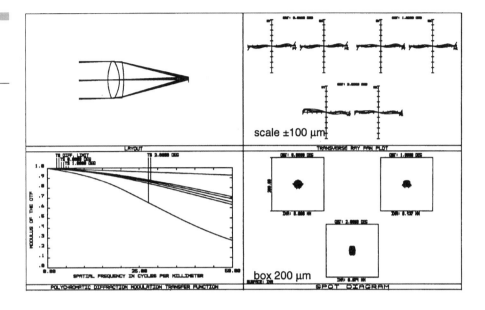

Figure 21.37
$f/4$ Achromatic
Doublet with
Aspheric Surface

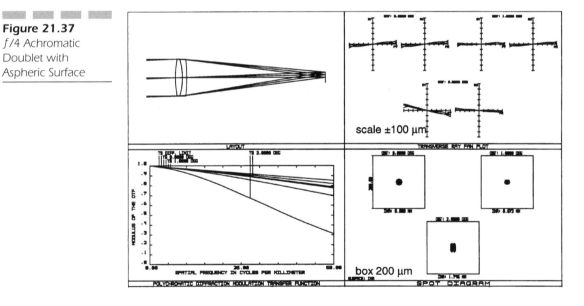

Figure 21.38
f/2 Single Element
with y^2 Diffractive
Kinoform Surface

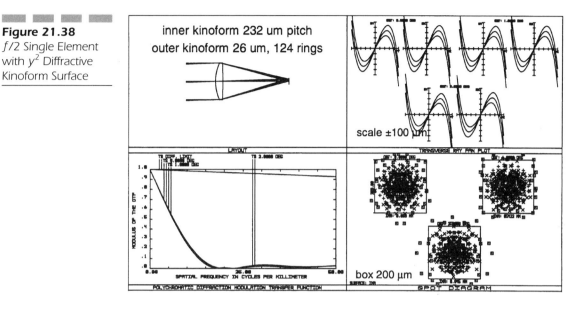

inner kinoform 232 um pitch
outer kinoform 26 um, 124 rings

scale ±100 μm

box 200 μm

Figure 21.39
f/4 Single Element
with y^2 Diffractive
Kinoform Surface

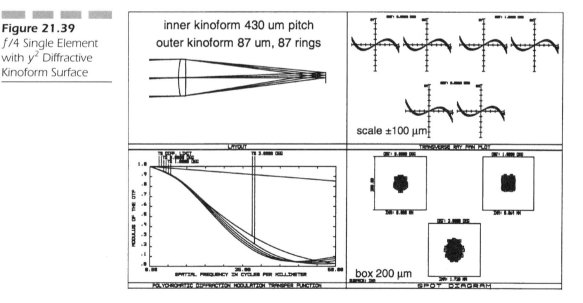

inner kinoform 430 um pitch
outer kinoform 87 um, 87 rings

scale ±100 μm

box 200 μm

Figure 21.40
f/2 Single Element
with $y^2 + y^4$
Diffractive Kinoform
Surface

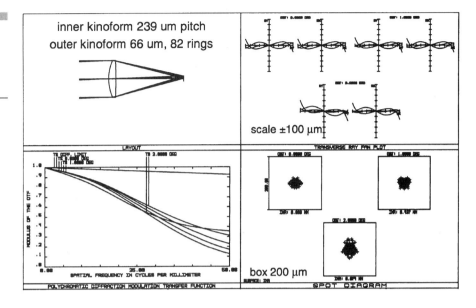

inner kinoform 239 um pitch
outer kinoform 66 um, 82 rings

scale ±100 μm

box 200 μm

Figure 21.41
f/4 Single Element
with $y^2 + y^4$
Diffractive Kinoform
Surface

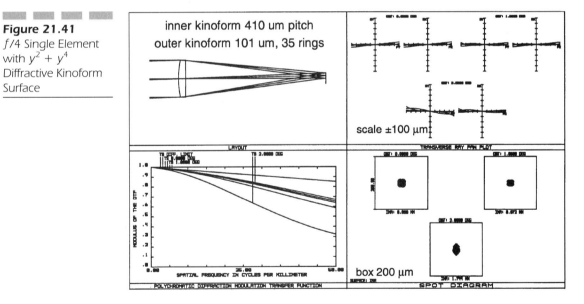

inner kinoform 410 um pitch
outer kinoform 101 um, 35 rings

scale ±100 μm

box 200 μm

Figure 21.42

f/2 Spherical Fresnel Lens

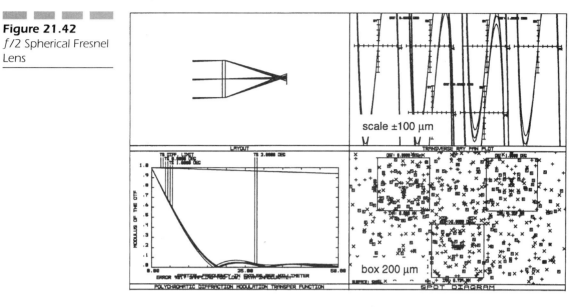

Figure 21.43

f/4 Spherical Fresnel Lens

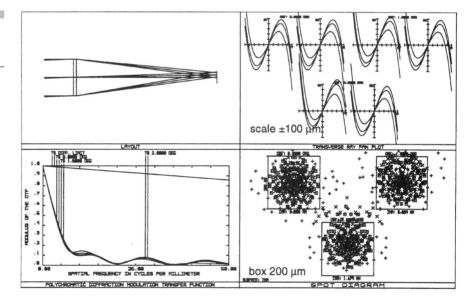

Figure 21.44
f/2 Aspheric Fresnel
Lens

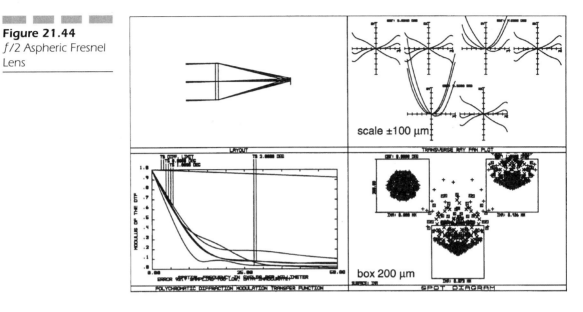

Figure 21.45
f/4 Aspheric Fresnel
Lens

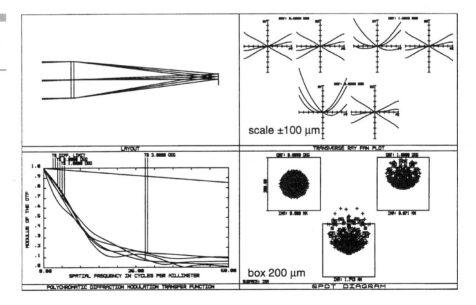

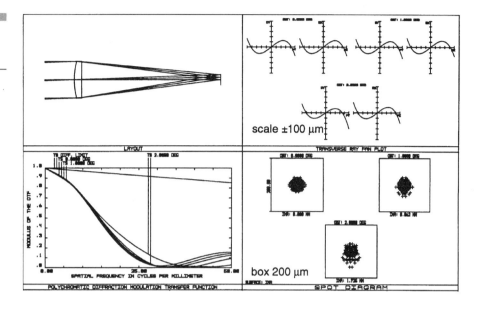

Figure 21.46
f/4 Plano BK7
Monochromatic

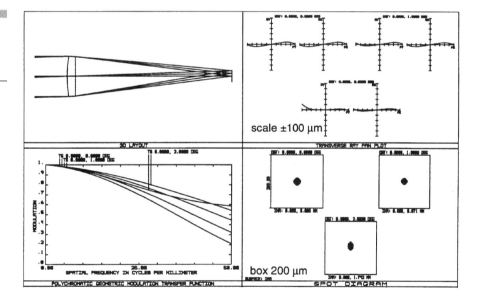

Figure 21.47
f/4 Plano
Axial-Gradient Index
Monochromatic

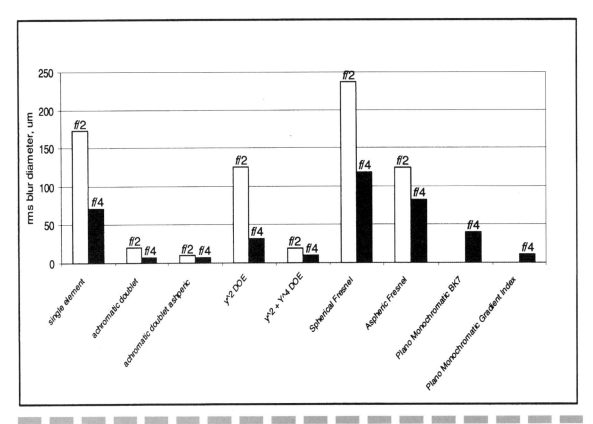

Figure 21.48 RMS Blur Diameter for Different Design Approaches, ƒ/2 and ƒ/4 Lenses (Diffractive, Fresnel, and Gradient Index Lenses Are Single Elements)

22

Bloopers and Blunders in Optics

This section is presented in the spirit that we all learn from our mistakes and/or the mistakes of others. The truth is that none of us is perfect, and from time to time even the best of us make mistakes. If we can share, in the right spirit, these mistakes, we will all learn, and our industry will improve. We are careful not to use names or affiliations in any of the following material.

Distortion in a 1:1 Imaging Lens

A lens that is fully symmetrical on both sides of a central aperture stop will be free of all orders of distortion, coma, and lateral color. This is because precisely equal and opposite amounts of these aberrations are introduced on each side of the central aperture stop, therefore, producing a net zero aberration at the image. Some years ago, a lens was required which imaged from a convex curved CRT onto a flat ground-glass image surface. The lens needed to have less than 0.25% of distortion. The lens was designed to be completely symmetrical about its central aperture stop. Only after the lens was assembled and tested did it become apparent that there was a residual distortion of several percent. This was never checked during the design effort because it was *assumed* that a fully symmetrical lens had zero distortion. The flaw in this assumption of symmetry was that the curved object surface immediately made the lens nonsymmetrical. This caused the distance from the edge of the field to the entrance pupil on the object side to differ from the corresponding distance from the edge of the field to the exit pupil, a clearly asymmetrical situation that will lead to distortion. If you can take advantage of symmetry, make sure that your system is fully and completely symmetrical!

We illustrate this in Figs. 22.1 to 22.3, where we show layouts and lens performance data for three different designs in which each of the lenses is fully symmetrical, except for the object radius which is shown as infinite, 100-mm convex, and 100-mm concave, respectively. The distortion is identically zero for the flat object design (the lens, object, and image are all completely symmetrical), and +1.4% for the 100-mm concave object and −1.2% for the 100-mm convex object.

Figure 22.1
Fully Symmetrical 1:1
Magnification Lens

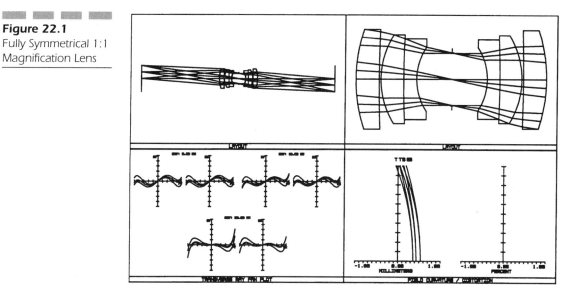

Figure 22.2
Symmetrical 1:1 Mag-
nification Lens with
100-mm Concave
Object and Flat
Image

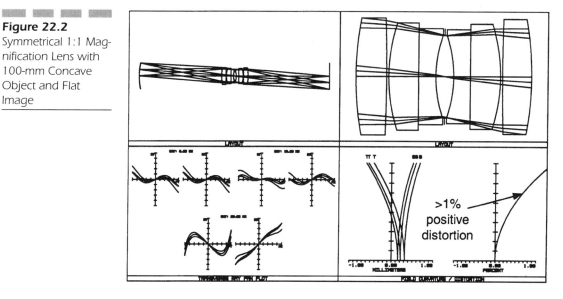

Figure 22.3
Symmetrical 1:1
Magnification Lens
with 100-mm
Convex Object and
Flat Image

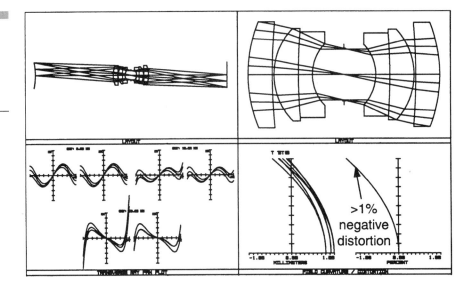

Zoom Periscope

A new periscope was designed for the U.S. Navy many years ago. There were various optical innovations in the system, including anomalous dispersion glasses for improved color correction. One innovation was to provide a continuous zoom lens to replace a 3× discrete field-of-view change. While technically a success, during sea trials the person at the helm became disoriented during docking procedures while zooming the periscope, and the submarine crashed into the dock, causing major damage. The decision quickly was made to freeze the zoom and revert back to a discrete field-of-view switch. While this is not an optical problem as such, it is indeed a human factors and human engineering issue. Generally, the optical designer is quite remote from the human factors issues; however, if you ever come across a similar situation in your future work, be bold and bring it up. After all, if you don't, perhaps no one else will either!

Sign of Distortion

Generally, ray tracing an optical system from one direction or the other will yield virtually the same results with respect to image quality (note that in complex systems the results may not be precisely identical). There are, however, some significant effects relating to distortion which can take on a totally different form, depending on which way light is traveling. This is a difficult concept to grasp, so we will illustrate it with a real design for an eyepiece.

Consider the Plössl form of eyepiece shown in Fig. 22.4. This eyepiece is designed to cover a full diagonal field of view of 40°, which is rather large for this design form. One result of this wide field of view is large distortion, at approximately 10%. If we ray trace from the eye to the image, we will predict negative or barrel distortion, as shown in Fig. 22.4. Now we will reverse the design in the computer and ray trace from the new object plane, which used to be our image, into the eye. The resulting distortion will be similar in magnitude; only it will be reversed in sign! This is very interesting indeed and is difficult to understand. The following explanation should be sufficient: Think of distortion as being analogous to spherical aberration of the chief ray. After all, it really is similar to this. Now if we are ray tracing from the eye to the image being viewed such as an LCD display in an HMD application, then the lens elements will bend the chief rays more severely than paraxial optics will dictate, meaning that the off-axis chief rays will end up closer to the

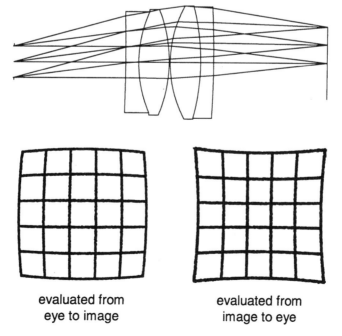

Figure 22.4
Plössl Eyepiece Showing Sign Reversal in Distortion Depending on Direction Lens Is Ray Traced

evaluated from
eye to image

evaluated from
image to eye

axis than their paraxial counterparts, resulting in negative or barrel distortion. Now let us reverse the eyepiece and ray trace from the image into the eye. The off-axis rays will bend more severely than their paraxial counterparts just like before; however, this will result in a *greater* ray angle entering the eye, and the resulting image will thus appear to the user as having positive or pincushion distortion.

Another way to explain and understand the opposite sign of distortion when tracing the rays backwards is the following: When we trace a rectangular object (*ABC*) from the eye as in Fig. 22.5, it will be imaged into *A'B'C'*, where *A'* and *C'* have smaller heights than they should have if there were no distortion. Now, if we trace the rays backwards, and if we take as the object the same points *A'B'C'*, we would end up with the angles *A, B, C* at the eye. However, when evaluating distortion, one always traces a rectangular object (or a regular-shaped object which is not distorted). In our case the object is *DEF*, where *E* is the same point as *B'*, and *D* and *F* are the points with larger height than *A'* and *C'*. Therefore, the angles after ray tracing to the eye corresponding to points *D* and *F* are *D'* and *F'*, and they are larger angles than *A* and *C*. *E'* is, of course, the same as *B*. Therefore, the key is that we do *not* take the same conjugate points in two ray traces.

The message here is to be extremely careful in assessing your performance, especially distortion. It would be quite disturbing if you predicted a given amount of negative distortion, only to find that the hardware produces the opposite sign.

Lens Elements That Are Not Necessary

An extremely weight-sensitive lens system was designed. The housing was manufactured of titanium and every gram needed to be accounted for. During the final design phase, one element became nearly flat with very long radii on each side. This element was about 10 mm thick. During the testplate fit, first the longest radius was made flat, and then the remaining radius of the element was made flat. This weight-sensitive lens system had, in effect, a flat window in its middle! In order not to look too foolish in front of the customer, one side was coated with a bandpass filter, which was originally planned to be coated onto one of the other elements. In the final design, this flat element was labeled

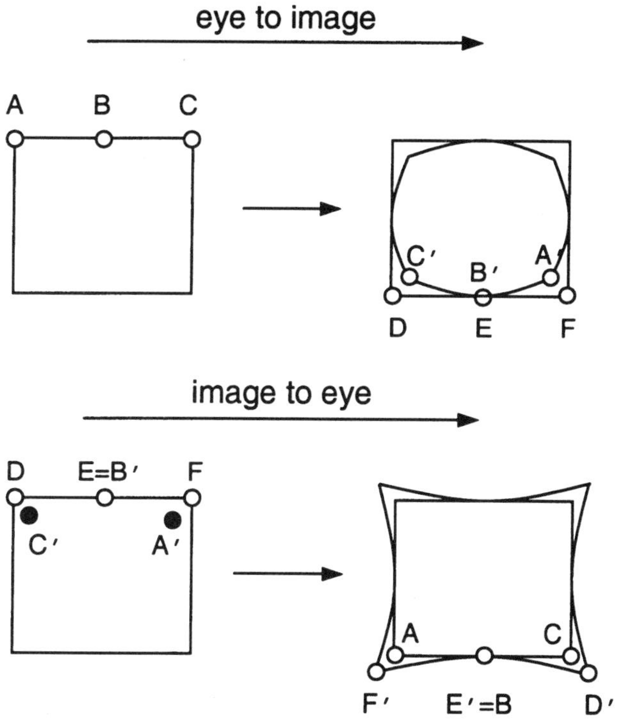

Figure 22.5
Explanation of
Distortion Sign
Reversal Illusion

"bandpass filter," and everyone was happy. If elements are not serving a real function in a design, remove them.

Pupil Problems

Many years ago, a lens system was designed to reimage high-resolution film onto a rear projection screen for viewing. As part of the system specifications it was necessary to provide an approximately 3× zoom of the central area of the film onto the screen. In order to "save money," it was decided to use all off-the-shelf optics. The optics consisted of a zoom lens and at least one relay lens group. After months of mechanical design and system integration, the initial imagery proved to be excellent. Unfortunately, when the zoom was initiated, the image became darker and darker until it was totally black prior to reaching the high magnification. The problem was that as zoom lenses are zoomed, their entrance and exit pupils translate axially, sometimes by large amounts. What had happened was that, at the high magnification, the light from the exit pupil of the first lens module or group simply did not get through, or even into, the entrance pupil of the next lens module. It took many months and was quite costly to remedy the problem.

Not Enough Light

A machine vision system was designed some years ago to provide a $50\times$ magnification from the object to the image, with a CCD chip located at the image. The specifications called for a relatively long working distance and for a working f/number of f/3 at the object in order to be able to sense the z location of the surface under test (in the focus direction). First-order optics tells us that the final f/number at the CCD will be f/150, which is quite high. The customer was informed that there may be an illumination problem and that there may not be enough light. The reply was "no problem...we have been there before, and we can simply turn up the rheostat." Eight months and several hundred thousand dollars later they did not have enough light and the project was cancelled!

How can you prevent this from happening to you? We have two suggestions: First, carefully work through the radiometry and derive the required irradiance in watts per square centimeter on the CCD chip. These data, along with the data sheet for your CCD device, should allow you to compute your signal-to-noise ratio, which will give you a level of confidence in your having enough light. Another approach is to perform an empirical experiment. Set up your object and your illumination just as you plan to implement it in your system. Now take a normal CCD camera lens and use it to reimage your object from some reasonable distance such as 0.5 to 1.0 m or thereabouts. The most important factor here is to now place a small circular aperture (a round hole in a piece of black paper is fine) in front of your lens in order to create the f/150 that you will have in your real system. If your camera lens has a focal length of 20 mm, for example, the aperture will need to be 20/150 = 0.133 mm in diameter, a very small diameter! If this is difficult to obtain, you can attenuate the light to the required level by using neutral density filters along with, or instead of, a small aperture. It is important to emulate the ultimate irradiance on the sensor. If you then determine that you have enough light, you can proceed ahead with your system design. If not, you have work to do! Empirical tests, such as described here, are extremely valuable, and often they are simple to execute.

Athermalization Using Teflon

Athermalization can be a serious problem, especially, but not limited to, thermal infrared systems where the change in refractive index with temperature (dn/dt) is large, such as for germanium where the value is 0.000396/C.° A system was built some years ago for the near infrared (IR) (just below 1-μm wavelength), and a bimetallic housing structure was utilized in order to maintain acceptable imagery as a function of temperature. Unfortunately, the required motion was larger than could be accomplished with typical housing materials, and Teflon was used as a spacer material in order to control one of the critical airspaces for athermalization. Initially, the system worked perfectly; however, it was later found that Teflon had hysteresis in its expansion characteristics, and when ambient temperature was restored, the system was out of focus. Ultimately, a more complex bimetallic housing using different metals solved the problem. If polymer materials are used for athermalization, do so with extreme care and do not ignore the hysteresis factor.

Athermalization Specifications

In a thermal infrared MWIR system, the airspace between a zinc sulfide and a zinc selenide element was very accurately controlled using a bimetallic housing structure in order to maintain focus through a wide temperature range. Initial tests in a thermal chamber showed the focus to be perfectly maintained. Several weeks afterward, it was discovered that from the outset of the project, refocus was permitted, and athermalization was not at all required. Read your specifications carefully!

Bad Glass Choice

A very compact telephoto lens was designed for a production system. Because of the demanding packaging and optical performance, SF58 glass was used. The first problem was that the optical shop could only find $^3/_4$ ft^3 of the glass in the world, not enough for full production. However, another problem arose after the first several hundred systems were completed—they all failed their MTF specification, especially on axis where astigmatism was present. After several weeks of intense study, it turned out that the SF58 elements had slumped during the coating operation. The lenses were supported in the coating chamber on rails, and the technician had been instructed to "set the temperature gauge to the red mark," since that is the temperature where they coated all of their elements. Unfortunately, SF58 has the second lowest transformation temperature in the entire catalog, and at the temperature in the chamber, the elements softened just enough to slump a little, and that was enough to introduce the astigmatism. Fortunately, the elements could be fine ground and repolished prior to recoating at a lower temperature. A short postscript: the design was reoptimized using SF6, a much better glass, and the performance was virtually the same as with SF58.

Elements in Backwards

This problem is far more common than it should be! At one of our short courses, 40% of course attendees' hands shot up in the air in response to the question "who has had the experience of elements being mounted backwards?" Figure 22.6 shows a scale drawing of two 25-mm diameter elements, one with radii of 40-mm convex on the left side and 42-mm convex on the right side, the other with both radii identical at 41 mm. The two lenses clearly look identical. Which has the nonequal radii? The answer is that the element on the left has the nonequal radii and the element on the right is perfectly equiconvex. During the assembly operation, the technician or assembly person will typically look at the reflection from each side from an overhead light source, and the side with the smaller reflected virtual image is the shorter radius. Unfortunately, the reflected imagery will look virtually identical from these two radii. What should be done in this case?

■ The best thing to do is either make the radii equal (both convex or both concave with the same radii), or make them sufficiently different so as to easily determine the correct orientation.

■ If you cannot make the radii the same, perhaps the best thing to do is to place an intentional bevel of a size or face width which can

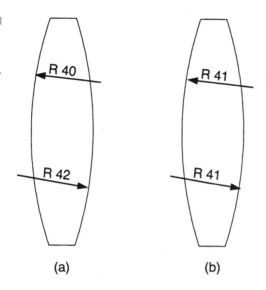

Figure 22.6
Two Nearly Identical
Elements

(a)   (b)

easily be distinguished on either S1 or S2. The reason this is a good idea is that the shop will most certainly place the bevel on the correct surface.

▪ Another approach is to request that the shop put an arrow pointing to the second surface showing the direction of light. This is a reasonable idea; however, you may be dealing with a shop whose practice is to put the marks following a different convention, and then you have a really serious problem!

Insufficient Sampling of Fields of View or Aperture

Computer programs, no matter how sophisticated, do only what they are told to do. If you specify, for example, semifields of view of 0, 7, and 10° off axis, the optimization algorithm will work specifically on those fields of view, and all other fields will be totally ignored as if they were nonexistent. If you have a system with higher-order aberrations and/or aspherics, it is very likely that the performance may degrade at field positions between these fields. Thus, for example, you may ultimately experience poor performance at 3 to 4° off axis. This was found to be the situation in our double Gauss case study in Chap. 21.

An MWIR infrared system with at least one aspheric surface was initially designed over five equally spaced fields of view. Unfortunately, during the final testplate fit, only three fields were used during the optimization. When the lens was tested, it performed superbly on axis and at its full field. However, at intermediate field positions the performance failed its specifications miserably. It is important to assure that the fields of view are sampled sufficiently during all phases of the lens design optimization. It is wise to evaluate performance of the system over 5 to 10 equally spaced field positions. It should be noted that sampling in pupil space is also important and not to be ignored.

Images Upside Down or Rotated

In visual systems the imagery must be both erect and right handed. In systems used for imagery onto a CCD or similar sensor, the image inversion and/or handedness can often be taken care of in the electronics. The orientation of the image can be verified by tracing a nonsymmetrical object through the system, which was described in Chap. 8.

Some years ago we had a panoramic system which used a prism to scan a wide azimuth field of regard. This form of system introduces image rotation which must be canceled by using another rotating prism subassembly such as a Pechan prism. Just after the machine shop started manufacturing the first components for the prototype, one of us in the team decided to carefully check once again if the direction of rotation of the Pechan prism was correct. Indeed, our original design had the direction of a Pechan prism rotation set incorrectly. Luckily there were not a lot of parts machined yet. If we did not discover this mistake on time, it would have been a major disaster, since these types of systems are generally very expensive.

This type of panoramic system has to be designed so that the image is properly oriented in the nominal (zero angle scan position), including the rotating element with a correctly determined axis of rotation. The second issue is the direction of rotation and the magnitude of angle of rotation. Generally, the magnitude of compensating element rotation is one-half of the scanning element angle of rotation. However, the direction of rotation has to be studied carefully in each specific case, since this is poorly covered in literature, and mistakes are very easily made.

The Hubble Telescope Null Lens Problem

As many of us know, a 1.3-mm error in the spacing of the null optics caused a significant error in the aspheric shape of the primary mirror of the Hubble telescope. The basic interferometric null test is shown in Fig. 22.7. An interferometer with a diverger lens similar to the system shown in Fig. 15.7 was used. Following the diverger focus, the diverging light is reflected from a concave spherical mirror, and after forming an intermediate image, it is reflected from a second concave spherical mirror which forms an image to the right of the first spherical mirror. A field lens is located at the last intermediate image as shown. The light then proceeds to the surface under test. The whole purpose of this test setup is to create a wavefront, which matches exactly the nominal mirror surface at the nominal location of the mirror surface. In other words, the null test creates a wavefront that precisely and perfectly nests into the nominal mirror surface under test. The extremely weak field lens appears to be doing nothing; however, since it is located at a highly aberrated image position, there is a significant difference between the paraxial rays and the real rays transmitting through the element, and this, in effect, allows the field lens to successfully balance the higher orders of spherical aberration. In the model we have developed, shown in Fig. 22.8, we have a residual double-pass optical path difference of 0.002 wave rms.

In initially setting up the test, it is imperative that the two mirrors and the field lens be properly positioned with respect to each other. In order to accomplish this, diverging light from the interferometer is first

Figure 22.7
Basic Setup of
Hubble Telescope
Null Optics

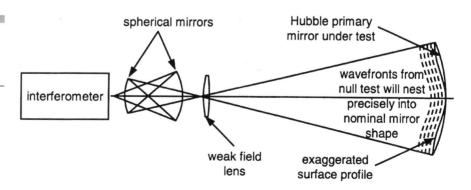

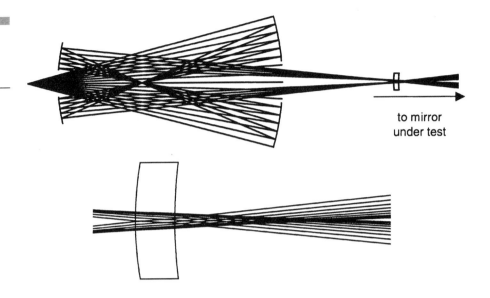

to mirror
under test

retroreflected from the right concave mirror back into the interferometer, as in Fig. 22.9a. When a null condition or straight fringes are seen,
this establishes a precisely known spacing between the diverger focus
and the mirror. The field lens now needs to be positioned, and a low-
expansion invar metering rod of a precisely known length with slightly
convex polished ends is now located so that the light from the interferometer focuses on the left-hand end of the rod and returns into the
interferometer, as in Fig. 22.9b. Once again, when a null fringe or straight
fringes are seen, we can be confident that the focus of the light is on the
end of the metering rod. Now the field lens is just barely touched to
the opposite end of the metering rod and bonded in place. The metering rod is now removed, and we are left with the right-hand mirror and
the field lens properly spaced with respect to each other. A further procedure is then used to locate the left-hand spherical mirror.

In order to assure that the light was properly incident onto the metering rod, a cap with a small aperture was placed onto the end of the rod,
as shown in Fig. 22.9c. The cap was painted flat black so that if the system were misaligned and the laser light were to strike the cap and not
pass through the aperture, then no fringes would be seen in the interferometer. The aperture plane was approximately 1.3 mm from the end
of the metering rod. After the system was initially tested, a piece of
masking tape was placed over the aperture in the cap in order to keep
dust out. Prior to testing, the tape was removed and a small piece of the

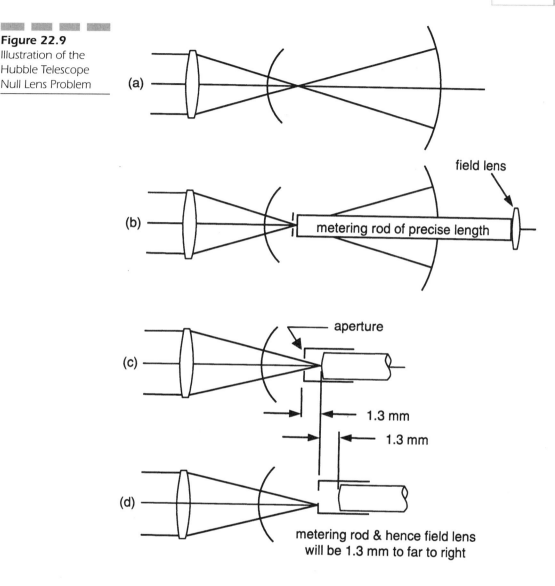

Figure 22.9
Illustration of the
Hubble Telescope
Null Lens Problem

(a)

field lens

(b)

metering rod of precise length

aperture

(c)

1.3 mm

1.3 mm

(d)

metering rod & hence field lens
will be 1.3 mm to far to right

flat black paint flaked off leaving the bare metal of the cap. During the setup of the null test components as outlined earlier, the diverger was inadvertently focused on the area of the cap where the paint had flaked off, and this caused the metering rod to be 1.3 mm too far to the right, as shown in Fig. 22.9d. This resulted in the field lens also being 1.3 mm to the right, and this was the problem. We would normally think that a bare metal surface like the top of a tin can would hardly be good

Figure 22.10
Image Point Spread
Function of the Hubble Telescope Null
Lens if Components
Were Manufactured
and Aligned Perfectly

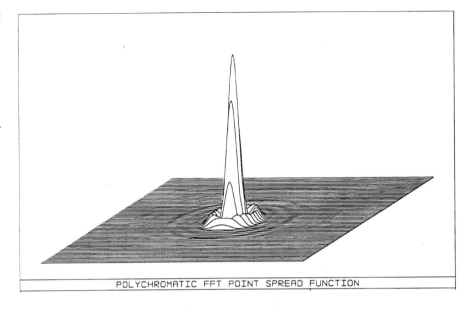

POLYCHROMATIC FFT POINT SPREAD FUNCTION

Figure 22.11
Optical Path Difference with Field Lens
Despaced 1.3 mm

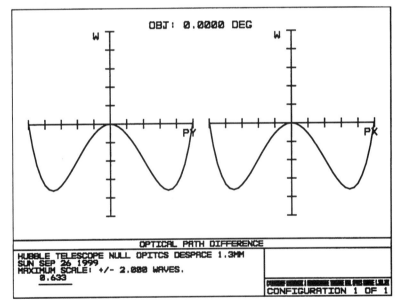

Figure 22.12
Image Point Spread
Function with Field
Lens Despaced
1.3 mm

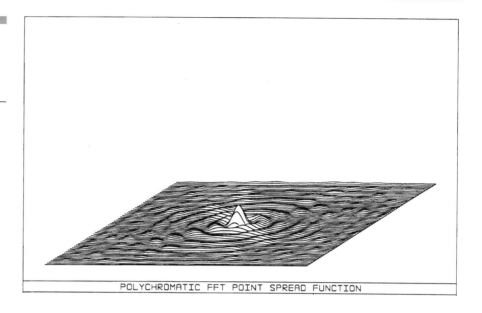

POLYCHROMATIC FFT POINT SPREAD FUNCTION

enough to produce straight fringes in an interferometric test. However, remember that the diameter of the focused spot is in the order of 2 μm, and the surface is quite likely to be good over this diameter.

Figure 22.10 shows the image-point spread function for the nominal (perfect) null test. This is, in effect, representative of the imagery predicted for the telescope at its Cassegrain focus if everything were manufactured perfectly. Figure 22.11 shows a plot of the OPD for the null test with the field lens axially shifted by 1.3 mm, and Fig. 22.12 shows the resulting point spread function. These data are representative of what the imagery would be like for the system with the primary mirror manufactured with the null test including the shifted field lens. The Hubble telescope was ultimately repaired and was a great success as evidenced by spectacular imagery.

## CHAPTER 23

Rules of Thumb
and
Hints

There are many "rules of thumb" in optics, and we will summarize the most useful ones here. While they may be discussed elsewhere in this book, this chapter is a compendium of the most important and useful rules of thumb and hints.

- Diffraction-limited Airy disk diameter = $2.44\lambda\ f/\#$ in units of λ.
  - In the visible the Airy disk diameter is approximately equal to the $f/\#$ expressed in micrometers.

- Diffraction-limited angular Airy disk diameter = $2.44\lambda/D$ rads, where D is the entrance pupil diameter in the same units as λ.

- Resolution of a diffraction-limited system in the visible is approximately $136/D$ in arc seconds, where D is the diameter of the entrance pupil in millimeters.

- A system will provide image quality, which is nearly indistinguishable from perfect, if the optical path difference from a nearest reference spherical wavefront reaching the image departs from sphericity by one-quarter of the wavelength of the light or radiation. This is the Rayleigh Criteria.

- The depth of focus for one-quarter-wave peak-to-valley optical path difference = $\pm(2\lambda\ f/\#)_2$ in units of λ.
  - In the visible, the depth of focus is approximately equal to the $(f/\#)_2$ in micrometers.

- The spatial frequency where the MTF goes to zero = $1/(\lambda\ f/\#)$ = 1000 line pairs/mm for an $f/2$ lens at a wavelength of 0.5 μm.

- The Nyquist frequency is the highest spatial frequency that a pixelated sensor can successfully record. It is 1/(2 pixel period) expressed in line pair per millimeter.

- The MWIR wavelength (3 to 5 μm) is approximately 8 times the visible and the LWIR wavelength (8 to 12 μm) is approximately 20 times the visible.

- RMS wavefront error is approximately 1/5 to 1/3.5 of the peak-to-valley optical path difference.

- "Clip it at the bud"—correct aberrations as close to where they are introduced in your system as possible. This will be easier to do and yield better overall performance.

- Always consider the possible effects of stray light, and provide suitable baffles and low-reflectivity interior system finish to attenuate it.

- Always consider the possible effect of ghost images due to multiple reflections from lens surfaces in your system. While ghost images are, for the most part, not a problem, it is best to analyze the situation.

- Make sure you tolerance your optical system and use realistic tolerances, otherwise your system will not be producible and/or will be very costly.

- Tolerances do not lie!

- The effect of tolerances is somewhat like standing at the edge of the Grand Canyon (well, not really, but this analogy is thought provoking). If someone pushes you a little on the shoulder, you will be fine. However, if they push very hard, you will tumble in rather fast. Tolerances are similar; small tolerances are generally fine, but as soon as they increase beyond a certain critical level, the performance will get bad very quickly.

- If your goal is to provide near to diffraction-limited performance, as your tolerances and the errors introduced by them increase as you error budget your system, the predicted MTF will degrade slowly at first and then faster and faster.

- If possible, try and assure that your specifications are based on what is *functionally* required for your system.

- Talk with your lens manufacturer to assure that the lens elements and other optical components are producible and your tolerances are reasonable. The same holds for the mechanics.

- Avoid very small, nearly concentric airspaces. This may lead to tight tolerances. You may be able to cement the two elements, which will also eliminate two antireflection coatings.

- Wherever possible, use good glasses which are easy to manufacture and low in cost. While the use of anomalous dispersion and other nonconventional glasses are sometimes valuable, they are not always required.

- Do not use aspheric surfaces unless they are mandatory, and if you do use them, make sure they are producible.

- The best rule of thumb regarding element thickness seems to be "if it looks good, it probably is good."

- Avoid elements with very thin edges and also very thin center thicknesses. They may warp during manufacturing, be costly, and/or be difficult to mount.

- For low-order aspherics use either a conic or a fourth-order asphericity, not both. While you can use both a y^4 and a conic, they often have a similar effect and can "beat" against one another.
  - If your surface is reasonably curved, a conic can be used.
  - If your surface is nearly flat, use only the fourth-order asphericity.
  - When you do use aspherics, work with the lower orders of asphericity first, and only use as high an order as you really need.
- Only use diffractive surfaces if you really need them.
  - Assure through discussions with your manufacturer that the diffractive profiles are producible.
  - You should be very concerned about diffraction efficiency and scattering, especially in the visible or at lower wavelengths.
  - If you use binary optics, make sure you have sufficient phase steps to minimize scattering.
- If possible, use only good quality glasses, which have low sensitivity to stain, bubbles, and other parameters.
  - Check the cost and availability of the glasses.
  - Avoid glasses with any parameters that are nonstandard such as transformation temperature and stain characteristics.
- Pay attention to possible polarization thin-film coating issues.
- If your system is to be used in a polarized light environment, stress birefringence should be considered.
- Make sure your image is right side up and right handed, especially in visible systems.
- Always try to match 100% of the radii to existing vendors' testplates.
- Remember, scattering increases with decreasing wavelength. You will thus have significantly more scattering in the blue and UV portions of the spectrum.
- When evaluating the effects of stray light and scattering in visible or IR systems (UV too), put your eye figuratively at the sensor and look outward and ask yourself "what do you see." The answer will be very revealing with respect to stray light and scattering, as well as to the means for controlling the stray light.
- When working with asymmetrical systems in which you have small tilted surfaces or decentered elements, you can validate your

setup and the sign convention by increasing the tilt or decentration to a very large value so it shows up clearly on a layout.

■ Make sure you consider the thermal and other environmental requirements for your system. If athermalization is required, do not wait until the last minute to determine how to accomplish the task.

■ Be extremely careful in working with and interpreting the terminology and conventions used in lens design programs. A good example is the use of the term *spot size;* is this a spot *diameter* or a spot *radius?*

■ Finally, remember that hindsight is diffraction limited!

GLOSSARY

Abbe number (also *V* number) $(n_d - 1)/(n_F - n_C)$, where n_d is the refractive index at wavelengths $d = 0.5876$ μm, $F = 0.4861$ μm, and $C = 0.6563$ μm. Also called *V* number.

aberration Geometrical errors in imagery whereby a perfect (or stigmatic) image is not formed. Typical aberrations include spherical aberration, astigmatism, coma, and chromatic aberration. The aberration of distortion does not affect the image quality but rather the image position. Similarly, field curvature creates an image on a curved image surface. Lens bendings, locations, powers, glass types, as well as number of lenses and stop position, are all used to minimize aberrations.

achromatic lens A lens using two or more glass types which brings two colors to a common focus. Refractive along with diffractive optics can also lead to achromatic designs.

afocal lens A lens system which takes collimated light input and produces collimated light out, such as a pair of binoculars, a telescope, and a beam expander.

Airy disk The central maximum of the diffraction pattern from a perfect optical system with a circular unobscured aperture. The diameter of the Airy disk is 2.44 λ *f*/number.

anomaly, image False or "ghostlike" images in thermal infrared systems which are caused by the detector "seeing," or sensing energy from, portions of the system interior (rather than scene energy). If the system interior is at a different temperature and emissivity than the scene, bright (or dark) areas will appear on the display. Sometimes this can be serious enough to make the system nonfunctional.

aperture stop The location within a lens system where the chief, or principal or central, ray passes through and crosses the optical axis. This is the location within the lens where the ray bundles appear to pivot about. The presence of a mechanical limiting aperture typically creates the limiting size.

aplanatic lens A lens which is free of third-order spherical aberration and coma.

apochromatic lens A lens in which three colors have been brought to a common focus.

539

apparent field of view   The field of view that the eye sees when looking through an eyepiece. The ratio of the apparent field of view to the real field of view in object space is the "magnification."

aspheric surface   A lens or mirror surface which departs from a spherical shape. Conic surfaces (paraboloidal, hyperboloidal, etc.) as well as higher-order aspheric departures are often required for aberration reduction.

astigmatism   An aberration in which light in one plane (the "plane of the paper" or meridional plane) focuses at a different location from the orthogonal plane. Astigmatism varies in proportion to the aperture and quadratic with field of view.

axial color   The aberration whereby different colors focus at different distances from a lens. Primary axial color is the residual between the upper and lower wavelengths. Secondary axial color is the residual between the upper/lower wavelengths and the central wavelength.

back focal distance   The distance from the last lens vertex to the image.

binary optics   Diffractive optics where a staircase approximation to a kinoform is used for the surface profile.

blocking, lens   A support whereby many spherical lenses can be mounted and optically ground and polished at one time.

boresight error   An error of alignment of the optical axis of two related systems parallel to each other. It is also an error of parallelism of an optical and a related mechanical axis. This is generally expressed as an angle.

Cassegrain telescope   A reflecting system consisting of a concave primary mirror and a convex secondary mirror. The image is located behind the vertex of the primary. The mirrors are typically aspheric in shape (paraboloidal/hyperboloidal) for the "classical Cassegrain," and both hyperbolic for the coma-free Ritchey–Chrétien Cassegrain.

catadioptric system   An optical system consisting of both lenses and mirrors with optical power.

chief ray   The ray passing through the center of the aperture stop of a lens or a mirror system. Also called a "principal" or "central" ray.

chromatic aberration   An axial or off-axis aberration whereby different colors have different focus positions, magnifications, spherical aberration, or other aberration forms.

clear aperture   The element diameter required for complete imagery over the full field of view. The term "optical clear aperture" refers to the lens (or mirror) aperture and the term "mechanical clear aperture" refers to the aperture created by mechanical features.

clipping   The effect in an infrared optical system whereby the beam is vignetted or clipped by an aperture within the system. This often produces undesirable cosmetic effects on to the imagery.

cold finger   A cooling device mounted directly behind a detector in thermal infrared cameras.

cold shield   A cold aperture within a dewar in an infrared system inside of which there is image-forming radiation. Radiation outside of the imaging cones and inside of the cold shield can be seen by the detector and is typically interior system structures.

cold stop   A cold aperture within a dewar in an infrared system which is also the aperture stop of the system. It allows for the detector to only see scene energy and no system interior.

cold-stop efficiency   In thermal infrared systems, the ratio of the imaging cone solid angle onto the sensor to the solid angle subtended by the cold aperture (cold shield) within the dewar assembly.

collimated light   Light where the rays are all parallel from a given object. A point source at infinity will yield collimated light. This also means that the wavefront is plane.

coma   An off-axis aberration where the outer periphery of a lens has a higher or a lower magnification than the central portion of the lens. The resulting image of a point object looks like a small comet.

coma, axial   Coma occurring at the center of the field of view introduced by element tilts, decentrations, and/or wedges. This coma generally carries over the field of view.

computer-aided optical design   The process of using a lens design computer program to optimize the performance of a lens system and then to evaluate its performance.

concave surface   A lens or mirror surface which is inward curving.

conjugate   A location which is at an image of another location.

convex surface   A lens or mirror surface which is outward curving.

cosmetic effects   Those defects in/on a lens or mirror which appear undesirable but which may have little or no functional impact on performance. Scratches and digs are often classified in this way. Also refers to video anomalies in infrared system imagery.

crown glass One of two main types of glasses, the other being flint glasses. Crown glass is harder than flint and has a lower index of refraction and a lower dispersion.

curvature 1/ radius of a surface.

depth of field The maximum change in axial position of an object which produces an acceptable image quality. This is typically looser than the Rayleigh criteria, and relates to the optics and the detector. This is a common term in photography.

depth of focus The focus shift corresponding to plus or minus one-quarter of wavefront error. This corresponds to a wavefront error which just meets the Rayleigh criteria and produces imagery which is essentially perfect. The depth of focus is $d = \pm\lambda/(2 \sin^2\theta) = \pm 2\ \lambda(f/\text{number})^2$, where θ is the half angle of the final image cone.

dewar A vacuum bottle which is cooled to cryogenic temperatures and holds an infrared detector.

diamond turning A process of ultraprecision machining whereby optical surfaces can be directly produced using a diamond-tipped tool and air bearings, air slides, and, as appropriate, numerical control. Surfaces accurate to within a few tenths of a micrometer are achievable. Nearly all nonferrous metals, as well as several of the infrared transmitting material, can be diamond turned.

diffraction A spreading of light after a wavefront of light passes by an opaque edge, due to the wave nature of the light or electromagnetic radiation. This spreading causes the formation of the classical Airy disk pattern when a perfect unobscured optical system images a point object.

diffractive optics The use of holographic, kinoform, and binary surfaces which use diffraction in controlling wavefronts. Often assists in introducing optical power to a system as well as helping in cancellation of monochromatic and chromatic aberrations.

dispersion The change in refractive index of glass or other refracting materials with color or wavelength.

distortion An aberration which is a change in magnification with field of view. Distortion is typically cubic with field of view and is a mapping error. It does not affect resolution or image quality.

entrance pupil The position along the optical axis of a lens or mirror system where the chief ray would intersect the optical axis if it were not redirected by the lenses or mirrors. Also, the location and size of the image of the aperture stop when looking into the front of the system.

etendue Product of the area of the light beam and the solid angle that the beam includes. Etendue represents the conservation of radiance, and is maintained throughout a given optical system.

exit pupil The position along the optical axis of a lens or mirror system where the chief ray exiting the system appears to have crossed the optical axis. Also, the location and size of the image of the aperture stop when looking into the rear of the system.

f/**number** The ratio of the focal length to the clear aperture diameter. First-order optics says that f/number $= 1/(2 \tan \theta)$, and for an aplanatic system f/number $= 1/(2 \sin \theta)$, where θ is the half angle of the final image cone.

field lens A lens or group of lenses located at or near an intermediate image which images the exit pupil of an objective lens on the entrance side of the field lens into the entrance pupil of an objective lens on the exit side of the field lens so as to minimize vignetting and increase light throughput.

flint glass One of two main types of glasses, the other being crown glasses. Flint glass is softer than crown, has a higher index of refraction and a higher dispersion.

focal length The distance measured along the optical axis from the image to the plane where the backwards-extended axial imaging cone of light intersects the extended input light bundle.

focal plane array (FPA) A linear or two-dimensional matrix of individual detector elements, typically used at the focus of an imaging system.

fringes Dark (and light) bands on an interferogram formed by interference of two light beams. After reflection from a surface with one wave of irregularity, we see two fringes. After transmission through a material with one wave of irregularity on one surface, we see $(n - 1)(1$ fringe) or about 0.5 fringe for glass of refractive index $n = 1.5$. A net system error of one fringe of wavefront error is one wave of optical path difference.

gaussian beam A beam of light whose intensity profile is gaussian in cross section. Lasers typically emit gaussian beams.

ghosting An effect in infrared systems where a facet of a polygon mirror adjacent to a facet actually being used is imaging onto a detector.

interferometer An instrument whereby a test wavefront is caused to interfere with a reference wavefront. Any difference between the reference and the test will show as light and dark fringes, which are typically photographed or viewed using a vidicon, CCD, or other sensor.

kinoform The surface profile of a diffractive surface which is like a saw tooth to direct the diffraction to the desired order.

lateral color A chromatic aberration which is a change in magnification with color. Sometimes called *color fringing.*

left-handed image After reflection in a mirror, an image is upside down, and even if it is rotated around the direction of image propagation, it cannot be oriented as the input object.

magnification, lateral The ratio of the image height to the object height. This is also equal to the ratio of the image distance to the object distance, or to the ratio of the image side f/number to the object side f/number.

magnification, longitudinal The ratio of the image motion along the optical axis to the corresponding object motion along the optical axis. The longitudinal magnification is the square of the lateral magnification.

modulation transfer function (MTF) The ratio of the modulation in the image to the modulation in the object for a sinusoidal object as a function of spatial frequency. The MTF is affected by diffraction and geometrical aberrations. For a perfect system, the maximum resolvable frequency is $1/(\lambda \; f/\text{number})$ in line pairs per millimeter, and this is where the MTF goes to zero.

narcissus The effect in an infrared system where the detector "sees" a reflection of itself. If this reflected radiation changes or modulates through scan and/or over a field of view, then less cold radiation is reflected back into the detector off axis and a dark central region or "porthole" appears in the display.

objective lens The primary lens, which takes light from an object and forms an image. An objective lens generally consists of multiple elements in order to minimize aberrations.

optical path difference (OPD) The difference between a reference (or perfect) wavefront and a real wavefront. If the OPD is one-quarter of the wavelength, then the system just meets the *Rayleigh criteria* and the system will be essentially diffraction limited.

paraxial The region where the angles between the rays and the optical axis are small, and the approximation that the sines and tangentsof angles can be represented by their value, in radians, is valid. This makes computations fast and easy, and provides for a convenient means of locating, for example, the image position without regard to aberrations.

partial dispersion Difference in index of refraction for two wavelengths. Main dispersion is $(n_F - n_C)$ where $F = 486.13$ nm and $C = 653.27$ nm.

power fit to testplate The number of fringes of power seen by the optician when placing in close contact the surface being fabricated and a surface of known radius (the testplate). Each fringe represents one-half wave of sag or difference between the two surfaces.

principal planes The locations within an optical system where incoming collimated light intersects the light directed to the image (if each are extended until they meet).

Rayleigh criteria The rule of thumb developed by Lord Rayleigh that if the difference between the longest and shortest paths leading to a selected focus is less than or equal to one-quarter of the wavelength, then the imagery is nearly indistinguishable from perfect. If a system meets the Rayleigh criteria, the optical path difference is approximately $\frac{1}{4}\lambda$ or less, and the imagery is essentially perfect.

refractive index The ratio of the velocity of the radiation (light) in a vacuum to the velocity of the radiation (light) in a material. The higher the refractive index, the more the radiation "bends" or is refracted at the air-material surface. Radiation incident on a surface obeys Snells law which says $n \sin \theta = n' \sin \theta'$, where n and n' are the refractive indices on each side of the interface and θ and θ' are the angles of incidence and refraction measured from the normal to the surface.

relay lens A lens or lens group which relays a finite object to a remote location at a magnification of unity or some other value.

right-handed image After reflections and refraction in the system, the image is oriented such that if it is rotated around the direction of image propagation, it can be brought to the same orientation as the input object.

sag The height of the curve measured from the chord. Often used to specify an aspheric surface.

scan noise An effect in infrared systems whereby the cooled detector "sees" more radiation or energy (from within the system) at one position in its field of view or scan than another position, therefore producing undesirable cosmetic effects on the system display.

secondary spectrum   In the systems where the primary color is corrected, and the blue and red focus brought to the same point, secondary spectrum or secondary color is the distance between the green focus and the red-blue focus.

spherical aberration   The axial aberration where rays from the outer periphery of the lens focuses closer (or further) from the lens than the rays closer to the axis. Spherical aberration is typically proportional to the cube of the aperture.

spherochromatism   Variation of spherical aberration with wavelength.

stigmatic   Perfect imagery, in the geometrical sense.

Strehl ratio   The ratio of the peak intensity in the diffraction pattern of an aberrated point image to the peak intensity in the diffraction pattern of the same aberration free point image.

vignetting   A clipping or truncation of the off-axis ray bundles by elements distant from the aperture stop. Vignetting is usually intentional in visible systems, as elements can be made smaller and lighter in weight while producing better imagery (by eliminating severely aberrated rays). For visible systems, vignetting of 30% to even as much as 50% can typically be tolerated. Vignetting is usually not tolerable in scanning infrared systems.

BIBLIOGRAPHY

Buralli, Dale A., "Optical Design with Diffractive Lenses," Sinclair Optics, design notes, 1991.

Glatzel, Erhard (1980) "New Lenses for Microlithography," *1980 International Lens Design Conference Proceedings*, SPIE.

Handbook of Optics, vols. 1 and 2 (1995), New York: McGraw-Hill.

Holland, L. (1956), *Vacuum Deposition of Thin Films*, London: Chapman & Hall.

Jacobson, M. (1986), *Deposition and Characterization of Optical Thin Films*, New York: Macmillan.

Kingslake, R. (1965), *Applied Optics and Optical Engineering*, New York: Academic Press.

Kingslake, R. (1978), *Lens Design Fundamentals*, New York: Academic Press.

Kingslake, R. (1983), *Lens Design Fundamentals*, New York: Academic Press.

Laikin, Milton (1991), *Lens Design*, New York: Marcel Dekker.

Londono, Carmina (1992), *Design and Fabrication of Surface Relief Diffraction Optical Elements, or Kinoforms, with Examples for Optical Athermalization*, Tufts University.

Lotmar, W. (1971), "Theoretical Eye Model with Aspherics," *Journal of the Optical Society of America*, **61**:1522–1529.

Macleod, H. A. (1986), *Thin-Film Optical Filters*, New York: Macmillan.

Melzer and Moffitt (1997), *Head Mounted Displays*, New York: McGraw-Hill.

Morris, G. Michael and Dale A. Buralli (1992), "Diffractive and Binary Optics," OSA Short Course.

Mouroulis, Pantazis (1999), *Visual Instrumentation*, New York: McGraw-Hill.

ORA Seminar Notes, April 1999, "Design of Efficient Illumination Systems," ORA, Pasadena, CA

O'Shea, D. C. (1985), *Elements of Modern Optical Design*, New York: John Wiley.

Pulker, H. K. (1984), *Coatings on Glass*, Amsterdam: Elsevier.

Rancourt, James D. (1987), *Optical Thin Films Users' Handbook*, New York: McGraw-Hill.

549

Ray, Sidney F. *Applied Photographic Optics, Lenses and Optical Systems for Photography, Film, Video and Electronic Imaging,* 2d ed., Focal Press.

Rhyins, R., *Laser Focus,* June 1974, p. 55.

Riedel, Max J., "Optical Design Fundamentals for Infrared Systems," tutorial texts in optical engineering; v.TT20, SPIE.

Shannon, R. (1980), "Aspheric Surfaces," *Applied Optics and Optical Engineering,* R. Kingslake, ed. vol. 8, New York: Academic Press.

Shannon, R., and James C. Wyant (1980), *Applied Optics and Optical Engineering,* R. Kingslake, ed., vol. 8, New York: Academic Press.

Siegman, A.E. (1986), *Lasers,* University Science Books.

Smith, Warren J. (1990), *Modern Optical Engineering,* New York: McGraw-Hill.

Smith, Warren J. (1992), *Modern Lens Design,* New York: McGraw-Hill.

Smith, Warren J. (1997), *Practical Optical System Layout,* New York: McGraw-Hill.

Stover, John C. (1990), *Optical Scattering Measurements and Analysis,* New York: McGraw-Hill.

Strong, John (1958), *Concepts of Classical Optics,* New York: Freeman.

The Photonics Directory, Laurin Publishing Co. (annual).

Thelen, Alfred (1989), *Design of Optical Interference Coatings,* New York: McGraw-Hill.

U.S. Military Handbook for Optical Design (1987), republished by Sinclair Optics, Fairport, NY.

Walker, Bruce (1995), *Optical Engineering Fundamentals,* New York: McGraw-Hill.

Wetherell, W. (1980), "The Calculation of Image Quality," *Applied Optics and Optical Engineering,* R. Kingslake, ed., vol. 8, New York: Academic Press.

Williams, C.S. (1989), *Introduction to the Optical Transfer Function,* New York, John Wiley & Sons.

Yoder, P. (1986), *Opto-Mechanical System Design,* New York: Marcel Dekker.

"A Chart Demonstrating Variation in Acuity with Retinal Position," letter to the editor, *Vision Research,* vol. 14, 1974, p. 589.

INDEX

553

About the Authors

Robert E. Fischer is the president and founder of OPTICS 1, Incorporated, of Westlake Village, California. He has a BS and MS in Optics from the Institute of Optics at the University of Rochester. He has been involved in optical design and engineering since 1967 when he joined the Itek Corporation. Prior to founding OPTICS 1, he was with Hughes Aircraft. His primary areas of technical interest and expertise include lens design, optical engineering, optical system manufacturing and testing, illumination systems, and related engineering technologies. In addition to chairing many conferences with the International Society for Optical Engineering (SPIE), Fischer has held several positions with the society, including president in 1984. His popular short course "Practical Optical System Design" has been widely attended over the years and forms the basis for this book. Mr. Fischer is a Fellow of SPIE and the Optical Society of America. He was awarded the Albert M. Pezzuto award from SPIE in 1986, and the Gold Medal of SPIE in 2000.

Biljana Tadic-Galeb is a Senior Staff Optical Engineer at OPTICS 1, Incorporated, of Westlake Village, California. She has a BS in Physics from the University of Sarajevo, Yugoslavia, and an MS in Optics from Reading University in England. She also holds an MS in Metrology from the University of Beograd, Yugoslavia. Ms. Tadic-Galeb has over 20 years of experience as an Optical Systems Engineer with specialties in the development of complex visible, IR, and UV optical systems, projections systems, laser systems, and hybrid systems with diffractive elements and fibers. She was recognized by the National Academy of Engineering in their Featured Women Engineers Program.